Theory and Applications
of Natural Language Processing

Series Editors:
Graeme Hirst (Textbooks)
Eduard Hovy (Edited volumes)
Mark Johnson (Monographs)

Aims and Scope

The field of Natural Language Processing (NLP) has expanded explosively over the past decade: growing bodies of available data, novel fields of applications, emerging areas and new connections to neighboring fields have all led to increasing output and to diversification of research.

"Theory and Applications of Natural Language Processing" is a series of volumes dedicated to selected topics in NLP and Language Technology. It focuses on the most recent advances in all areas of the computational modeling and processing of speech and text across languages and domains. Due to the rapid pace of development, the diversity of approaches and application scenarios are scattered in an ever-growing mass of conference proceedings, making entry into the field difficult for both students and potential users. Volumes in the series facilitate this first step and can be used as a teaching aid, advanced-level information resource or a point of reference.

The series encourages the submission of research monographs, contributed volumes and surveys, lecture notes and textbooks covering research frontiers on all relevant topics, offering a platform for the rapid publication of cutting-edge research as well as for comprehensive monographs that cover the full range of research on specific problem areas.

The topics include applications of NLP techniques to gain insights into the use and functioning of language, as well as the use of language technology in applications that enable communication, knowledge management and discovery such as natural language generation, information retrieval, question-answering, machine translation, localization and related fields.

The books are available in printed and electronic (e-book) form:

* Downloadable on your PC, e-reader or iPad
* Enhanced by Electronic Supplementary Material, such as algorithms, demonstrations, software, images and videos
* Available online within an extensive network of academic and corporate R&D libraries worldwide
* Never out of print thanks to innovative print-on-demand services
* Competitively priced print editions for eBook customers thanks to MyCopy service http://www.springer.com/librarians/e-content/mycopy

For other titles published in this series, go to
www.springer.com/series/8899

Thierry Poibeau • Horacio Saggion
Jakub Piskorski • Roman Yangarber
Editors

Multi-source, Multilingual Information Extraction and Summarization

 Springer

Editors
Thierry Poibeau
LATTICE-CNRS
Ecole Normale Supérieure
Université Sorbonne Nouvelle
Paris
France

Jakub Piskorski
Institute for Computer Science
Polish Acadmey of Science
Warsaw
Poland

Horacio Saggion
Information & Communication
  Technologies
Universitat Pompeu Fabra
Barcelona
Spain

Roman Yangarber
Department of Computer Science
University of Helsinki
Helsinki
Finland

ISSN 2192-032X   ISSN 2192-0338 (electronic)
ISBN 978-3-642-28568-4   ISBN 978-3-642-28569-1 (eBook)
DOI 10.1007/978-3-642-28569-1
Springer Heidelberg New York Dordrecht London

Library of Congress Control Number: 2012945066

Printed on acid-free paper

Springer is part of Springer Science+Business Media (www.springer.com)

Preface

1 Introduction

Information extraction (IE) and text summarization (TS) are key technologies aiming at extracting relevant information from texts and presenting the information to the user in condensed form. The on-going information explosion makes IE and TS particularly critical for successful functioning within the information society. These technologies, however, face new challenges with the adoption of the Web 2.0 paradigm (e.g. blogs, wikis) because of their inherent multi-source nature. These technologies have to deal no longer with isolated texts or single narratives, but with large-scale repositories, or sources – possibly in several languages – containing a multiplicity of views, opinions, or commentaries on particular topics, entities or events. There is thus a need to adapt and/or develop new techniques to deal with these new phenomena.

Recognising similar information across different sources and/or in different languages is of paramount importance in this multi-source, multi-lingual context. In information extraction, merging information from multiple sources can lead to increased accuracy relative to extraction from a single source. In text summarization, similar facts found across sources can inform sentence scoring algorithms. In question answering, the distribution of answers in similar contexts can inform answer ranking components.

Often, it is not the similarity of information that matters, but its complementary nature. In a multi-lingual context, information extraction and text summarization can provide solutions for cross-lingual access: key pieces of information can be extracted from different texts in one or many languages, merged, and then conveyed in many natural languages in concise form. Applications need to be able to cope with the idiosyncratic nature of the new Web 2.0 media: mixed input, new jargon, ungrammatical and mixed-language input, emotional discourse, etc. In this context, synthesizing or inferring opinions from multiple sources is a new and exciting challenge for NLP. On another level, profiling of individuals who engage in the new social Web, and identifying whether a particular opinion is appropriate/relevant in a given context are important topics to be addressed.

It is therefore important that the research community address the following issues:

- What methods are appropriate to detect similar/complementary/contradictory information? Are hand-crafted rules and knowledge-rich approaches suitable?
- What methods are available to tackle cross-document and cross-lingual entity and event co-reference?
- What machine learning approaches are most appropriate for this task—supervised, unsupervised, semi-supervised? What type of corpora are required for training and testing?
- What techniques are appropriate to synthesize condensed synopses of the extracted information? What generation techniques are useful here? What kind of techniques can be used to cross domains and languages?
- What techniques can improve opinion mining and sentiment analysis through multi-document analysis? How do information extraction and opinion mining connect?
- What tools exist for supporting multi-lingual/multi-source access to information? What solutions exist beyond full document translation to produce cross-lingual summaries?

This volume contains a series of recent papers covering most of the above challenges. Some of them also bridge the gap between IE and TS and show that these are complementary technologies that can be valuably integrated in real-world natural language applications.

2   Content of this volume

Part I of the volume contains two background chapters describing the state of the art in Text Summarization (Saggion and Poibeau) and Information Extraction (Piskorski and Yangarber). These are intended to provide a broad overview of the field for the reader, and to define the context for the subsequent technical chapters.

Part II contains four chapters on named entity analysis in a multilingual context. Named entity recognition plays a prominent role both for IE and TS. Named entities carry major information that can help determine what is the text about. It is a major component of any IE system, which consists in large part in identifying relations between named entities. Lastly, named entity also plays a role in determining what are the most important sentences of a text, which is obviously crucial for TS. This Part of the volume provides different studies on named entity recognition, addressing issues such as variation across languages (Mani et al.) or in one language (Driscoll). The last two chapters address the complex problem of relating various pieces of information to one referring entity despite language variation (Rao et al.) and co-reference resolution across documents (Saggion).

The first chapter is "Learning to Match Names Across Languages" by Inderjeet Mani, Alex Yeh and Sherri Condon. The authors report on research on matching names in different scripts across languages. They explore two trainable approaches based on comparing pronunciations. The first, a cross-lingual approach, uses an automatic name-matching program that exploits rules based on phonological comparisons of the two languages carried out by humans. The second, monolingual approach, relies only on automatic comparison of the phonological representations of each pair. Alignments produced by each approach are fed to a machine learning algorithm. Results show that the monolingual approach results in machine-learning based comparison of person-names in English and Chinese at an accuracy of over 97.0 F-measure.

The following chapter is "Computational Methods for Name Normalization Using Hypocoristic Personal Name Variants" by Patricia Driscoll. A growing body of research addresses name normalization as part of co-reference and entity resolution systems, but the problem of hypocoristics has not been systematically addressed as a component to such systems. In many languages, these name variants are governed by morphological and morpho-phonological constraints, providing a dataset rich in features which may be used to train and run matching systems. This chapter gives a full treatment to the phenomenon of hypocoristics and presents a supervised learning method that takes advantage of their properties to untangle the relationships between hypocoristics and corresponding full form names.

The following chapter, "Entity Linking: Finding Extracted Entities in a Knowledge Base", by Delip Rao, Paul McNamee and Mark Dredze, deals with named entity disambiguation as well as connecting various pieces of information from different texts to the same real-world object. Entity linking is a new task that has recently drawn much attention in NLP research. Entity Linking, also referred to as record linkage or entity resolution, involves aligning a textual mention of a named-entity to an appropriate entry in a knowledge base, which may or may not already contain the entity. This has manifold applications, ranging from linking patient health records to maintaining personal credit files, prevention of identity crimes, and supporting law enforcement. The authors discuss key challenges of this task, and present a high-performing system that links entities using max-margin ranking. The chapter also summarizes recent work in this area and describes several open research problems.

Once named entities have been analyzed, they can be used to identify related documents, which involve co-reference resolution. This is the topic addressed by the last chapter of Part II: "A Study of the Effect of Document Representations in Clustering-based Cross-document Co-Reference Resolution" by Horacio Saggion. Finding information about people on huge text collections or on-line repositories on the Web is a common activity. The author describes experiments aiming at identifying the contribution of semantic information (e.g. named entities) and summarization (e.g. sentence extracts) in a cross-document co-reference resolution algorithm. Its cross-document co-reference system is a clustering-based algorithm which groups documents referring to the same entity. Clustering uses vector representations created by summarization and semantic tagging analysis components. The author investigates different configurations achieving state of the art performance demonstrating the potential of the applied techniques. He shows that selection of the type of summary and the type of term to be used for vector representation is important to achieve good performance.

Part III contains three chapters on Information Extraction. The chapters address various aspects of IE, both practical and theoretical. IE typically refers to filling a predefined template with information extracted from text related to a specific domain. The chapters address IE is less favorable environments: for example, when no domain or no template have been defined, or when the system faces a large number of different domains, or different sources (Neumann and Schmeier). In this context, it can be useful to predict the utility of the extracted information (Huttunen et al.). The output of IE can also serve as a basis for the generation of summaries (Ji et al.).

The first chapter is "Interactive Topic Graph Extraction and Exploration of Web Content" by Günter Neumann and Sven Schmeier. The authors consider IE when no domain has been a priori defined. It is then necessary to let the user dynamically explore the corpus and define templates on the fly. In their framework, the initial information request (in the form of a query topic description) is issued by a user online to the system using a search engine. A topic graph is then constructed using collocations identified in snippets returned by the search engine. This graph allows the user to dynamically explore the results, refine his query and identify additional relevant knowledge usingn the topic graph. The authors conclude their chapter with a user-oriented evaluation, which shows that the approach is especially helpful for finding new interesting information on topics about which the user has only a vague idea or no idea at all.

The following chapter is "Prediction of Utility in Event Extraction", by Silja Huttunen, Arto Vihavainen, and Roman Yangarber. The goal of the chapter is to estimate the relevance of the results of an Information Extraction system to the end-user's needs. Traditional criteria for evaluating the performance of IE focus on correctness of the extracted information, e.g., in terms of recall, precision and F-measure. Here, the authors introduce subjective criteria for evaluating the quality of the extracted information: utility of results to the end-user. They use methods from text mining and linguistic analysis to identify features that are good predictors of the relevance of an event or a document to a user. They report on experiments in two real-world news domains: business activities and epidemics of infectious disease.

In "Open-Domain Multi-document Summarization via Information Extraction: Challenges and Prospects", Heng Ji, Benoit Favre, Wen-Pin Lin, Dan Gillick, Dilek Hakkani-Tur and Ralph Grishman propose ideas to bridge the gap between IE and TS. The authors observe that IE ad TS share the same goal of extracting and presenting relevant information in a document. They show that while IE was a primary element of early abstractive summarization systems, it has been left out in more recent extractive systems. However, extracting facts, recognizing entities and events should provide useful information to those systems and help resolve semantic ambiguities that they cannot tackle. The chapter explores novel ways of taking advantage of cross-document IE for multi-document summarization. The authors propose several approaches to IE-based summarization and analyze their strengths and weaknesses. One of them, re-ranking the output of a high performing summarization system with IE-informed metrics, leads to improvements in manually-evaluated content quality and readability.

The final part, Part IV of this volume, concerns multi-document summarization. There is a clear link between this and the previous parts, since recognition of named entities and identification of key information in text are among the main components of any automatic summarization system. The five chapters of Part IV address recent trends in automatic summarization, such as production of update summaries, containing only new information after a first set of documents has already been summarized (Bossard) and production of summaries in a highly multilingual environment (Kabadjov et al.). An important issue in this context is coherent ordering of the information, especially when information is extracted from multiple sources (Bollegala et al.). The two last contributions consider TS techniques in multimedia environments, namely in relation to speech (Ribeiro and Martins de Matos) and images (Aker et al.).

The first chapter in Part IV is "Generating Update Summaries: Using an Unsupervised Clustering Algorithm to Cluster Sentences" by Aurélien Bossard. The author presents an original approach based on clustering techniques: sentence clustering makes it possible to group sentences conveying the same event or the same idea. In the first step, sentences of an initial set of documents are clustered based on their content. Sentences contained in new documents are then projected onto the result of the clustering of the initial set of documents, making it possible to distinguish new information (merged in existing clusters) from already known information (forming new clusters). The system is evaluated on the TAC 2009 "Update task" and shows promising results.

In "Multilingual Statistical News Summarisation", Mijail Kabadjov, Josef Steinberger and Ralf Steinberger present a generic approach for summarizing multilingual news clusters, such as those produced by the Europe Media Monitor (EMM) system. The authors use robust statistical techniques to summarize news from different sources and languages. A multilingual entity disambiguation system is used to build the source representation. The authors show that their system obtains good performances on the TAC datasets. Lastly, the authors have run a small-scale evaluation on languages other than English, providing interesting evidence that contradicts the pervasive assumption "if it works for English, it works for any language."

The following chapter, "Coherent Ordering of Information Extracted from Multiple Sources" by Danushka Bollegala, Naoaki Okazaki and Mitsuru Ishizuka, in creating a summary of information extracted from multiple sources, deals with the problem of deciding on the order in which information must be presented in the summary. Incorrect ordering can lead to misunderstandings. In this chapter, the authors discuss the challenges involved when ordering information selected from multiple sources and present several approaches to overcome those challenges. They also introduce several semi-automatic evaluation measures to empirically evaluate an ordering of sentences created by an algorithm.

The two last chapters establish connection between TS and other media, speech and images. Concerning the former, Ricardo Ribeiro and David Martins de Matos present a chapter entitled "Improving Speech-to-Text Summarization by Using Additional Information Sources". Speech-to-text summarization systems usually take as input the output of an automatic speech recognition (ASR) system that is affected by speech recognition errors, disfluencies, or difficulties in identification of sentence boundaries. The authors propose the inclusion of related, solid background information to cope with the difficulties of summarizing spoken language and the use of multi-document summarization techniques in single document speech-to-text summarization. They explore the possibilities offered by phonetic information to select the background information and conduct a perceptual evaluation to assess the relevance of the inclusion of that information. Results show that summaries generated using this approach are better than those produced by a state-of-the-art latent semantic analysis (LSA) summarization method and suggest that humans prefer summaries restricted to the information conveyed in the input source.

The last chapter, "Towards automatic image description generation using multi-document summarization techniques" by Ahmet Aker, Laura Plaza, Elena Lloret, and Robert Gaizauskas, reports an initial study of the viability of multi-document summarization techniques for automatic captioning of geo-referenced images. The automatic captioning procedure requires summarizing multiple web documents that contain information related to the images' location. The authors use different state-of-the art summarization systems to generate generic and query-based multi-document summaries and evaluate them using ROUGE metrics relative to human-generated summaries. Results show that query-based summaries perform better than generic ones and are thus more appropriate for the task of image captioning, or generation of short descriptions related to the location/place captured in the image.

Montrouge, France   Thierry Poibeau
Barcelona, Spain   Horacio Saggion
Warszawa, Poland   Jakub Piskorski
Helsinki, Finland   Roman Yangarber

Acknowledgements

This volume contains original papers, as well as a selection of extended contributions to the two MMIES workshops—Multi-source, Multilingual Information Extraction and Summarization. The first workshop of these series was held during RANLP 2007, in Borovets (Bulgaria), and was organized by Thierry Poibeau and Horacio Saggion. The second workshop was part of the COLING 2008 conference, held in Manchester, and was organized by Sivaji Bandyopadhyay, Thierry Poibeau, Horacio Saggion and Roman Yangarber. We would like to thank all the people who helped organizing these workshops, especially those who served in the program committees.

We would also like to thank the following people, who helped review the papers for this volume:

- Nathalie Colineau, CSIRO, Australia
- Nigel Collier, National Institute of Informatics, Japan
- Hercules Dalianis, Stockholm University, Sweden
- Patricia Driscoll, Columbia University, USA
- Robert Gaizauskas, University of Sheffield, UK
- Michel Généreux, University of Lisbon, Portugal
- Brigitte Grau, LIMSI-CNRS, France
- Ralph Grishman, New York University, USA
- Heng Ji, City University of New York, USA
- Guy Lapalme, University of Montreal, Canada
- Inderjeet Mani, The MITRE Corporation, USA
- Diana Maynard, University of Sheffield, UK
- Jean-Luc Minel, Université Paris Ouest Nanterre La Défense, France
- Günter Neumann, German Research Center for Artificial Intelligence GmbH (DFKI), Germany
- Constantin Orasan, University of Wolverhampton, UK
- Cecile Paris, CSIRO, Australia
- Maria-Teresa Pazienza, University Roma 2 Tor Vergata, Italy
- Ricardo Ribeiro, Universitário de Lisboa, Portugal

- Agnes Sandor, Xerox Research Centre Europe, France
- Satoshi Sekine, New York University, USA
- Josef Steinberger, Joint Research Centre, Italy
- Stan Szpakowicz, University of Ottawa, Canada
- Lucy Vanderwende, Microsoft research, USA
- Sumithra Velupillai, Stockholm University, Sweden
- José-Luis Vicedo, University of Alicante, Spain

Contents

Contributors

Ahmet Aker University of Sheffield, Sheffield, UK, a.aker@dcs.shef.ac.uk

Aurélien Bossard Laboratoire d'Informatique de Paris-Nord, Villetaneuse, France, aurelien.bossard@lipn.univ-paris13.fr

Danushka Bollegala Graduate School of Information Science and Technology, The University of Tokyo, Bunkyo-ku, Tokyo, Japan, danushka@iba.t.u-tokyo.ac.jp

Sherri Condon The MITRE Corporation, McLean, VA, USA, scondon@mitre.org

Mark Dredze Department of Computer Science and Human Language Technology Center of Excellence, Johns Hopkins University, Baltimore, USA, delip@cs.jhu.edu

Patricia Driscoll Columbia University, New York City, NY, USA, pdriscoll@gmail.com

Benoit Favre LIF, Aix-Marseille Université, Marseille, France, benoit.favre@lif.univ-mrs.fr

Robert Gaizauskas University of Sheffield, Sheffield, UK, r.gaizauskas@dcs.shef.ac.uk

Dan Gillick Computer Science Department, University of California, Berkeley, CA, USA, dgillick@berkeley.edu

Ralph Grishman Computer Science Department, New York University, New York, NY, USA, grishman@cs.nyu.edu

Dilek Hakkani-Tur Speech Labs, Microsoft, Mountain View, CA, USA, dilek@ieee.org

Silja Huttunen Department of Modern Languages, University of Helsinki, Helsinki, Finland, first.last@helsinki.fi

Mitsuru Ishizuka Department of Information and Communication Engineering, Graduate School of Information Science and Technology, The University of Tokyo, Bunkyo-ku, Tokyo, Japan, ishizuka@i.u-tokyo.ac.jp

Heng Ji Computer Science Department, Queens College and Graduate Center, City University of New York, New York, NY, USA, hengji@cs.qc.cuny.edu

Mijail Kabadjov EC Joint Research Centre, Ispra (VA), Italy, mijail.kabadjov@jrc.ec.europa.eu

Wen-Pin Lin Computer Science Department, Queens College and Graduate Center, City University of New York, New York, NY, USA

Elena Lloret University of Alicante, Alicante, Spain, elloret@dlsi.ua.es

Inderjeet Mani The MITRE Corporation, Bedford, MA, USA, imani@mitre.org

David Martins de Matos L^2F – INESC ID/IST, Rua Alves Redol, Lisboa, Portugal, david.matos@inesc-id.pt

Paul McNamee Human Language Technology Center of Excellence, Johns Hopkins University, Baltimore, USA, paul.mcnamee@jhuapl.edu

Günter Neumann German Research Center for Artificial Intelligence GmbH (DFKI), Saarbrücken, Germany, neumann@dfki.de

Naoaki Okazaki Department of System Information Sciences, Graduate School of Information Sciences, Tohoku University, Sendai, Japan, okazaki@ecei.tohoku.ac.jp

Jakub Piskorski Institute for Computer Science, Polish Academy of Science, Warszawa, Poland, jakub.piskorski@frontex.europa.eu

Laura Plaza Universidad Complutense de Madrid, C/ José García Santesmases, Madrid, Spain, lplazam@fdi.ucm.es

Thierry Poibeau Laboratoire LaTTiCe-CNRS, École Normale Supérieure and Université Sorbonne-Nouvelle, Montrouge, France, thierry.poibeau@ens.fr

Delip Rao Department of Computer Science, Johns Hopkins University, Baltimore, USA, delip@cs.jhu.edu

Ricardo Ribeiro L^2F – INESC ID/ISCTE – Instituto, Universitário de Lisboa, Lisboa, Portugal, ricardo.ribeiro@inesc-id.pt

Horacio Saggion Department of Information and Communication Technologies, Barcelona, Spain, horacio.saggion@upf.edu

Sven Schmeier German Research Center for Artificial Intelligence GmbH (DFKI), Saarbrücken, Germany, schmeier@dfki.de

Josef Steinberger EC Joint Research Centre, 21027, Ispra (VA), Italy, josef.steinberger@jrc.ec.europa.eu

Ralf Steinberger EC Joint Research Centre, 21027, Ispra (VA), Italy, ralf. steinberger@jrc.ec.europa.eu

Roman Yangarber Department of Computer Science, University of Helsinki, Helsinki, Finland, first.last@cs.helsinki.fi

Alexander Yeh The MITRE Corporation, Bedford, MA, USA, asy@mitre.org

Acronyms

ACE	Automatic Content Extraction
ACL	Association for Computational Linguistics
AFP	Agence France Presse
APP	Audio Preprocessing Module
ASR	Automatic Speech Recognition
CCP	Correct Conditional Probability
CE	Concept Extractor
CLEF	Cross Language Evaluation Forum
CNF	Conjunctive Normal Form
CRF	Conditional Random Fileds
DUC	Document Understanding Conference
EM	Expectation Maximization
EMM	European Media Monitor
IE	Information Extraction
HMM	Hidden Markov Models
HSD	Honest Significant Difference
HTML	Hypertext Markup Language
HVS	Hub Vertices Sets
ILP	Integer Linear Program
IPA	International Pronunciation Alphabet
IR	Information Retrieval
KB	Knowledge Base
KBP	Knowledge Base Population
LDC	Linguistic Data Consortium
LSA	Latent Semantic Analysis
MeSH	Medical Subject Heading
MIL	Multiple Instance Learning
MLN	Markov Logic Network
MLP	Multi-layer Perceptron
MMR	Maximal Marginal Relevance
MT	Machine Translation

NE	Named Entity
NIST	National Institute of Standards and Technology
NLG	Natural Language Generation
NLP	Natural Language Processing
NP	Noun Phrase
NT	Negative Threshold
NTCIR	National Institute of Informatics Test Collection for IR Systems?
NYT	New York Times
OT	Optimality Theory
POS	Part of Speech
PT	Positive Threshold
QA	Question Answering
RBF	Radial Basis Function
RDF	Resource Description Framework
RE	Relation Extractor
ROUGE	Recall-Oriented Understudy for Gisty Evaluation
RSS	Really Simple Syndication
RST	Rhetorical Structure Theory
SCU	Single Content Unit
SRL	Semantic Role Labelling
SU	Sentence-like Unit
SVD	Singular Value Decomposition
SVM	Support Vector Machine
TAC	Text Analysis Conference
TREC	Text Retrieval Conference
TS	Text Summarization
TSC	Text Summarization Challenge
UI	User interface
URL	Unified Resource Locator
WePS	Web People Search
WER	Word Error Rate
WIE	Web Information Extraction
WFST	Weighted Finite State Transducer
XML	eXtended Markup Language

Part I
Background and Fundamentals

Chapter 1
Automatic Text Summarization: Past, Present and Future

Horacio Saggion and Thierry Poibeau

Abstract Automatic text summarization, the computer-based production of condensed versions of documents, is an important technology for the information society. Without summaries it would be practically impossible for human beings to get access to the ever growing mass of information available online. Although research in text summarization is over 50 years old, some efforts are still needed given the insufficient quality of automatic summaries and the number of interesting summarization topics being proposed in different contexts by end users ("domain-specific summaries", "opinion-oriented summaries", "update summaries", etc.). This paper gives a short overview of summarization methods and evaluation.

1.1 Introduction

Automatic Text Summarization, the reduction of a text to its essential content, is a very complex problem which, in spite of the progress in the area thus far, poses many challenges to the scientific community. It is also a relevant application in todays information society given the exponential growth of textual information online and the need to promptly assess the contents of text collections.

Research in automatic summarization started to attract the attention of the scientific community in the late 1950s [35], when there was a particular interest

H. Saggion
Department of Information and Communication Technologies, Universitat Pompeu Fabra,
C/Tanger 122, Barcelona, 08018, Spain
e-mail: horacio.saggion@upf.edu

T. Poibeau (✉)
Laboratoire LaTTiCe-CNRS, École Normale Supérieure and Université Sorbonne-Nouvelle,
1 rue Maurice Arnoux, Montrouge, 92120, France
e-mail: thierry.poibeau@ens.fr

T. Poibeau et al. (eds.), *Multi-source, Multilingual Information Extraction
and Summarization 11*, Theory and Applications of Natural Language Processing,
DOI 10.1007/978-3-642-28569-1__1, © Springer-Verlag Berlin Heidelberg 2013

in the automation of summarization for the production of abstracts of technical documentation. The interest in the area declined for a few years until Artificial Intelligence started to show interest in the topic [11].

It has long been assumed that summarization presupposes to understand the input text, which means that an explicit (semantic) representation of the text must be calculated so as to be able to identify its essential content. Therefore, text summarization became an interesting application to test the understanding capabilities of artificial systems. However, the interest for this approach to text summarization decreased rapidly given the complexity of the task, and text understanding became an open research area in itself.

There was a resurge of interest in summarization in the nineties with the organization of a number of relevant scientific events [71] and a peak of interest from the year 2000 with the development of evaluation programs such as the Document Understanding Conferences (DUC) [48] and the Text Analysis Conferences (TAC) [49] in the United States. These frameworks will be further detailed in Sect. 1.4 dedicated to evaluation.

Two fundamental questions in text summarization are; (i) how to select the essential content of a document, and (ii) how to express the selected content in a condensed manner [69, 71]. Text summarization research has many times concentrated more on the product: the summary, and less on the cognitive basis of text understanding and production that underly human summarization. Some of the limitations of current systems would benefit from a better understanding of the cognitive basis of the task. However, formalizing the content of open domain documents is still a research issue, so most systems are only based on a selection of sentences from the set of original documents.

Where the transformation of the original text content into a summary is concerned, there are two main types of summaries: an *extractive* summary, a set of sentences from the input document and an *abstractive* summary (i.e., an abstract), a summary in which some of its material is not present in the input document [36]. Summaries can also be classified into *indicative* or *informative* depending on their intended purpose to *alert* or to *inform* respectively.

Early research in summarization concentrated on summarization of single documents (i.e., *single document summarization*), however in the current Web context, many approaches focus on summarization of multiple, related documents (i.e., *multi-document summarization*). Most summarization algorithms today aim at the generation of extracts given the difficulties associated to the automatic generation of well formed texts in arbitrary domains. It is generally accepted that there are a number of factors that determine the content to select from the source document and the type of output to produce [70], for example factors such as the audience (e.g., expert vs. non-expert reader) certainly influences the material to select.

This short introduction overviews some classic works on content selection and realization, summarization tools and resources, and summarization evaluation. For the interested reader, there are a number of papers and books that provide extensive overviews of text summarization systems and methods [23, 33, 36].

1.2 Overview of Summarization Methods

Most summarization today is based on a sentence-extraction paradigm where a list of features believed to indicate the relevance of a sentence in a document or set of documents [36] is used as text interpretation mechanism. The approaches to sentence selection can be driven by statistical information or based on some informed summarization theory, which takes into account linguistic and semantic information.

1.2.1 Superficial Techniques

Basic statistical approaches to content selection have relied on the use of frequency computation to identify relevant document keywords [35]. The basic mechanism used is to measure the frequency of each word in a document and to adjust the frequency of words with additional corpus evidence such as the inverse document frequency of the word in a general document collection. This helps boost the score of words which are not too frequent in a general corpus and moderate the score of otherwise too frequent general words. These methods assume that the relevance of a given concept in a text is proportional to the number of times the concept is "mentioned" in the document, assuming that each distinct word corresponds to a different concept. However, counting concept occurrences in text is not a trivial task given the presence of synonymy (e.g., "dog" and "puppy" could refer to the concept dog) and coreferential expressions (e.g., "Obama" and "the President" could refer to the same individual) which contribute to text cohesion. Once keywords are identified in the document, sentences containing those keywords could be selected using various sentence scoring and ranking mechanisms (e.g., the relevance of a sentence could be proportional to the number of keywords it contains). More sophisticated techniques than using simple word counts also exists: for example topic signatures were proposed in [32] as a way to model relevant keywords in particular domains and as interpretation and sentence relevance determination mechanism.

The position of sentences in text [12, 31] is also deemed indicative of sentence relevance. For example in news stories, the first or leading paragraph usually contains the main information about the event reported in the news, while the rest of the text gives details as well as background information about the event, therefore selecting sentences from the beginning of the text could be a reasonable strategy. However, in scientific texts the introduction usually gives background information, while the main developments are reported in the conclusions, so the position strategy needs to be adapted to the text type to be summarized.

Another superficial, easy-to-implement method consists in measuring the relevance of sentences by comparing them with the title of the document to be summarized, a sentence containing title words could be regarded as relevant and the more title words a sentence has the more relevant the sentence would be [12].

Yet another method looks at the presence of very specific cue-words or expressions in sentences [12, 50] such as "in conclusion", "to summarize", or "the objective" which could be regarded as pointing to relevant information out-of-context.

These now classical features are usually applied following an Edmundsonian empirical approach [12] or a machine learning framework. In the Edmundsonian approach a kind of linear combination of sentence features to score sentences is used. In a machine learning approach [26], sentence extraction is seen as sentence classification where sentences in the document have to be classified as either "summary sentence" or "non summary sentence". Superficial methods such as frequency, position, title, and cue-words are incorporated in one way or another in text summarization systems, even though the summaries they produce are far from those a human would produce.

Superficial techniques are also those based on graph representations [14, 43] which are so popular nowadays. These approaches compute sentence similarity values and use graph algorithms (e.g., random walks or PageRank) to weight word or sentence relevance to make decisions about sentence centrality. Graph-based approaches were also explored earlier for single document summarization of encyclopædic articles [68].

1.2.2 Knowledge-Based Approaches

Knowledge-rich approaches either extend basic methods by the incorporation of sophisticated, yet general lexical resources or apply discourse organization theories in generic or specific contexts. Few approaches bring domain knowledge to the summarization enterprise. One of the best known artificial intelligence approaches to summarization was DeJong's FRUMP system [11] which was based on text understanding and generation mechanisms. More specifically, the system was supposed to map chunks of text to rich conceptual structures such as sketchy scripts, which contained information about the key elements of specific story types (e.g., earthquakes, demonstrations). A main drawback of early artificial intelligence approaches was that they relied on manual, intensive coding of world knowledge which made such approaches impractical.

The situation is nowadays being corrected to some extent with studies into knowledge induction from text corpora. For example, in [28] clustering is applied to generate templates for specific entity types (actors, companies, etc.) and patterns are automatically produced that describe the information in the templates. In [8] narrative schemas are induced from corpora using coreference relations between participants in texts. Participants' roles are identified and typical verb sequences in stereotypical situations are extracted. Induction results of these approaches have still to be incorporated and tested in summarization systems.

Lexical cohesion has long been considered a key component in assessing content relevance in text summarization. In [5] cohesive links between sentences were identified, based on the use of a thesaurus, and the resulting linked structure was used to select sentences. In [2], lexical chains that connect related words were created based on WordNet relations [15], then sentences were selected depending on which chains sentences' words belong to.

A text is a complex linguistic unit, therefore many works rely on discourse structure or text organization theories for text interpretation and "sound" sentence selection. For example, in scientific domains, it is well known that texts follow a more or less predefined semantic or rhetorical organization [73]. Various works have identified categories of information such as "purpose", "method", "results", "conclusions" which make up the structure of scientific/technical documents and which are also present in research abstracts. In [29], a full model of summaries of empirical research was developed with 36 information types organized in a three-level hierarchy. The top of the hierarchy contains types of information which are very frequent in abstracts while the other two levels contain less frequent types of information. Automatic identification of some information types in textual input is essential to produce good quality abstracts. In [74], a Naïve Bayes machine learning algorithm was used to identify semantic information such as "Aim", "Background", "Contrast", etc. The method is similar to [26] although the problem is more challenging. In [63] a semantic dictionary and manually developed syntactic/conceptual patterns were used to identify sentences, instantiate templates, and generate indicative-informative abstracts.

In restricted domains, Information Extraction [16], the mapping of textual input into predefined template structures, can be used to extract information and then generate a summary. The approach has been applied in [51] where the objective was to produce short indicative informative abstracts (i.e., non-extractive summaries) and in [55] that uses templates instantiated from different articles referring to the same event to generate a coherent multi-document summary.

General discourse organization theories have also been used in summarization. Rhetorical Structure Theory (RST) [39] establishes that a text can be represented as a tree structure (i.e., rhetorical tree) linking text spans by using a set of predefined discourse relations. Because the rhetorical relations link text spans with different informational status (i.e., nucleus and satellite), text elements could be selected depending on their role in a relation. In [47] the rhetorical tree has been used to select sentences by pruning the tree while in [40] text units were promoted based on their status. However, analyzing discourse organization is still an open research issue and these kinds of techniques still need to be more robust in order to be more widely used.

1.2.3 Non-extractive Methods

In text summarization research, most of the attention has been paid to the problem of selecting information, whilst the problem of generating a new cohesive and coherent text has somehow received less attention. Non-extractive summarization in the current context refers to problems such as sentence compression, headline generation, cut-and-paste summarization, and sentence regeneration.

In headline generation [76] the objective is to produce a short title for a news story. This problem has been addressed following the statistical machine translation paradigm, where words have been selected from the document and then have been combined using a language model. In [77], two ways of generating headlines have been proposed, one based on the pruning of a syntactic tree which is supposed to generate well-formed headlines, a second one based on a Hidden Markov Model which is more robust.

Together with headline generation, sentence compression has received considerable attention, the objective here is for example to reduce sentence size by eliminating components which might be unnecessary. In [22], compression was carried out by removing sentence components using manually developed rules operating on syntactic trees which consider syntactic restrictions (e.g., obligatory verb arguments) and contextual information. In [25], a noisy-channel model which learns based on an aligned corpus how to translate uncompressed to compressed sentences. The method only dealt with removal of elements and used a very small corpus of compressed sentences, it was therefore extended in [75]. In professional abstracting settings [13], it was observed that in order to produce abstracts, textual fragments are usually combined to create new sentences and sometimes new linguistic material is included. Simulation of text abstracting operations have been implemented using rule-based and machine learning approaches in [21, 60, 61].

1.2.4 Multi-Document Summarization

A multi-document summary is a brief representation of the essential contents of a set of related documents. The relation between the documents can be of various types, for example documents can be related because they are about the same entity, or because they discuss the same topic, or because they are about the same event or the same event type.

Fundamental problems when dealing with multi-source input in summarization are the detection and reduction of redundancy as well as the identification of contradictory information. In [55], the multi-document summarization problem is studied in the context of multi-source information extraction in specific domains. Here templates instantiated from various documents are merged using specific operators aiming at detecting identical or contradictory information.

In an information retrieval context, where multi-document summaries are required for a set of documents retrieved from a search engine in response to the query, the Maximal Marginal Relevance (MMR) method [7] can be applied. The method scores text passages (e.g., paragraphs) iteratively taking into consideration the relevance of each passage to a user query and the redundancy of the passage with respect to summary content already selected. In the case of generic summarization, computing similarity between sentences and the centroid of the documents to summarize has resulted in competitive summarization solutions [54, 62]. Domain specific multi-document summarization techniques are explored in a chapter on summarizing images in this volume (cf. Chap. 14).

Sentence ordering is also an issue for multi-document summarization. In single-document summarization it is assumed that presenting the information in the order this information appears in the input document would generally produce an acceptable summary. By contrast, in multi-document summarization particular attention has to be paid to how sentences extracted from multiple sources are going to be presented. Various techniques exist for dealing with sentence ordering, for example if sentences are timestamped by publication date, then they could be presented in chronological order. However this is not always possible because recognizing the date of a reported event is not trivial and not all documents contain a publication date.

Sentence ordering can also be conceived as to represent the different topics to be addressed in the summary. For example, a clustering algorithm can be used to identify topics in the set of input documents and discover in what order the topics are presented in the input documents [3], this in turn could be used to present sentences in an order similar to that observed in the input set. A probabilistic approach to sentence ordering seeks to estimate the likelihood of a sequence of sentences. It tries to find a locally optimal order by learning ordering constraints for pairs of sentences [27]. An entity-grid model [1] tries to represent coherent texts by modelling entity roles (e.g., subject, object) in consecutive sentences, the model is able to discriminate coherent and incoherent texts. Advanced techniques improving on these methods are presented in this volume Chap. 12.

1.2.5 Multilingual Summarization

While most summarization research so far has been carried out for the English language probably because of the availability of data and evaluation resources, summarization in languages other than English is not rare.

Activities to promote summarization in Japanese have been undertaken in the Text Summarization Challenges [46], a series of evaluations of text summarization systems with tasks such as single document summarization and topic focused multi-document summarization. Summarization in a cross-lingual environment was studied in the 2001 Johns Hopkins research workshop [64] where evaluation resources and summarization algorithms for English and Chinese were developed [53]. The 2005 Multilingual Summarization Evaluation concentrated on summarization from mixed input in Arabic and English, where the challenge was to generate output from automatic translations [57].

Various research projects have produced multilingual summarization technology based on features already used for the English language: the SUMMARIST research project has produced a summarizer available for Korean and Spanish, the SweSum summarizer works for Scandinavian languages and the MUSE system is a language independent summarizer which has been tested for Arabic and Hebrew. The 2011 Text Analysis Conference in its text summarization evaluation program has included a pilot task on Multilingual Summarization which objective is to evaluate the application of language independent summarization algorithms. In the multilingual task ten clusters with ten documents (in Arabic, Czech, English, French, Greek, Hebrew, and Hindi) were produced for evaluation, and each system had to produce summaries in at least four different languages. Moreover, various researchers argue their summarizers to be language independent [21, 24, 42].

1.3 Summarization Resources

1.3.1 Text Summarization Tools

From a commercial point of view, there are a number of software products offering summarization capabilities (e.g., Microsoft Autosummarize option) as well as a number of companies offering standalone summarization applications (e.g., Copernic, Pertinence). From the research point of view, a few tool-kits exist, probably the best known text summarization software for research available today is MEAD [53] a set of Perl components for summarization of English as well as other languages such as Chinese. MEAD implements classical sentence relevance features such as position and term frequency, but also implements multi-document summarization features such as centroid. MEAD has been used many times for comparison purposes in text summarization research. SUMMA [58, 59] is another available tool which relies on the GATE framework [41] for document processing and text representation. It is based mainly on statistical approaches such as frequency computation, cue-word identification, position, title, centroid, etc. It also exploits the vector space model to represent documents, sets of documents, sentences, and other textual units to be able to compute text relevance based on similarity measures between text units. SUMMA is implemented in Java and used within GATE as a plug-in or as a Java library for standalone applications. Both MEAD and SUMMA systems need to be adjusted for optimal performance therefore requiring training data to set up parameters.

1.3.2 Text Summarization Datasets

Data is fundamental in any scientific activity, and it is of paramount relevance in text summarization. Without data it will be impossible to formulate working hypotheses, verify them, and adjust system parameters. Over the past few years, various datasets have been produced for the study of text summarization, among them datasets pertaining to the various evaluation frameworks we mention in Sect. 1.4 (mainly The Document Understanding Conference and the Text Analysis Conference).

SummBank is a dataset of parallel English and Chinese documents containing, in addition to source documents, multi-document summaries for sets of related stories (i.e., 40 clusters), relative utility judgements for sets of sentences (e.g., an indication of how valuable the sentence is for a summary), and automatic summaries [65]; it has been used in large scale evaluation experiments [56]. The CAST corpus [18] contains a set of documents where each sentence and sentence fragment has been annotated as either essential or non essential; this kind of dataset could be of help for developing sentence selection and sentence reduction algorithms. The Ziff-Davis corpus [17] which has been partially used for experiments in cut-and-paste summarization [20] and sentence reduction [25], contains newspaper articles and their human written abstracts. The abstracts were automatically sentence aligned to their source documents therefore being of value for studying non-extractive summarization.

1.4 Evaluation

A good summary must be easy to read and give a good overview of the content of the source text. Since summaries tend to be more and more oriented towards specific needs, it is necessary to tune existing evaluation methods accordingly. Unfortunately these needs do not give a clear basis for evaluation and the definition of what is a good summary remains to a large extent an open question.

Therefore, the evaluation of human or automatic summaries is known to be a difficult task. It is difficult for humans, which means the automation of the task is even more challenging and hard to assess. However, because of the importance of the research effort in automatic summarization, a series of proposals have been made to partially or fully automate the evaluation [38, 72]. It is also useful to note that in most cases automatic evaluations already correlate positively with human evaluations.

1.4.1 Evaluating Automatically Produced Summaries

A series of evaluation campaigns have been organized since the late 1990 in the US, which provided a forum for evaluation and discussion. These campaigns are essentially SUMMAC (1996–1998) [37], DUC (the Document Understanding Conference, 2000–2007) [48] and more recently TAC (the Text Analysis Conference, 2008-) [45]. Evaluation in these conferences is based on human as well as automatic scoring of the summaries proposed by participants. Therefore, these conferences have played a major role in the design of evaluation measures; they also play a role in the meta-evaluation of scoring methods since it is possible to check to what extend the scores obtained automatically correlate with human judgements.

Broadly, one can say that there are three main difficulties in the automatic evaluation of summaries: (i) it is necessary to determine what are the most important pieces of information that should be kept from the initial text; (ii) evaluators must be able to automatically recognize these pieces of information in the candidate summary since this information can be expressed using various expressions; (iii) lastly, the readability (including grammaticality and coherence) of the summary should be evaluated.

In this section we mainly refer to evaluation methods that can be applied to extractive summaries. These summaries are made of extracts from the original text, which means it is possible to evaluate their quality by comparing their content (i.e. sequences of words) to reference summaries. If this comparison is not possible (for example in the case of abstractive summaries), then manual methods are the only reliable solution within the current state of the art.

Even for extractive summaries, evaluation methods range from purely manual approaches to purely automatic ones, and there are of course a lot of possibilities in between. Manual approaches refer to methods where a human evaluates a candidate summary from different points of view, for example coverage, grammaticality or style; this kind of evaluation is necessary but is known to be highly subjective. Automatic approaches compare segments of texts from the candidate summary with one or several reference abstracts; this approach is easy to reproduce but cannot be applied when the system uses reformulation techniques. Mixed approaches allow one to manually analyse and score the most important pieces of information and rank the candidate summaries according to these (the most important pieces of information must be contained in the candidate summary, independently of their linguistic formulation).

A lot of diverse approaches for summary evaluation have been proposed in the last two decades. This prevents us from being comprehensive. We will instead focus on three methods that have been widely used during recent evaluation campaigns (esp. during the last Text Analysis Conferences organized by NIST [10]): ROUGE (a fully automatic method), PYRAMID (a mixed method) and a series of indicators resulting from a manual evaluation. Lastly, we will consider a recent body of research aiming at evaluating summaries with no human reference.

1.4.2 Manual Methods

The most obvious and simple way to evaluate a summary is to have assessors evaluating its quality. For example for DUC, the judges had to evaluate the coverage of the summary, which means they had to give a global score assessing to what extent the candidate summary covers the text given as input. In more recent frameworks, and especially in TAC, query-oriented summaries have to be produced: judges then have to evaluate the responsiveness of the summary, that is to say to what extent a given summary answers the query given in input.

Manual evaluation can also provide some indicators to assess the quality and readability of a text. A good summary is supposed to be:

- Syntactically accurate;
- Semantically coherent;
- Logically organized
- Without redundancy.

These different points are too complex to be fully automatically calculated, especially semantic coherence and logical organization. In order to get a reliable evaluation of these different aspects, it is necessary to get human judgements. For the DUC [48] and TAC [10] campaigns, human experts had to give different scores to each candidate summary, using the following indicators:

- Grammaticality;
- Non redundancy;
- Focus (integration of the most important pieces of information of the original text);
- Structure and coherence.

Experts had to give a grade between 0 (void) and 10 (perfect) for each of these indicators. For TAC 2009, reference summaries written by human experts got an average grade of 8.8/10 (TAC did not provide the figure for each of the criteria in isolation). Thus, this grade can be seen as the upper bound score reachable by candidate summaries.

1.4.3   Automatic Methods Based on a Comparison with a Manual Reference

Since the early 2000s, a series of measures have been proposed to automate the evaluation of summaries. Most of these measures are based on a direct comparison of the produced and the reference summaries [56, 66]. We detail here the two most popular measures: Rouge and Pyramid.

1.4.3.1   ROUGE

The ROUGE measures (*Recall-Oriented Understudy for Gisty Evaluation*) have been introduced by Lin [30]. These measures are based on the comparison of n-grams (i.e. a sequence of n elements) between the candidate summary (the summary to be evaluated) and one or several reference summaries.[1] Most of the time, several

---

[1] Rouge was inspired by BLEU, a measure used in the evaluation of machine translation also based on the comparison of n-grams [52].

reference summaries are used for comparison, which allows more flexibility and a fairer evaluation. There are several variants of ROUGE and we present the most widely used of them below.

- **ROUGE-n**: This measure is based on a simple comparison of n-grams (most of the time a sequence of two or three elements, more rarely 4). A series of n-grams (hence series of sequences of n consecutive words) is extracted from the reference summaries and the candidate summary. The score is the ratio between the number of common n-grams (between the candidate summary and the reference summary) and the number of n-grams extracted from the reference summary only.
- **ROUGE-L**: In order to overcome some of the shortcomings of ROUGE-n, more precisely the fact that the measure may be based on too small sequences of text, ROUGE-L takes into account the longest common sequence between two sequences of text divided by the the length of one of the text. Even if this method is more flexible than the previous one, it still suffers from the fact that all n-grams have to be continuous.
- **ROUGE-SU**: Skip-bigram and uni-gram ROUGE takes into account bigrams as well as unigrams. However, the bi-grams, instead of being just continuous sequences of words, allows insertions of words between their first and last element. The maximal distance between the two elements of the bi-gram corresponds to a parameter (n) of the measure (often, the measure is instantiated with $n = 4$). During TAC 2008, it has been shown that ROUGE-SUn was the most correlated measure with human judgements.

ROUGE has been very useful for comparing different summaries based on extractive methods, but the use of n-grams only is a strong limitation since it requires an exact match of different units of text. More recently, other evaluation measures have been developed to better capture the semantics of texts.

1.4.3.2   Pyramid

As we have seen, ROUGE measures are based on the discovery of perfect matches between some sequences of the candidate summary and some of the reference summaries. These methods are thus inefficient if the candidate summary has been produced using reformulation techniques. PYRAMID [44] is supposed to overcome some of these issues.

First, the most important pieces of information to be included in reference summaries are extracted.[2] This can be done by a group of experts or directly by analyzing reference summaries. These pieces of information are called SCU (*Summarization Content Units*). They are then weighted: if a SCU only appears in one of the reference summaries it will get a low score, but if it is included in all

---

[2]Hovy et al. [19] also proposed an approach of this kind with the notion of *Basic Units*.

the reference summaries, it will get a high score. The analogy with a pyramid comes here: the result of this process is a lot of SCU with a low score as the basis of the pyramid, and a few SCU with a high score at its top.

A list of linguistic expressions is associated with each SCU. It is then possible to map sequences of text with SCU. The last step is then obvious: the evaluation identifies all the SCU contained in the candidate summary and calculates a score for the candidate summary, based on the number and the weight of the SCU it contains.

This evaluation is more precise than ROUGE but requires some manual work to identify the SCU, associate linguistic expressions to them and calculate the weights necessary for the evaluation. However, according to TAC 2008, this method seems to better correlate with human judgements than ROUGE probably because it takes into account some of the semantics of the text.

1.4.4   Automatic Evaluation with no Manual Reference

Even if the previous automatic methods have proven useful for evaluation, they require that reference summaries be available. Several authors have observed that this is not always the case and reference summaries are costly to produce when they are not directly available. Moreover, even when reference summaries are available, the definition and weighting of SCU remains a difficult and time-consuming task. On the other hand, the input text used to produce the summary is of course always available and contains valuable information to evaluate the summaries derived from it.

The fact that information from the original text can be used to directly evaluate summaries with no manual reference was first explored in [34]. The authors proposed to use four classes of easily computable features that are supposed to capture aspects of the input text: input and output text size (coverage is generally proportional to the length of the summary), information-theoretic properties of the distribution of words in the input, presence of descriptive (topic) words and similarity between the documents in multi-document inputs. Then they directly compared a set of candidate summaries using these features and the Jensen-Shannon measure. The authors showed on different tasks (query-focused and update summaries, using different test sets from different evaluation campaigns) that their approach correlates favourably with Pyramid.

Finally, [67] showed that this method is effective in most cases but not always: they specifically pointed out that the method does not perform so well for biographical information or for the summarization of opinions. They propose a new method, mainly based on n-grams, skip bi-grams and the Jensen-Shannon measure. Further, their approach provides interesting results on different summarization tasks for different languages.

1.5   Conclusion and Perspectives

This introductory chapter provided a quick insight into recent trends in automatic summarization methods. However, despite more than 50 years of research, there is still room for improvement.

The most effective and versatile methods used so far in automatic summarization rely on extractive methods: they aim at selecting the most relevant sentences from the collection of original documents in order to produce a condensed text rendering important pieces of information. As we have seen, these kinds of methods are far from being ideal: in multi-document summarization, the selection of sentences from different documents leads to redundancy, which in turn must be eliminated. Moreover, most of the time only a part of a sentence is relevant, but extracting only sub-sentences is still far from being operational. Lastly, extracting sentences from different documents may produce an inconsistent and/or hard-to-read summary.

These limitations suggest a number of desirable improvements. We detail here three very active research trends, some of them being illustrated in the papers in this volume.

- In order to overcome the redundancy problem, researchers are actively working on a better representation of the text content and, more interestingly, are now trying to provide summaries tailored towards specific user needs. Evaluation tasks proposed in TAC reflect this trend since recent evaluations concerned, among other things, the production of opinion-based summaries and the production of update summaries (where only new information should be selected), see for example [10]. In this new context, even if the Edmundsonian approach (based on the recognition of cues and key phrases from the set of original documents) is still widely used, the integration of new methods and new modules has proven necessary (e.g., the integration of an opinion mining module for opinion-based summarization [6]), leading to a more fine grained representation of the original set of documents.
- As explained above, sentence compression is another very active domain of research. Rather than just selecting relevant sentences from original documents, sentence compression aims at keeping only the part of the sentence that is really meaningful and of interest for the abstract. Most of the time, compression rules are defined manually [22] even if recent experiments tried to automatically learn these rules from a set of examples [9]. The improvement brought by this method in readability and summarization quality still needs to be assessed: most approaches in sentence compression require an accurate analysis of the input sentence in order to provide reliable results, which is not always possible given the state of the art in parsing. So, sentence compression is a promising source of improvement but its application still needs to be validated
- One of the main drawback of extractive methods is often the lack of readability of the text produced. In most systems, sentence ordering is based on simple heuristics (e.g. location of the sentence in the original documents) that are not enough to produce a coherent text. Recent research aims at finding new methods

for producing more coherent texts. For example [4] suggest that calculating the local coherence between candidate sentences may lead to more readable summaries. Developing a global document model may also help.

Finally, the evaluation of automatic summarization is still an open issue. Recent automatic methods (like Pyramid, see above) have proven more consistent that human-based evaluation. However, the lack of consensus between humans when evaluating summaries is a real problem. The development of more focused summaries may lead to a more consistent evaluation and to a better convergence between human and automatic evaluation methods, which is highly desirable. However, automatic summarization evaluation is still a very promising research area with many challenges ahead.

The tenet of this research goes a lot further than just evaluation, since evaluating a summary involves having an accurate description of the content of the documents. Hence, automatic summarization is not *just* another natural language application: it raises important issues related to artificial intelligence and cognitive science. Beyond practical applications, research in the domain may lead to a better understanding of human comprehension.

Acknowledgements Horacio Saggion is grateful to a fellowship from Programa Ramón y Cajal, Ministerio de Ciencia e Innovación, Spain. Thierry Poibeau is supported by the "Empirical Fundations of Linguistics" labex, Sorbonne-Paris-Cité. We acknowledge the support from the editors of this volume.

References

1. Barzilay, R.: Modeling local coherence: an entity-based approach. In: Proceedings of ACL 2005, Michigan, pp. 141–148. Association for Computational Linguistics, Stroudsburg (2005)
2. Barzilay, R., Elhadad, M.: Using lexical chains for text summarization. In: Proceedings of the ACL/EACL'97 Workshop on Intelligent Scalable Text Summarization, Madrid, pp. 10–17 (1997)
3. Barzilay, R., Elhadad, N., Mckeown, K.R.: Inferring strategies for sentence ordering in multidocument news summarization. J. Artif. Intell. Res. **17**, 2002 (2002)
4. Barzilay, R., Lapata, M.: Modeling local coherence: an entity-based approach. Comput. Linguist. **34**(1), 1–34 (2008)
5. Benbrahim, M., Ahmad, K.: Text summarisation: the role of lexical cohesion analysis. In: The New Review of Document and Text Management, pp. 321–335. Taylor Graham Pub., London, UK (1995)
6. Bossard, A., Généreux, M., Poibeau, T.: Cbseas, a summarization system – integration of opinion mining techniques to summarize blogs. In: Proceedings of the 12th Meeting of the European Association for Computational Linguistics (system demonstration), EACL '09, Athens. Association for Computational Linguistics, Stroudsburg (2009)
7. Carbonell, J.G., Goldstein, J.: The use of MMR, diversity-based reranking for reordering documents and producing summaries. In: Research and Development in Information Retrieval, pp. 335–336. The Association for Computing Machinery, New York (1998)
8. Chambers, N., Jurafsky, D.: Unsupervised learning of narrative schemas and their participants. In: ACL/AFNLP, Singapore, pp. 602–610. Association for Computational Linguistics, Stroudsburg (2009)

9. Cohn, T., Lapata, M.: Sentence compression as tree transduction. J. Artif. Intell. Res. (JAIR) **34**, 637–674 (2009)
10. Dang, H.T., Owczarzak, K.: Overview of the tac 2008 opinion question answering and summarization tasks. In: Proceedings of the TAC 2008 Workshop, Notebook Papers and Results, Gaithersburg, MD, USA. NIST, Gaithersburg, MD, USA (2008)
11. DeJong, G.: An overview of the FRUMP system. In: Lehnert, W., Ringle, M. (eds.) Strategies for Natural Language Processing, pp. 149–176. Lawrence Erlbaum Associates, Hillsdale (1982)
12. Edmundson, H.: New methods in automatic extracting. J. Assoc. Comput. Mach. **16**(2), 264–285 (1969)
13. Endres-Niggemeyer, B.: SimSum: an empirically founded simulation of summarizing. Inf. Process. Manag. **36**, 659–682 (2000)
14. Erkan, G., Radev, D.: Lexrank: graph-based lexical centrality as salience in text summarization. J. Artif. Intell. Res. (JAIR) **22**, 457–479 (2004)
15. Fellbaum, C. (ed.): WordNet: An Electronic Lexical Database. MIT, Cambridge (1998)
16. Grishman, R.: Information extraction: techniques and challenges. In: Pazienza, M.T. (ed.) Information Extraction. A Multidisciplinary Approach to an Emerging Information Technology. Lecture Notes in Artificial Intelligence, vol. 1299. Springer, Berlin/New York (1997)
17. Harman, D., Liberman, M.: Tipster Complete. Technical Report, University of Pennsylvania, Philadelphia, USA (1993)
18. Hasler, L., Orãsan, C., Mitkov, R.: Building better corpora for summarisation. In: Proceedings of Corpus Linguistics, Lancaster, pp. 309–319 (2003)
19. Hovy, E., Lin, C.Y., Zhou, L., Fukumoto, J.: Automated summarization evaluation with basic elements. In: Proceedings of the Fifth Conference on Language Resources and Evaluation (LREC), Genoa, Italy. ELDA, Paris, France (2006)
20. Jing, H.: Using hidden markov modeling to decompose human-written summaries. Comput. Linguist. **28**, 527–543 (2002)
21. Jing, H., McKeown, K.: The decomposition of human-written summary sentences. In: Hearst, M., Gey, F., Tong, R. (eds.) Proceedings of SIGIR'99 – 22nd International Conference on Research and Development in Information Retrieval, University of California, Berkeley, pp. 129–136 (1999)
22. Jing, H., McKeown, K.: Cut and paste based text summarization. In: Proceedings of the 1st Meeting of the North American Chapter of the Association for Computational Linguistics, Seattle, pp. 178–185. Association for Computational Linguistics, Stroudsburg (2000)
23. Jones, K.S.: Automatic summarising: the state of the art. Inf. Process. Manage. **43**(6), 1449–1481 (2007)
24. Kabadjov, M.A., Atkinson, M., Steinberger, J., Steinberger, R., der Goot, E.V.: Newsgist: a multilingual statistical news summarizer. In: ECML/PKDD (3), Barcelona, pp. 591–594. Springer, Berlin/New York (2010)
25. Knight, K., Marcu, D.: Statistics-based summarization – step one: sentence compression. In: Proceedings of the 17th National Conference of the American Association for Artificial Intelligence, Austin. AAAI, Palo Alto, CA, USA (2000)
26. Kupiec, J., Pedersen, J., Chen, F.: A trainable document summarizer. In: Proceedings of the 18th ACM-SIGIR Conference, Seattle, pp. 68–73. ACM, New York (1995)
27. Lapata, M.: Probabilistic text structuring: experiments with sentence ordering. In: Proceedings of the 41st Meeting of the Association of Computational Linguistics, Sapporo, pp. 545–552. Association for Computational Linguistics, Stroudsburg (2003)
28. Li, P., Jiang, J., Wang, Y.: Generating templates of entity summaries with an entity-aspect model and pattern mining. In: Proceedings of ACL, Uppsala. Association for Computational Linguistics, Uppsala (2010)
29. Liddy, E.D.: The discourse-level structure of empirical abstracts: an exploratory study. Inf. Process. Manag. **27**(1), 55–81 (1991)
30. Lin, C.Y.: Rouge: a package for automatic evaluation of summaries. In: Proceedings of the Workshop on Text Summarization Branches Out (WAS 2004), Barcelona (2004)

31. Lin, C., Hovy, E.: Identifying topics by position. In: Fifth Conference on Applied Natural Language Processing, Washington, DC, pp. 283–290. Association for Computational Linguistics, Stroudsburg (1997)
32. Lin, C.Y., Hovy, E.: The automated acquisition of topic signatures for text summarization. In: Proceedings of the COLING Conference, Saarbrumlcken. Association for Computational Linguistics, Stroudsburg (2000)
33. Lloret, E., Palomar, M.: Text summarisation in progress: a literature review. Artif. Intell. Rev. **37**(1), 1–41 (2011)
34. Louis, A., Nenkova, A.: Automatically evaluating content selection in summarization without human models. In: Proceedings of EMNLP'09, Singapore, pp. 306–314. Association for Computational Linguistics, Stroudsburg (2009)
35. Luhn, H.P.: The automatic creation of literature abstracts. IBM J. Res. Dev. **2**(2), 159–165 (1958)
36. Mani, I.: Automatic Text Summarization. John Benjamins, Amsterdam/Philadelphia (2001)
37. Mani, I., Klein, G., House, D., Hirschman, L., Firmin, T., Sundheim, B.: Summac: a text summarization evaluation. Nat. Lang. Engin. **8**, 43–68 (2002). DOI 10.1017/S1351324901002741. URL http://portal.acm.org/citation.cfm?id=973860.973864
38. Mani, I., Maybury, M.T.: Advances in Automatic Text Summarization. MIT, Cambridge (1999)
39. Mann, W., Thompson, S.: Rhetorical structure theory: towards a functional theory of text organization. Text **8**(3), 243–281 (1988)
40. Marcu, D.: From discourse structures to text summaries. In: The Proceedings of the ACL'97/EACL'97 Workshop on Intelligent Scalable Text Summarization, Madrid, pp. 82–88 (1997)
41. Maynard, D., Tablan, V., Cunningham, H., Ursu, C., Saggion, H., Bontcheva, K., Wilks, Y.: Architectural elements of language engineering robustness. J. Nat. Lang. Engin. Spec. Issue Robust Methods Anal. Nat. Lang. Data **8**(2/3), 257–274 (2002)
42. Mihalcea, R.: Language independent extractive summarization. In: AAAI, Pittsburgh, Pennsylvania, pp. 1688–1689. Association for Computational Linguistics, Stroudsburg (2005)
43. Mihalcea, R., Tarau, P.: TextRank: Bringing order into texts. In: Proceedings of EMNLP-04and the 2004 Conference on Empirical Methods in Natural Language Processing, Barcelona (2004)
44. Nenkova, A., Passonneau, R., McKeown, K.: The pyramid method: incorporating human content selection variation in summarization evaluation. ACM Trans. Speech Lang. Process. **4**(2), 1–23 (2007)
45. Hoa Trang Dang (ed.): NIST: Proceedings of the Text Analysis Conference. NIST, Gaithesburg (2008)
46. Okumura, M., Fukusima, T., Nanba, H., Hirao, T.: Text summarization challenge 2 text summarization evaluation at ntcir workshop 3. SIGIR Forum **38**(1), 29–38 (2004)
47. Ono, K., Sumita, K., Miike, S.: Abstract generation based on rhetorical structure extraction. In: Proceedings of the International Conference on Computational Linguistics, Kyoto, Japan, pp. 344–348. ACL, Stroudsburg, USA (1994)
48. Over, P., Dang, H., Harman, D.: DUC in context. Inf. Process. Manag. **43**, 1506–1520 (2007). DOI 10.1016/j.ipm.2007.01.019. URL http://portal.acm.org/citation.cfm?id=1284916.1285157
49. Owczarzak, K., Dang, H.: Overview of the tac 2010 summarization track. In: Proceedings of TAC 2010, NIST, Gaithersburg, MD, USA (2010)
50. Paice, C.D.: Constructing literature abstracts by computer: technics and prospects. Inf. Process. Manag. **26**(1), 171–186 (1990)
51. Paice, C.D., Oakes, M.P.: A Concept-Based Method for Automatic Abstracting. Technical Report 27, Library and Information Commission, Wetherby (1999)
52. Papineni, K., Roukos, S., Ward, T., Zhu, W.J.: Bleu: a method for automatic evaluation of machine translation. In: Proceedings of the 40th Annual Meeting on Association for Computational Linguistics, ACL '02, Philadelphia, pp. 311–318 (2002)
53. Radev, D., Allison, T., Blair-Goldensohn, S., Blitzer, J., Çelebi, A., Dimitrov, S., Drabek, E., Hakim, A., Lam, W., Liu, D., Otterbacher, J., Qi, H., Saggion, H., Teufel, S., Topper,

M., Winkel, A., Zhang, Z.: MEAD — A platform for multidocument multilingual text summarization. In: Conference on Language Resources and Evaluation (LREC), Lisbon (2004)
54. Radev, D.R., Jing, H., Budzikowska, M.: Centroid-based summarization of multiple documents: sentence extraction, utility-based evaluation, and user studies. In: ANLP/NAACL Workshop on Summarization, Seattle (2000)
55. Radev, D.R., McKeown, K.R.: Generating natural language summaries from multiple on-line sources. Comput. Linguist. **24**(3), 469–500 (1998)
56. Radev, D.R., Teufel, S., Saggion, H., Lam, W., Blitzer, J., Qi, H., Çelebi, A., Liu, D., Drabek, E.: Evaluation challenges in large-scale document summarization. In: Proceedings of the 41st Annual Meeting on Association for Computational Linguistics - Vol. 1, ACL '03, Sapporo, Japan, pp. 375–382. ACL, Stroudsburg, USA (2003)
57. Saggion, H.: Multilingual multidocument summarization tools and evaluation. In: Proceedings of LREC 2006, Genoa, Italy. ELDA, Paris, France (2006)
58. Saggion, H.: Experiments on semantic-based clustering for cross-document coreference. In: Proceedings of the Third Joint International Conference on Natural Language Processing, AFNLP, Hyderabad, pp. 149–156 (2008)
59. Saggion, H.: SUMMA: a robust and adaptable summarization tool. Traitement Automatique des Langues **49**(2), 103–125 (2008)
60. Saggion, H.: A classification algorithm for predicting the structure of summaries. In: UCNLG+Sum '09: Proceedings of the 2009 Workshop on Language Generation and Summarisation, pp. 31–38. Association for Computational Linguistics, Morristown (2009)
61. Saggion, H.: Learning predicate insertion rules for document abstracting. In: CICLing, Tokyo, pp. 301–312. Springer, Berlin/New York (2011)
62. Saggion, H., Gaizauskas, R.: Multi-document summarization by cluster/profile relevance and redundancy removal. In: Proceedings of the Document Understanding Conference 2004, NIST, Boston (2004)
63. Saggion, H., Lapalme, G.: Generating indicative-informative summaries with sumUM. Comput. Linguist. **28**, 497–526 (2002)
64. Saggion, H., Radev, D., Teufel, S., Lam, W.: Meta-evaluation of summaries in a cross-lingual environment using content-based metrics. In: Proceedings of COLING 2002, Taipei, pp. 849–855. Association for Computational Linguistics, Stroudsburg (2002)
65. Saggion, H., Radev, D., Teufel, S., Wai, L., Strassel, S.: Developing infrastructure for the evaluation of single and multi-document summarization systems in a cross-lingual environment. In: LREC 2002, Las Palmas, pp. 747–754 (2002)
66. Saggion, H., Teufel, S., Radev, D., Lam, W.: Meta-evaluation of summaries in a cross-lingual environment using content-based metrics. In: Proceedings of the 19th international conference on Computational linguistics - Vol. 1, COLING '02, Taipei, pp. 1–7. Association for Computational Linguistics, Stroudsburg (2002)
67. Saggion, H., Torres-Moreno, J.M., da Cunha, I., SanJuan, E., Velazquez-Morales, P.: Multilingual summarization evaluation without human models. In: In Proceedings of COLING, Beijing (2010)
68. Salton, G., Allan, J., Singhal, A.: Automatic text decomposition and structuring. Inf. Process. Manag. **32**(2), 127–138 (1996)
69. Sparck Jones, K.: What might be in a summary? In: K. Knorz, Womser-Hacker (eds.) Information Retrieval 93: Von der Modellierung zur Anwendung (1993)
70. Sparck Jones, K.: Automatic summarizing: factors and directions. In: Mani, I., Maybury, M. (eds.) Advances in Automatic Text Summarization. MIT, Cambridge (1999)
71. Sparck Jones, K., Endres-Niggemeyer, B.: Automatic summarizing. Inf. Process. Manag. **31**(5), 625–630 (1995)
72. Spärck Jones, K., Galliers, J.R.: Evaluating Natural Language Processing Systems. Springer, Berlin (1996)
73. Swales, J.: Genre Analysis: English in Academic and Research Settings. Cambridge University Press, Cambridge (1990)

74. Teufel, S., Moens, M.: Argumentative classification of extracted sentences as a first step towards flexible abstracting. In: Mani, Maybury, M. (eds.) Advances in Automatic Text Summarization, pp. 155–171. MIT, Cambridge (1999)
75. Turner, J., Charniak, E.: Supervised and Unsupervised Learning for Sentence Compression. In: ACL, Michigan, Ann Arbor, USA. ACL, Stroudsburg, USA (2005)
76. Witbrock, M.J., Mittal, V.O.: Ultra-summarization: a statistical approach to generating highly condensed non-extractive summaries. In: In SIGIR99, Berkeley, pp. 315–316. ACM, New York (1999)
77. Zajic, D., Dorr, B., Lin, J., Schwartz, R.: Multi-candidate reduction: sentence compression as a tool for document summarization tasks. In: Information Processing and Management Special Issue on Summarization, p. 43. Elsevier, Amsterdam, The Netherlands (2007)

Chapter 2
Information Extraction: Past, Present and Future

Jakub Piskorski and Roman Yangarber

Abstract In this chapter we present a brief overview of Information Extraction, which is an area of natural language processing that deals with finding factual information in free text. In formal terms, *facts* are structured objects, such as database records. Such a record may capture a real-world entity with its attributes mentioned in text, or a real-world event, occurrence, or state, with its arguments or actors: who did what to whom, where and when. Information is typically sought in a particular target setting, e.g., corporate mergers and acquisitions. Searching for specific, targeted factual information constitutes a large proportion of all searching activity on the part of information consumers. There has been a sustained interest in Information Extraction for over two decades, due to its conceptual simplicity on one hand, and to its potential utility on the other. Although the targeted nature of this task makes it more tractable than some of the more open-ended tasks in NLP, it is replete with challenges as the information landscape evolves, which also makes it an exciting research subject.

2.1 Introduction

The recent decades witnessed a rapid proliferation of textual information available in digital form in a myriad of repositories on the Internet and intranets. A significant part of such information—e.g., online news, government documents, corporate reports, legal acts, medical alerts and records, court rulings, and social media

J. Piskorski
Institute for Computer Science, Polish Academy of Sciences, Warsaw, Poland
e-mail: Jakub.Piskorski@ipipan.waw.pl

R. Yangarber (✉)
Department of Computer Science, University of Helsinki, Finland
e-mail: Roman.Yangarber@cs.helsinki.fi

T. Poibeau et al. (eds.), *Multi-source, Multilingual Information Extraction and Summarization 11*, Theory and Applications of Natural Language Processing, DOI 10.1007/978-3-642-28569-1__2, © Springer-Verlag Berlin Heidelberg 2013

"Three bombs have exploded in north-eastern Nigeria, killing 25 people and wounding 12 in an attack carried out by an Islamic sect. Authorities said the bombs exploded on Sunday afternoon in the city of Maiduguri."

⇓

TYPE:	Crisis
SUBTYPE:	Bombing
LOCATION:	Maiduguri
DEAD-COUNT:	25
INJURED-COUNT:	12
PERPETRATOR:	Islamic sect
WEAPONS:	bomb
TIME:	Sunday afternoon

Fig. 2.1 Example of automatically extracted information from a news article on a terrorist attack

communication—is transmitted through *unstructured*, free-text documents and is thus hard to search in. This resulted in a growing need for effective and efficient techniques for analyzing free-text data and discovering valuable and relevant knowledge from it in the form of *structured* information, and led to the emergence of Information Extraction technologies.

The task of Information Extraction (IE) is to identify a predefined set of concepts in a specific domain, ignoring other irrelevant information, where a domain consists of a corpus of texts together with a clearly specified information need. In other words, IE is about deriving structured factual information from unstructured text. For instance, consider as an example the extraction of information on violent events from online news, where one is interested in identifying the main actors of the event, its location and number of people affected. Figure 2.1 shows an example of a text snippet from a news article about a terrorist attack and a structured information derived from that snippet. The process of extracting such structured information involves identification of certain small-scale structures like noun phrases denoting a person or a person group, geographical references and numeral expressions, and finding semantic relations between them. However, in this scenario some domain-specific knowledge is required (understanding the fact that terrorist attacks might result in people being killed or injured) in order to correctly aggregate the partially extracted information into a structured form.

Even in a limited domain, IE is a non-trivial task due to the complexity and ambiguity of natural language. There are many ways of expressing the same fact, which can be distributed across multiple sentences [26], documents, or even knowledge repositories. Further, a significant amount of relevant information might be implicit, which may be difficult to discern and an enormous amount of background knowledge is needed to infer the meaning of unrestricted natural language. However, the scope of IE is narrower than the scope of full text understanding—computing all possible interpretations and grammatical relations in natural language text—whose realization is, as of today, still impossible from the technical point of view. As a consequence, the use of considerably less sophisticated linguistic analysis tools for solving IE tasks may be beneficial, since it might be sufficient for the extraction and assembly of relevant pieces of information and it requires less knowledge engineering. In particular, recent advances in the field of Natural Language Processing (NLP) in robust, efficient and high-coverage shallow text processing techniques, as opposed to deep linguistic analysis, have contributed to the spread of deployment of IE techniques in real-world applications for processing of vast amount of textual data.

Information Extraction has not received as much attention as Information Retrieval (IR) and is often confounded with the latter. The task of IR is to select from a collection of textual documents a subset which are relevant to a particular query, based on key-word search and possibly augmented by the use of a thesaurus. The IR process usually returns a ranked list of documents, where the rank corresponds to the relevance score that the system assigned to the document in response to the query. However, the ranked document list does not provide any detailed information on the content of those documents. The goal of IE is not to rank or select documents, but to extract from the documents salient facts about pre-specified types of events, entities, or relationships, in order to build more meaningful, rich representations of their semantic content, which can be used to populate databases that provide structured input for mining more complex patterns (e.g., trends, summaries) in text collections. To sum up, IE aims to process textual data collections into a shape that facilitates searching and discovering knowledge in such collections.

IE systems are in principle more difficult and knowledge-intensive to build than IR systems. However, IE and IR techniques can be seen as complementary and can potentially be combined in various ways. IR is often used in IE for pre-filtering a very large document collection to a manageable subset, to which IE techniques could be applied. Alternatively, IE could be used as a subcomponent of an IR system to identify structures for intelligent document indexing.

Recent advances in IE provide dramatic improvements in the conversion of the flow of raw textual information into structured data and are increasingly being deployed in commercial application, for example in the financial, medical, legal, and security domains. Furthermore IE can constitute a core component technology in many other NLP applications, such as Machine Translation, Question Answering, Text Summarization, Opinion Mining, etc.

This chapter provides a brief survey of Information Extraction, including past advances, current trends and future challenges, and is organized as follows. Section 2.2 introduces basic definitions, typical IE tasks, evaluation of IE systems, and IE competitions and challenges. Section 2.3 presents a generic architecture of an IE system. Early IE systems and approaches are addressed in Sect. 2.4. An overview of supervised and unsupervised machine-learning based techniques to develop IE systems is given in Sect. 2.5. Finally, we mention recent trends and topics in Information Extraction in Sect. 2.6, which covers multi-linguality, cross-document and cross-lingual IE, extraction from social media, and open-domain IE.

The reader may also refer to other, more in-depth surveys, such as [4, 32].

2.2 Information Extraction Tasks

2.2.1 Definitions

The task of Information Extraction is to identify instances of a particular pre-specified class of entities, relationships and events in natural language texts, and the extraction of the relevant properties (arguments) of the identified entities, relationships or events. The information to be extracted is pre-specified in user-defined structures called *templates* (or objects), each consisting of a number of *slots* (or attributes), which are to be instantiated by an IE system as it processes the text. The slots fills are usually: strings from the text, one of a number of pre-defined values, or a reference to a previously generated object template. One way of thinking about an IE system is in terms of database population, since an IE system creates a structured representation (e.g., database records) of selected information drawn from the analyzed text.

2.2.2 IE Task Types

Applying information extraction on text aims at creating a structured view—i.e., a representation of the information that is machine understandable. The classic IE tasks include:

- **Named Entity Recognition (NER)** addresses the problem of the identification (detection) and classification of predefined types of named *entities*, such as organizations (e.g., *'World Health Organisation'*), persons (e.g., *'Muammar Kaddafi'*), place names (e.g., *'the Baltic Sea'*), temporal expressions (e.g., *'1 September 2011'*), numerical and currency expressions (e.g., *'20 Million Euros'*), etc. NER task can additionally include extracting descriptive information from the text about the detected entities through filling of a small-scale template. For example, in the case of persons, it may include extracting the title, position, nationality, sex, and other attributes of the person. It is important to note that NER also involves lemmatisation (normalisation) of the named entities, which is particularly crucial in highly inflective languages. For example in Polish there are six inflected forms of the name *'Muammar Kaddafi'* depending on grammatical case: *'Muammar Kaddafi'* (nominative), *'Muammara Kaddafiego'* (genitive), *Muammarowi Kaddafiemu* (dative), *'Muammara Kaddafiego'* (accusative), *Muammarem Kaddafim* (instrumental), *Muammarze Kaddafim* (locative), *Muammarze Kaddafi* (vocative).
- **Co-reference Resolution (CO)** requires the identification of multiple (co-referring) mentions of the same entity in the text. Entity mentions can be:
  - (a) Named, in case an entity is referred to by name; e.g., *'General Electric'* and *'GE'* may refer to the same real-world entity,

(b) Pronominal, in case an entity is referred to with a pronoun; e.g., in *'John bought food. But he forgot to buy drinks.'*, the pronoun *he* refers to *John*,

(c) nominal, in case an entity is referred to with a nominal phrase; e.g., in *'Microsoft revealed its earnings. The company also unveiled future plans.'* the definite noun phrase *The company* refers to *Microsoft*, and

(d) Implicit, as in case of using zero-anaphora[1]; e.g., in the Italian text fragment *'[Berlusconi]$_i$ ha visitato il luogo del disastro. ϕ_i Ha sorvolato, con l'elicottero.'* (Berlusconi has visited the place of disaster. [He] flew over with a helicopter.) the second sentence does not have an explicit realisation of the reference to *Berlusconi*.

- **Relation Extraction (RE)** is the task of detecting and classifying predefined relationships between entities identified in text. For example:

  - `EmployeeOf(Steve Jobs, Apple)`: a relation between a person and an organisation, extracted from *'Steve Jobs works for Apple'*
  - `LocatedIn(Smith, New York)`: a relation between a person and location, extracted from *'Mr. Smith gave a talk at the conference in New York'*,
  - `SubsidiaryOf(TVN, ITI Holding)`: a relation between two companies, extracted from *'Listed broadcaster TVN said its parent company, ITI Holdings, is considering various options for the potential sale.*

  Note, although in general the set of relations that may be of interest is unlimited, the set of relations within a given task is predefined and *fixed*, as part of the specification of the task.

- **Event Extraction (EE)** refers to the task of identifying events in free text and deriving detailed and structured information about them, ideally identifying *who did what to whom, when, where, through what methods (instruments), and why*. Usually, event extraction involves extraction of several entities and relationships between them. For instance, extraction of information on terrorist attacks from the text fragment *'Masked gunmen armed with assault rifles and grenades attacked a wedding party in mainly Kurdish southeast Turkey, killing at least 44 people.'* involves identification of perpetrators (*masked gunmen*), victims (*people*), number of killed/injured (*at least 44*), weapons and means used (*rifles and grenades*), and location (*southeast Turkey*). Another example is the extraction of information on new joint ventures, where the aim is to identify the partners, products, profits and capitalization of the joint venture. EE is considered to be the hardest of the four IE tasks.

While in the early years of IE the focus was on solving the IE tasks listed above at document level, research has shifted to cross-document information extraction, which will be addressed in Sect. 2.6.

---

[1]Zero-anaphora are typical in many languages—including the Romance and Slavic languages, Japanese, etc.—in which subjects may not be explicitly realized.

2.2.3   Evaluation in Information Extraction

Given an input text, or a collection of texts, the expected output of an IE system can be defined very precisely. This facilitates the evaluation of different IE systems and approaches. In particular, the *precision* and *recall* metrics were adopted from the IR research community for that purpose. They measure the system's effectiveness from the user's perspective, i.e., the extent to which the system produces all the appropriate output (recall) and only the appropriate output (precision). Thus, recall and precision can bee seen as measure of completeness and correctness, respectively. To define them formally, let *#key* denote the total number of slots expected to be filled according an *annotated* reference corpus, representing ground truth or a "gold-standard", and let *#correct* (*#incorrect*) be the number of correctly (incorrectly) filled slots in the system's response. A slot is said to be filled incorrectly either if it does not align with a slot in the gold standard (*spurious slot*) or if it has been assigned an invalid value. Then, precision and recall may be defined as follows:

$$precision = \frac{\#correct}{\#correct + \#incorrect} \qquad recall = \frac{\#correct}{\#key} \qquad (2.1)$$

In order to obtain a more fine-grained picture of the performance of IE systems, precision and recall are often measured for each slot type separately.

The *f-measure* is used as a weighted harmonic mean of precision and recall, which is defined as follows:

$$F = \frac{(\beta^2 + 1) \times precision \times recall}{(\beta^2 \times precision) + recall} \qquad (2.2)$$

In the above definition β is a non-negative value, used to adjust their relative weighting ($\beta = 1.0$ gives equal weighting to recall and precision, and lower values of β give increasing weight to precision).

Other metrics are used in the literature as well, e.g., the so called *slot error rate*, SER [40], which is defined as follows:

$$SER = \frac{\#incorrect + \#missing}{\#key} \qquad (2.3)$$

where *#missing* denotes the number of slots in the reference that do not align with any slots in the system response. It reflects the ratio between the total number of slot errors and the total number of slots in the reference. Depending on the particular needs, certain error types, (e.g., spurious slots) may be weighted in order to deem them more or less important than others.

2.2.4 Competitions, Challenges and Evaluations

The interest and rapid advances in the field of IE has been essentially influenced by the DARPA-initiated series of Message Understanding Conferences (MUCs).[2] They were conducted with the intention to coordinate multiple research groups and US government agencies seeking to improve IE and IR technologies [23]. The MUCs defined several types of IE tasks and invited the scientific community to tune their systems to complete these tasks. MUC participants were initially given a detailed description of an IE task scenario, along with annotated training data to adapt their systems to this scenario within a limited time-period, of 1–6 months. During the testing phase, each participant received a new set of test documents, applied their systems on these documents, and returned the extracted templates to MUC organizers. These results were then compared to a set of templates manually filled by human annotators, similarly to how it is done in evaluation in other fields of NLP, including summarization, described in Chap. 1.

The IE tasks defined in MUC conferences were intended to be prototypes of extraction tasks that will arise in real-world applications. The first two MUC conferences (1987–1989) focused on automated analysis of military messages containing textual information about naval sightings and engagements, where the template to be extracted had ten slots. Since MUC-3 [35] the task shifted to the extraction from newswire articles, i.e., information concerning terrorist activities, international joint-venture foundations, corporate management succession events, and space vehicle and missile launches. Over time, template structures became more complex. Since MUC-5, nested template structure and multilingual IE were introduced. The later MUC conferences distinguished several different IE subtasks in order to facilitate the evaluation and identification of IE sub-component technologies which could be immediately useful. The generic IE tasks defined in MUC-7 (1998) provided progressively higher-level information about texts, which, basically correspond to the IE tasks described in Sect. 2.2.2.

The ACE (Automated Content Extraction) Program,[3] started in 1999, is a continuation of the MUC initiative, which aims to support the development of automatic content extraction technologies for automatic processing of natural language in text from various sources [15]. The ACE Program has defined new and harder-to-tackle IE tasks centered around extraction of entities, relations and events. In particular, the increased complexity resulted from: (a) inclusion of various information sources (e.g., news, web-logs, newsgroups) and quality of input data (e.g., telephone speech transcripts), (b) introduction of more fine-grained entity types (e.g., facilities, geopolitical entities, etc.), template structures and relation types, and (c) widening the scope of the core IE tasks, e.g., the classic NER task has been converted to Entity

---

[2]http://www-nlpir.nist.gov/related_projects/muc/.
[3]http://www.itl.nist.gov/iad/mig/tests/ace/.

Detection and Tracking task, which includes detection of all mentions of entities within a document, whether named, nominal or pronominal.

The Linguistic Data Consortium[4] (LDC) develops annotation guidelines, corpora and other linguistic resources to support the ACE Program, which are used to evaluate IE systems in a similar manner to conducting MUC competitions. Effort has been put into preparing data for languages other than English, i.a., Spanish, Chinese, and Arabic. In the recent years, corpora for the evaluation of cross-document IE tasks (e.g., global entity detection and recognition task which requires cross-document co-reference resolution of certain types of entities) have been created.

Both MUC and ACE initiatives are of central importance to the IE field, since they provided a set of corpora that are available to the research community for the evaluation of IE systems and approaches.

IE systems and approaches are evaluated in other, broader venues as well. The Conference on Computational Natural Language Learning (CoNLL) has organized some "shared tasks" (competitions) on language-independent NER.[5] As part of the Text Analysis Conference (TAC)[6] a dedicated track, namely the Knowledge Base Population track, comprises evaluation of tasks centering around discovering information about entities (e.g., entity attribute extraction) as found in a large corpus and incorporating this information into a given knowledge base (*entity linking*), i.e., deciding whether a new entry needs to be created. Some IE-related tasks are being evaluated in the context of the Senseval[7] initiative, whose goal is to evaluate semantic analysis systems. For instance, the Web People Search Task [5] focused on grouping web pages referring to the same person, and extracting salient attributes for each of the persons sharing the same name. The task on Co-reference Resolution in Multiple Languages [55] focused on the evaluation of co-reference resolution for six different languages to provide, i.e., better insight into porting co-reference resolution components across languages.

2.2.5 *Performance*

The overall quality of extraction depends on multiple factors, including, i.e.: (a) the level of logical structures to be extracted (entities, co-reference, relationships, events), (b) the nature of the text source (e.g., online news, corporate reports, database textual fields, short text messages sent through mobile phones and/or social media), (c) the domain in focus (e.g., medical, financial, security, sports), and (d) the language of the input data (e.g., English vs. morphologically rich languages like Russian). To give an example, the precision/recall figures of the

---

[4]http://projects.ldc.upenn.edu.

[5]http://www.clips.ua.ac.be/conll2002/ner/.

[6]http://www.nist.gov/tac/.

[7]http://www.senseval.org.

top scoring IE systems for extracting information related to aircraft accidents from online English news evaluated at MUC-7[8] oscillated around: (a) 95/95% for NER task (b) 60/80% for co-reference, (c) 85/70% for relations, and (d) 70/50% for events. These figures reflect the relative complexity of the main IE tasks, with best performance achievable on NER[9] and event extraction being the most difficult task. It is important to note that a direct comparison between precision and recall figures obtained by IE systems in various competitions and challenges is difficult and not straightforward, due to the factors listed above, i.e., different set-up of the evaluation tasks w.r.t. template structure, domain, quality of input texts and language to be processed. In general, IE from English texts appears to be somewhat easier than for other languages which exhibit more complex linguistic phenomena, e.g., richer morphology and freer word order. Similarly for the extraction from grammatically correct texts, such as online news, versus extraction from short and often ungrammatical short messages, such as tweets and blog posts.

2.3 Architecture: Components of IE Systems

Although IE systems built for different tasks may differ from each other significantly, there are certain core components shared by most IE systems. The overall IE processing chain can be analyzed along several dimensions. The chain typically includes core linguistic components—which can be adapted to or be useful for NLP tasks in general—as well as IE-specific components, which address the core IE tasks. On the other hand, the chain typically includes domain-independent vs. domain-specific components. The domain-independent part usually consists of language-specific components that perform linguistic analysis in order to extract as much linguistic structure as possible. Usually, the following steps are performed:

- **Meta-data analysis:** Extraction of the title, body, structure of the body (identification of paragraphs), and the date of the document.
- **Tokenization:** Segmentation of the text into word-like units, called tokens and classification of their type, e.g., identification of capitalized words, words written in lowercase letters, hyphenated words, punctuation signs, numbers, etc.
- **Morphological analysis:** Extraction of morphological information from tokens which constitute potential word forms—the base form (or *lemma*), part of speech, other morphological tags depending on the part of speech: e.g., verbs have features such as tense, mood, aspect, person, etc. Words which are ambiguous with respect to certain morphological categories may undergo disambiguation. Typically part-of-speech disambiguation is performed.

---

[8]http://www-nlpir.nist.gov/related_projects/muc/proceedings/muc_7_toc.htmlconference.
[9]Industrial-strength solutions for NER for various languages exist.

- **Sentence/Utterance boundary detection:** Segmentation of text into a sequence of sentences or utterances, each of which is represented as a sequence of lexical items together with their features.
- **Common Named-entity extraction:** Detection of domain-independent named entities, such as temporal expressions, numbers and currency, geographical references, etc.
- **Phrase recognition:** Recognition of small-scale, local structures such as noun phrases, verb groups, prepositional phrases, acronyms, and abbreviations.
- **Syntactic analysis:** Computation of a dependency structure (parse tree) of the sentence based on the sequence of lexical items and small-scale structures. Syntactic analysis may be deep or shallow. In the former case, one is interested in computing all possible interpretations (parse trees) and grammatical relations within the sentence. In the latter case, the analysis is restricted to identification of non-recursive structures or structures with limited amount of structural recursion, which can be identified with a high degree of certainty, and linguistic phenomena which cause problems (ambiguities) are not handled and represented with under-specified structures.

The extent of the domain-independent processing may vary depending on the requirements of the particular application. The core IE tasks—NER, co-reference resolution, and detection of relations and events—are typically domain-specific, and are supported by domain-specific system components and resources.

Domain-specific processing is typically also supported on a lower level by detection of specialized *terms* in text. For example, in domains related to medicine, extensive specialized lexicons, ontologies and thesauri of medical terms will be essential, whereas for IE in business-related domains, these are unnecessary.

A typical architecture of an IE system is depicted in Fig. 2.2. In the domain-specific core of the processing chain, a NER component is applied to identify the entities relevant in a given domain. Patterns may then be applied[10] to: (a) identify text fragments, which describe the target relations and events, and (b) extract the key attributes to fill the slots in the template representing the relation/event. A co-reference component identifies mentions that refer to the same entity. Finally, partially-filled templates are fused and validated using domain-specific inference rules in order to create full-fledged relation/event descriptions. The last step is crucial since the relevant information might be scattered over different sentences or even documents. It is important to note that, in practice, the borders between domain-independent linguistic analysis components and core IE components may be blurred, e.g., there may be a single NER component which performs domain-independent and domain-specific named-entity extraction simultaneously.

There are several software packages, both freely available for research purposes and commercial use, which provide various tools that can be used in the process

---

[10]Various formalisms are used to encode patterns, ranging from character-level regular expressions, through context-free grammars to unification-based formalisms.

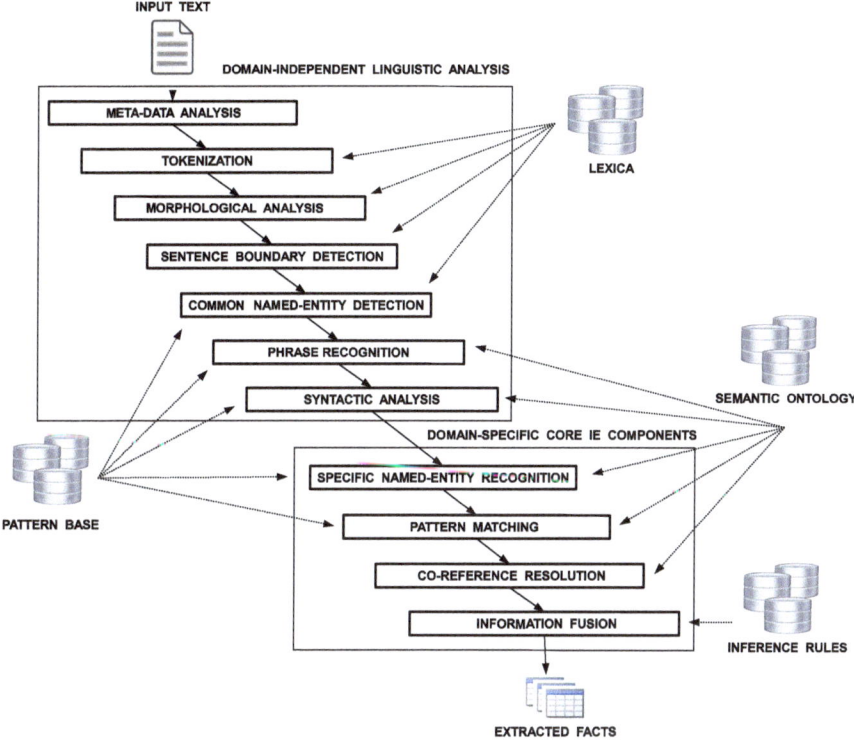

Fig. 2.2 Typical architecture of an information extraction system

of developing an IE system, ranging from core linguistic processing modules (e.g., language detectors, sentence splitters), to general IE-oriented NLP frameworks.[11]

2.4   Early Years: Knowledge Engineering Approaches

The strong application potential of IE was recognized already in the late 1980s. One of the first attempts to apply IE in the financial domain to extract information from messages regarding money transfers between banks was the ATRANS system [38], based on simple NLP techniques and script-frames approach. JASPER was an IE system, described in [1], which extracted information from reports on corporate earnings from small sentence fragments using robust NLP methods. SCISOR [28] was an integrated system incorporating IE for the extraction of facts related to corporate mergers and acquisitions from online news. Examples of IE system in

---

[11]Some examples of NLP tools which are relevant in the context of developing IE systems can be found at: http://alias-i.com/lingpipe/web/competition.html.

the security domain are described, for example, in [34, 35]. All the above systems and other early IE systems were developed using the Knowledge Engineering (KE) approach [4], where the creation of linguistic knowledge in the form of *rules*, or patterns, for detecting and extracting the target information from text is performed by human experts, through inspection of the test corpus and intuition. This is usually done in an iterative manner, starting with a small set of extraction rules which are tested on the available corpora and extended until a desired trade-off between precision and recall is reached. Most of the early IE systems had a major shortcoming: they exhibited a non-modular black-box character, were mono-lingual, and were not easily adaptable to new scenarios. However, they did demonstrate that relatively simple NLP techniques may be sufficient for solving real-world IE tasks that are narrow in scope.

The subsequent efforts in the area of KE-based approaches aimed at moving toward general-purpose IE systems and frameworks, which are modular and easier to adapt to new domains and languages. The FASTUS system described in [24] is an example of an efficient and robust general-purpose IE system developed at SRI International. FASTUS was able to process English and Japanese, and was among the top-scoring system in the MUC competitions. It was designed as a set of cascaded nondeterministic finite-state transducers, where each stage of the processing is represented as a finite-state device. The output structures of the successive stages serve as the input for the next stage. Many other IE systems developed by that time were implemented as cascades of finite-state devices, e.g., the SPPC system described in [45], that processes German text. The emergence of finite-state based IE systems was mainly motivated by two reasons. From the computational perspective, finite-state devices are time- and space-efficient due to their closure properties and the existence of efficient optimization and transformation algorithms. From the linguistic perspective, the vast majority of local linguistic phenomena encountered in IE can be easily expressed as finite-state devices, and in particular, they can be usually broken down into a number of autonomous finite-state sub-descriptions. In this manner, a finite-state cascade allows for a strong decomposition of the linguistic analysis and of the entire IE process into many subtasks, ordered by increasing complexity. The incremental character of this approach yields simplicity, higher modularity and ease of maintenance of the underlying linguistic resources. Although more powerful formalisms exist— e.g., context-free or unification grammars—the finite-state-based formalisms for extraction grammars became prevalent, due to the characteristics mentioned. In addition, finite-state approximation grammars [43] proved to provide surprisingly effective partial parsers. Finite-state based formalisms are used in the popular general-purpose text engineering platform GATE [14], which is deployed for the development of IE systems [42], as well as in other IE-oriented text engineering frameworks such as SPROUT [17] and EXPRESS [49].

Although IE systems based on shallow linguistic analysis implemented as finite-state cascades became popular, the later MUC conferences saw some integration of shallow and deep linguistic analysis in IE systems in order to obtain a better performance. For instance, deep linguistic analysis could be applied on text fragments deemed to be highly relevant or on those which could not have been processed successfully through shallow linguistic analysis. An example of a system that attempts to find a pragmatic middle way in the shallow vs. deep analysis debate is LASIE-II [25]. It deploys an eclectic mixture of techniques ranging from finite-state recognizers for detection of domain-specific lexical patterns to restricted context-free grammars for partial parsing. This system exhibited highly modularized architecture and was equipped with visual tools for selecting the control flow through different module combinations, which became an essential feature of many IE systems to enable one to obtain deeper insight into strengths and weaknesses of the particular subcomponents and their interactions. Two other examples of modular IE systems based on the KE approach, namely, IE2 and REES are described in [2] and [3], respectively. The former achieved the highest scores in almost all IE tasks in MUC-7, whereas the latter was a first attempt at constructing large-scale event and relation extraction system based on shallow text analysis methods, which covered more than 100 types of relations and events related to the area of business, finance and politics.

2.5   Emergence of Trainable IE Systems

Research in the late 1990s and the beginning of the twenty-first century resulted in significant advances in IE (and in NLP in general) in terms of emergence of KE-based, modular IE systems that were able to process vast amounts of textual data robustly and efficiently, including in languages other than English. By the turn of the century, the modular design of IE systems was widely accepted; the design consisted of a *generic core engine* on one hand, and language- and domain-specific components on the other, which may be called *knowledge bases*. The knowledge bases are further modularized according to the kind of knowledge needed to perform the IE tasks. Information contained in the different knowledge bases is of different forms and different levels of complexity. On the lowest-level are specialized lexicons and gazetteers. We may view ontologies, semantic networks and thesauri as being one level above, in that they specify inter-relationships among concepts and organize knowledge into more complex objects. Many of the described systems are pattern-based: they employ a pattern-matching mechanism to identify constituents in text that have semantic significance to the task—entities or events— and large parts of the pattern base are generally highly domain-specific. Patterns are much more easily constructed and understood when they are encoded in a declarative fashion, and languages for defining patterns emerged, in the systems mentioned above. Finally, certain *reasoning* operations that need to be performed on extracted facts are also highly domain-specific in nature—these operations can sometimes be encoded in declarative fashion, or as sub-programs; in either case, they can be viewed as a parameterization of the core processing chain.

Abstracting the domain-specific knowledge into knowledge bases and away from the core processing engine allows for much more flexible re-configuration of an existing IE engine from one task to another. However, the process of handcrafting language- and domain-specific resources and components remained a very time-consuming and difficult task. This stimulated research on trainable IE systems that deploy machine-learning (ML) techniques to offload some of the burden of customizing a general-purpose IE engine to a new domain or task.

The first "wave" in the shift away from KE-based approaches and toward trainable systems focused on supervised machine-learning approaches. The principal motivation behind this trend is to shift the human effort in customization of the IE knowledge bases away from knowledge engineering and toward annotating training data, which serves as input to machine-learning algorithms. This promotes further modularity in development, since knowledge engineering requires effort from both the system developer and the domain expert, whereas data annotation, in principle, requires mostly an effort on the part of the latter.

The broad array of tools and algorithms from supervised learning that were already in wide use in other areas of NLP came to be applied to IE tasks as well. This trend affected all IE-related tasks. For example, hidden Markov models (HMMs), at the core of many NLP tasks, including speech processing and part-of-speech tagging, found application in IE. An example of a system for named entity detection that uses HMMs is Nymble [9]. An HMM consists of states and probabilistic transitions between the states. The simple idea behind using the HMM for finding named entities is to associate one state with being inside a name, and another state with being outside a name, while scanning or generating the words of a sentence. This basic scheme is generalized by adding states to accommodate multiple *types* of names, etc. Given some amount of text annotated with labels that indicate which words constitute names, and what type of names, the HMM is trained and applied using standard algorithms well-known from the literature.

Supervised learning is similarly applied to other IE tasks, mentioned above.[12] As we may do with NER, we may also re-cast the task of relation or event extraction as a classification problem. This may be done as follows. Given a segment (or a window) of text, such as a sentence, or a paragraph, etc., pose the problem as: does this segment contain a relation of type R? The key to making the problem tractable as stated is in defining a set of independent variables, or *features*, of the text segment that may help us to determine the presence or absence of the relation. These features will typically include words and terms (possibly consisting of multiple words) present (or absent) in the segment, entities found in the segment, other relations, and other syntactic and semantic information extracted from the segment. Some of the earlier literature on application of supervised learning to IE tasks includes, e.g., [56, 57]. Conditional random fields (CRFs), another supervised learning method that has become popular in the recent years, has also been applied to IE tasks. CRFs are similar to HMMs in that they are good at modeling local dependencies, and they perform especially well on the "lower-level" tasks—e.g.,

---

[12]Standard references are available for application of supervised learning algorithms, e.g., [70], and in particular to NLP tasks, cf. [41].

named entity classification, [60], as they are well suited for sequence-labeling tasks in general. CRFs can be used to model certain variants of the relation detection task as well, as shown, e.g., by [7] which applies CRFs to the task of relation *labeling* in an open setting. More on this latter topic is mentioned in Sect. 2.6.2.[13]

The features used by a learning algorithm do not need to be limited to the simplest kinds of features of the text, such as "bags of words", but they do need to be more readily extractable from text than the classification problem we are trying to solve— i.e., they have to be handled by lower-level processing within the IE pipeline. This approach fits well with the overall cascaded design of the IE pipeline, where the text is first analyzed into small constituent pieces, and then successive stages extract higher-level information over the same text.

Co-reference resolution can be similarly re-cast as a classification problem. For example, for a given pair of entity mentions extracted from a text—entities that were extracted in an earlier processing phase by a component that precedes co-reference resolution in the IE pipeline—we may ask whether these mentions co-refer, i.e., whether they refer to the same real-world object. The features, in this case, may include bags of words found in the sentences containing the mentions, as well as other, more complex and abstract aspects of the text, including the distance between the entities, the words and/or parts of speech found along the path in the parse tree leading from one entity to the other (if the entities are in the same sentence, and the sentence has been fully parsed), the layout of the segments containing the two entities, etc.[14]

Once we have chosen a set of representative features, we can use supervised learning algorithms, such as Naive Bayes (NB), or Bayes network (BN), to train classifiers based on a set of annotated training data.

The choice of features also depends on the type of classifier we wish to train. Some of the simpler classifiers, such as NB, may suffer degraded performance if the features are not independent, while others, such as support-vector machines (SVM), may be are less sensitive to complex dependencies among feature, which may simplify feature engineering. Several tradeoffs need to be considered when choosing the features and the learning algorithms. On one hand we wish incorporate as many features as possible, including ones that are possibly not independent, and possibly not informative. On the other hand, we need to limit the computational cost. Methods for automatic feature selection (such as leave-one-out cross-validation) can be employed to reduce large sets of non-independent and possibly non-informative features, but then the feature selection process itself can become computationally expensive.

---

[13]This kind of labeling is actually more similar in nature to named entity classification, except that the information extracted is about not the nature of node in a semantic network—i.e., a (possibly named) entity—but the nature of an edge, i.e., a link between two nodes. The information extracted focuses mostly on what label the system should attach to the analyzed object.

[14]For an example of learning for anaphora resolution, cf. [46].

In this fashion, the burden on the human developer of the IE system shifts away from manual knowledge engineering, and toward feature engineering on one hand, and data annotation on the other. Although feature engineering is a complex task, to some extent we may expect that once appropriate features are chosen for one domain, they will work reasonably well for other domains as well, or at least for "similar" domains.

Thus, the effectiveness (quality of performance) of supervised learning depends on the complexity of the task and on the amount of available annotated training data. The larger the amount of annotated text, the better the performance of the learner. On the other hand, the more complex IE tasks, such as relation extraction, will require more annotated data than the simpler tasks (such as name entity recognition).[15] For the higher-level complex IE tasks, data sparseness becomes a serious problem, which requires still more annotated data. Similarly to supervised learning for syntactic parsing, (e.g., [11]) where tree-banks containing millions of annotated words are used for training statistical parsers, supervised learning for IE tasks can also require massive amounts of annotated data. To alleviate the data annotation bottleneck, recent research in IE has followed several approaches.

Active learning is an area of machine learning which aims to reduce the amount of annotation required from the human annotator, by "active" participation on the part of the learner in the learning process.[16] The idea is that the annotator initially provides a small set of examples, based on which the learner actively decides which other examples—from among a very large set of possible candidate examples—the human teacher/annotator should annotate next, for maximum gain. Intuitively, this means selecting the examples on which the learner's uncertainty is greatest. In practice, it means selecting such examples, on which the conditional distribution of the target class—e.g., a sought name type, or whether a relation is present or absent—has the highest entropy, given the feature values. These are examples from which the learner will benefit *most*, assuming that an annotator is willing to spend only a fixed amount of time on annotating data. Another way of viewing this is that the learner benefits less if the annotator marks up many examples which are very similar among themselves (in terms of the feature values), and provides redundant information. For examples of application of active learning in IE, cf., e.g., [31].

Active learning is a "human-in-the-loop" approach to learning, used to support supervised learning methods in IE, which results in a "hybrid" approach to building IE systems: the human developer is still involved in the knowledge engineering process but less directly and with substantial automated assistance.

Bootstrapping is another kind of learning that aims to reduce the amount of human involvement in knowledge engineering. Similarly to active learning, the

---

[15]Typically a substantial subset of the annotated data will need to be reserved for testing and evaluation purposes—otherwise we cannot estimate the quality of performance on unseen data. We cannot test on the same data on which the learning algorithm was trained.

[16]In contrast to a "passive" learner, which learns from a set of annotated examples—negative as well as positive—provided by the teacher.

system developer provides an initial set of annotated examples—the "seeds—but in bootstrapping the learning process proceeds without further supervision, until a convergence criterion is reached. Such *weakly supervised* methods,[17] require some amount of human supervision, in the initial phase—providing the seeds—as well as possibly in the final phase—checking the result of the learning, to remove candidates that appear "bad", or not useful. Methods for learning domain-specific names (such as names of companies for the domain of financial news, and names of medical conditions for the domain of epidemic surveillance) are described, e.g., in [13, 48, 74]. Application to learning domain-specific patterns for event extraction is described, e.g., in [63, 66, 71].

The main motivation for bootstrapping is that we expect that once the learner converges and produces a large set of candidate elements for the knowledge bases—e.g., for named entities or for relation extraction patterns—it is easier, or takes less time, for the system developer to go rank the set of candidates by their quality (even if the set is large), than to construct a set candidates from scratch. Thus we can hope to reduce the expense of knowledge engineering.

2.6   Current Trends, Challenges and the Future

We have presented IE in the "classic" setting, as it has been studied traditionally. We turn now to the challenges and directions that we anticipate will grow in importance and will dominate the IE research landscape in the coming years. In the context of currently emerging novel kinds of large-scale corpora, IE assumes new dimensions and reinvents itself.

For over a decade the bulk of research focused on information extraction in English. The growing amount of textual data available in other languages resulted in shifting of the focus to other non-English IE and language-independent (multilingual) IE techniques. While much of the attention in non-English IE focused on NER (cf. [44] which includes a collection of references to work on non-English NER), relatively little work has been reported on higher-level IE tasks, including relation and event extraction.

IE in languages other than English is, in general, harder, and the performance of non-English IE systems is usually lower. This is mainly due to lack of core NLP components and underlying linguistic resources for many languages, but most of all due to the various linguistic phenomena that are non-existent in English, which include, *inter alia*:

- Lack of whitespace, which complicates word boundary disambiguation (e.g., in Chinese [22]);
- Productive compounding, which complicates morphological analysis, as in German, where circa 10–15% words are compounds whose decomposition is

---

[17]These bootstrapping-type methods are sometimes called "unsupervised", but that is rather a misnomer; a more appropriate term would be *weakly* or *minimally* supervised learning.

crucial for higher-level NLP processing [45]; Finnic languages exhibit similar phenomena;

- Complex proper name declension, which complicates named-entity normalisation. Typical for Slavic languages and exemplified in [52], which describes techniques for lemmatisation and matching Polish person names;
- Zero anaphora, whose resolution is crucial in the context of CO task. This is typical for Japanese, Romance and Slavic languages, [27];
- Free word order and rich morphology, which complicates relation extraction. This is typical, e.g., for Germanic, Slavic and Romance languages, and exemplified in [76], which proposes more linguistically sophisticated extraction patterns for relation extraction than the classical surface-level linear extraction patterns applied in many English IE systems.

Though this is only a partial list of phenomena that complicate non-English IE, significant progress in non-English IE could be observed in the recent years.

Apart from tackling multi-linguality, one central characteristic theme that runs through current research in IE, as distinguished from classic IE presented so far, is the extraction of information from *multiple sources*—in contrast from finding information within a single piece of text. The clear motivation for this is that most given facts or stories typically do not exist in an isolated report, but are rather carried by many on-line sources, possibly in different languages. In particular, the information redundancy resulting from the access to multiple sources reporting can be exploited for the validation of facts. Further, the fact or story may undergo evolution through time, with elaborations, modifications, or possibly even contradictory related information emerging over time.

Examples of approaches to IE from multiple sources and existing IE systems that exploit such approaches may be found in [16, 21, 30, 33, 36, 47, 51, 62, 65, 69, 72, 73]. In the area of cross-lingual IE, experiments on cross-lingual bootstrapping of ML-based event extraction systems have been reported in [12, 33, 64], and on cross-lingual information fusion in [50]. An overview on the challenges in IE, in particular addressing cross-document IE, cross-lingual IE, and cross-media IE and fusion may be found, e.g., in [29].

There have been several workshops dedicated to research on the various aspects of IE from multiple sources, in particular [10, 53, 54]. The latter two include the body of work which was invited to be presented in this volume, as mentioned in the Preface. This is the primary purpose and focus of this volume—to detail the research in the direction of these current and future needs served and supported by IE; the subsequent chapters provide a wealth of references to a wide range of recent and current work.

We conclude this survey with an excursion into two more closely related aspects of state-of-the-art IE from large-scale text collections—applying IE to novel types of media and attempting to broaden the application of IE to any number of domains simultaneously.

2.6.1 Information Extraction from Social Media

The advent of social media resulted in new forms of individual expression and communication. At present, an ever-growing amount of information is being transferred through social media, including blogs, Web fora, and micro-blogging services like Facebook[18] or Twitter.[19] These services allow users to write large numbers of short text messages in order to communicate, comment and chat about current events of any kind, current events, politics, products, etc. The trend of using social media has been significantly boosted by the low-barrier cross-platform access to such media and global proliferation of mobile devices. It is worth noting that social media can in many situations provide information that is more up-to-date than conventional sources of information, such as online news—e.g., in the context of natural disasters, as the earthquakes in Japan in 2011, or the Arab Spring—which makes them a particularly attractive source of information in these settings. As a consequence, some work was reported by the research community on automated analysis of social media content, including also attempts to information extraction from social media.

Extraction of information from social media is more challenging than classic IE, i.e., extraction from trusted sources and well-formed grammatical texts. The major problems encountered when processing social media content are: (a) texts are typically very short—e.g., Facebook limits status updates to 255 characters, whereas Twitter limits messages ("tweets") to 140 characters, (b) texts are noisy and written in an informal setting, include misspellings, lack punctuation and capitalisation, use non-standard abbreviations, and do not contain grammatically correct sentences, and (c) high uncertainty of the reliability of the information conveyed in the text messages, e.g., compared to the news media. A straightforward application of standard NLP tools on social media content typically results in significantly degraded performance. The mentioned characteristics influenced a new line of research in IE, which focuses on developing methods to extract information from short and noisy text messages.

Some recent work reports on NER from Tweets. Locke and Martin [39] presents results of tuning a SVM-based classifier for classifying persons, locations and organisations names in Twitter, which achieves relatively poor results, i.e., precision and recall around 75% and 50% respectively. Ritter et al. [58] describes an approach to segmentation and classification of a wider range of names in tweets based on CRFs (using POS and shallow parsing features) and Labeled Latent Dirichlet Allocation respectively. The reported precision and recall figures oscillate around 70% and 60% respectively. Liu et al. [37] proposed NER (segmentation and classification) approach for tweets, which combines KNN and CRFs paradigms and achieves precision and recall figures around 80% and 70% respectively.

---

[18]http://www.facebook.com/.

[19]http://twitter.com/.

Higher level IE tasks such as event extraction from Twitter or Facebook are more difficult than NER since not all information on an event may be expressed within a single short message. Most of the reported work in the area of event extraction from short messages in social media focus on event detection, e.g., [59] presents a SVM-based approach to classify tweets on earthquakes and typhoons (80–90% recall and 60–65% precision) and a probabilistic spatio-temporal model to find the centre and the trajectory of the event, [67] reports on an approach of using linguistic features to detect tweets with content relevant to situational awareness during mass emergencies (emergency event detection), which achieves 80% accuracy. Apart from research on pure event detection from micro-blogs, some work reported on extracting structured information on events. For instance, [68] studies the content of tweets during natural hazards events in order to identify event features which can be automatically extracted. Benson et al. [8] presents a CRF-based approach to extracting entertainment events with 85% precision (including extraction of location, event name, artist name, etc.) through aggregating information across multiple messages, similar in spirit to the approaches referred to above.

Research on IE from social media is still in its early stages, and focuses mainly on processing English. Future research will likely focus on further adaptation of classic IE techniques to extract information from short messages used in micro-blogging services, tackling non-English social media, and techniques for aggregation and fusion of information extracted from conventional text documents (e.g., news) and short messages posted through social media, e.g., for using Twitter to enhance event descriptions extracted from classic online news with situational updates, etc.

2.6.2   Open-Domain Information Extraction

In the last two decades IE has moved from mono-lingual, domain-tailored, knowledge-based IE systems to multilingual trainable IE systems that deploy weakly supervised ML techniques which automate to a large extent the entire IE customisation process. In [6] the paradigm of *Open Information Extraction* (OIE) was introduced, that aims to facilitate domain-independent discovery of relations extracted from texts and to scale to heterogeneous and large-size corpora such as the Web. An OIE system takes as input only a corpus of texts without any a priori knowledge or specification of the relations of interest and outputs a set of all extracted relations. The main rationale behind introducing the concept of OIE was to move away from the classic IE systems, where the relations of interest have to be specified prior to posing a query to the system, since adapting such a system to the extraction of a new relation type requires a training phase. Instead, an OIE system should: (a) discover all possible relations in the texts, using time linear in the number of documents in the corpus, and (b) create an IE-centric index that would support answering a broad range of unanticipated questions over arbitrary relations. Transforming classic IE methods and systems into fully unsupervised OIE technologies that can handle large-size and diverse corpora such as Web would have various immediate applications, e.g., factual web-based Q/A systems capable of handling complex relational queries, and intelligent indexing for search engines.

A first step towards OIE, namely tackling automated customisation and handling heterogeneity, was presented with KNOWITALL [18] and similar systems. KNOWITALL is a self-supervised Web extraction system that uses a small set of domain-independent extraction patterns (e.g., `<A> is a <B>`) to automatically instantiate "reliable", relation-specific extraction rules (training data), which are than used to learn domain-specific extraction rules iteratively in a bootstrapping process. It relies solely on part-of-speech tagging, and requires neither a NER component nor a parser, i.e., no features which rely on syntactic or semantic analysis are required, which helps to better tackle the problem of heterogeneity and scaling to other languages. However, the system requires: (a) a large number of search engine queries in the process of assessing the reliability of the candidate relation-specific extraction rules, (b) providing the name of the relation of interest to the system, and (c) a new learning cycle in case of adding a new relation. An OIE system TEXTRUNNER, described in [75], addresses the aforementioned problems, i.e., it needs just one pass through the corpus without the need to name any relations of interest in advance. TEXTRUNNER first learns a general model of how relations are expressed in a particular language using CRF paradigm.[20] The creators of the system claimed that most relations in English can be characterized by a set of several lexico-syntactic patterns, e.g., `Entity-1 Verb Prep Entity-2`. In the second phase, the system scans each sentence and uses the model to assign to each word labels that denote the beginning/end of an entity or string describing the relation. Analogously to KNOWITALL, the model uses only low-level linguistic features such as part-of-speech, token type (e.g., capitalisation), closed-word classes, etc., which is attractive from the perspective of handling genre diversity and different languages. For each sentence the system returns one or more triples, each representing a binary relation between two entities (e.g., `(Paris, capital-of, France)`), which is accompanied by a probability of the triple (relation) being extracted correctly, relying primarily on information about the frequency of occurrence on the Web. An evaluation of TEXTRUNNER in [7] revealed that it achieves on average 75% precision, however, the output of the system is unnormalised to a large extent, e.g., problems of matching names referring to the same real-world entity or identifying synonymous relations were not handled. Some improvements of the system in terms of handling relation synonymy problem, which resulted in improving the overall recall, are also described in [7]. Related work on OIE, similar in spirit, is reported in [61], focusing on avoiding relation-specificity in the relation extraction task, but it does not scale well to Web size.

A comparison of the performance of an OIE system versus a traditional relation extraction system is given in [7]. In particular, variants of the aforementioned TEXTRUNNER system were compared against a relation extraction system that uses

---

[20]Trained on extractions heuristically generated from PennTreebank.

the same CRF-based extraction model, but which was trained from hand-labelled data and which used a richer feature set, including lexical features. Experiments on extracting binary relations such as "corporate acquisitions" or "inventors of products" revealed that both systems achieve comparable precision (circa 75%), whereas the traditional extraction system achieved significantly higher recall, i.e., circa 60% vs. 20% achieved by OIE system. A hybrid system, combining the two using stacked generalisation, was reported to achieve a 10% relative improvement in precision with slight deterioration of the recall over the classic extraction system. In case higher level of recall is more important, in the context of extraction of binary relations, as the ones mentioned above, using traditional relation extraction is still by far more effective, although an OIE system might be deployed to reduce the number of hand-labeled training data. If, however, precision is more important and the number of relations is large, then an OIE system might potentially constitute a better alternative. One of the most frequent errors of the early OIE systems such as TEXTRUNNER were: (a) incoherent extractions (i.e., no meaningful interpretation of the relation phrase), (b) uninformative extractions (critical information omitted in relation phrases), and (c) incorrect or improperly scoped arguments (early OIE systems did not handle complex arguments, e.g., complex NPs with relative clauses). In [19], new OIE systems were introduced which try to tackle the aforementioned problems based on more fine-grained linguistic analysis of the structure of strings/phrases denoting relations and their arguments in English, which lead to further improvements.

The area of OIE, briefly described here, and referred work indicate progress in tackling Web-scale IE and further reduction of human involvement in the customization process, however, it is important to note that most of the work on OIE exhibits certain limitations since it is mainly focused on the extraction of binary relations within the scope of sentence boundaries. Attempts at devising OIE methods to extract more complex structures have been reported; e.g., [20] addresses the application of OIE techniques to the extraction of ordered sequences from the Web—sequence name and a set of ordered pairs, where the first element is the string naming the member of the sequence and the second element represents the position of that member, e.g., ordered list of the presidents of a given country. OIE for events, which have time, location, and potentially involve extraction of more than one relation that go beyond sentence level, constitutes a challenging task to be studied in the future. Since most of the work on OIE focused on English, the portability of existing OIE algorithms to other languages also needs to be investigated. Furthermore, the applicability of OIE to corpora consisting of short text messages might constitute an interesting area for research. Finally, another line of research might focus on the integration of some inference mechanisms in OIE systems, that would allow reasoning based on the facts they extract from texts.

References

1. Andersen, P., Hayes, P., Huettner, A., Schmandt, L., Nirenburg, I., Weinstein, S.: Automatic extraction of facts from press releases to generate news stories. In: Proceedings of the 3rd Conference on Applied Natural Language Processing, ANLC '92, Trento, pp. 170–177. Association for Computational Linguistics, Stroudsburg (1992)
2. Aone, C., Halverson, L., Hampton, T., Ramos-Santacruz, M., Hampton, T.: SRA: description of the IE2 system used for MUC-7 In: Proceedings of MUC-7. Morgan Kaufmann, Columbia (1999)
3. Aone, C., Ramos-Santacruz, M.: REES: a large-scale relation and event extraction system. In: Proceedings of the 6th Conference on Applied Natural Language Processing, ANLP 2000, Seattle, pp. 76–83. Association for Computational Linguistics, Stroudsburg (2000)
4. Appelt, D.: Introduction to information extraction. AI Commun.**12**, 161–172 (1999)
5. Artiles, J., Borthwick, A., Gonzalo, J., Sekine, S., Amigï£¡, E.: WePS-3 evaluation campaign: overview of the web people search clustering and attribute extraction tasks. In: Braschler, M., Harman, D., Pianta, E. (eds.) CLEF (Notebook Papers/LABs/Workshops), Padua (2010)
6. Banko, M., Cafarella, M., Soderland, S., Broadhead, M., Etzioni, O.: Open information extraction from the web. In: Proceedings of the 20th International Joint Conference on Artifical intelligence, Hyderabad, pp. 2670–2676. Morgan Kaufmann, San Francisco (2007)
7. Banko, M., Etzioni, O.: The tradeoffs between open and traditional relation extraction. In: Proceedings of ACL-08: HLT, Columbus, pp. 28–36. Association for Computational Linguistics, Columbus (2008)
8. Benson, E., Haghighi, A., Barzilay, R.: Event discovery in social media feeds. In: Proceedings of the 49th Annual Meeting of the Association for Computational Linguistics: Human Language Technologies – Vol. 1, pp. 389–398. Association for Computational Linguistics, Stroudsburg (2011)
9. Bikel, D., Miller, S., Schwartz, R., Weischedel, R.: Nymble: a high-performance learning name-finder. In: Proceedings of the 5th Applied Natural Language Processing Conference, Washington. Association for Computational Linguistics, Washington, DC (1997)
10. Califf, M.E., Greenwood, M.A., Stevenson, M., Yangarber, R. (eds.): In: Proceedings of the Workshop on Information Extraction Beyond The Document. COLING/ACL, Sydney (2006)
11. Charniak, E.: Statistical Language Learning. MIT, Cambridge (1993)
12. Chen, Z., Ji, H.: Can one language bootstrap the other: a case study on event extraction. In: Proceedings of the NAACL HLT 2009 Workshop on Semi-Supervised Learning for Natural Language Processing, SemiSupLearn '09, Boulder, pp. 66–74. Association for Computational Linguistics, Stroudsburg (2009)
13. Collins, M., Singer, Y.: Unsupervised models for named entity classification. In: Proceedings of the Joint SIGDAT Conference on Empirical Methods in Natural Language Processing and Very Large Corpora. University of Maryland, College Park (1999)
14. Cunningham, H., Maynard, D., Bontcheva, K., Tablan, V.: GATE: a framework and graphical development environment for robust NLP tools and applications. In: Proceedings of the 40th Anniversary Meeting of the Association for Computational Linguistics, Philadelphia (2002)
15. Doddington, G., Mitchell, A., Przybocki, M., Ramshaw, L., Strassel, S., Weischedel, R.: Automatic content extraction (ACE) program – task definitions and performance measures. In: Proceedings of the 4th International Conference on Language Resources and Evaluation (LREC 2004) (2004)
16. Downey, D., Etzioni, O., Soderland, S.: A probabilistic model of redundancy in information extraction. In: Proceedings of the 19th International Joint Conference on Artificial Intelligence, IJCAI'05, Edinburgh, pp. 1034–1041. Morgan Kaufmann, San Francisco (2005)
17. Drożdżyński, W., Krieger, H.U., Piskorski, J., Schäfer, U., Xu, F.: Shallow processing with unification and typed feature structures – foundations and applications. Künstliche Intell. **1/04**, 17–23 (2004)

18. Etzioni, O., Cafarella, M., Downey, D., Popescu, A.M., Shaked, T., Soderland, S., Weld, D., Alexander, A.: Unsupervised named-entity extraction from the web: an experimental study. Artif. Intell. **165**, 91–134 (2005)
19. Etzioni, O., Fader, A., Christensen, J., Soderland, S., Mausam: Open information extraction: the second generation. In: Proceedings of IJCAI 2011, Barcelona, pp. 3–10 (2011)
20. Fader, A., Soderland, S., Etzioni, O.: Extracting sequences from the web. In: Proceedings of the ACL 2010 Conference Short Papers, Uppsala, pp. 286–290. Association for Computational Linguistics, Uppsala (2010)
21. Finkel, J.R., Grenager, T., Manning, C.: Incorporating non-local information into information extraction systems by gibbs sampling. In: Proceedings of the 43rd Annual Meeting on Association for Computational Linguistics, ACL '05, Michigan, pp. 363–370. Association for Computational Linguistics, Stroudsburg (2005)
22. Gao, J., Wu, A., Li, M., ning Huang, C.: Chinese word segmentation and named entity recognition: a pragmatic approach. Comput. Linguist. **31**, 574 (2005)
23. Grishman, R., Sundheim, B.: Message understanding conference – 6: a brief history. In: Proceedings of the 16th International Conference on Computational Linguistics (COLING), Kopenhagen, pp. 466–471. The Association for Computational Linguistics, Stroudsburg (1996)
24. Hobbs, J.R., Appelt, D., Bear, J., Israel, D., Kameyama, M., Stickel, M., Tyson, M.: FASTUS: A cascaded finite-state transducer for extracting information from natural-language text. In: Roche, E., Schabes, Y. (eds.) Finite State Language Processing. MIT, Cambridge (1997)
25. Humphreys, K., Gaizauskas, R., Huyck, C., Mitchell, B., Cunningham, H., Wilks, Y.: University of sheffield: description of the LaSIE-II system and used for MUC-7. In: Proceedings of MUC-7, Virginia. SAIC (1998)
26. Huttunen, S., Yangarber, R., Grishman, R.: Complexity of event structure in information extraction. In: Proceedings of the 19th International Conference on Computational Linguistics (COLING 2002). Taipei (2002)
27. Iida, R., Poesio, M.: A cross-lingual ILP solution to zero anaphora resolution. In: The 49th Annual Meeting of the Association for Computational Linguistics: Human Language Technologies, Proceedings of the Conference, The Association for Computer Linguistics, Portland, Oregon, 19–24 June 2011, pp. 804–813 (2011)
28. Jacobs, P., Rau, L.: SCISOR: extracting information from on-line news. Commun. ACM **33**, 88–97 (1990)
29. Ji, H.: Challenges from information extraction to information fusion. In: Proceedings of the 23rd International Conference on Computational Linguistics: Posters, COLING '10, pp. 507–515. Association for Computational Linguistics, Stroudsburg (2010)
30. Ji, H., Grishman, R.: Refining event extraction through cross-document inference. In: Proceedings of ACL-08: HLT, pp. 254–262. Association for Computational Linguistics, Columbus (2008)
31. Jones, R., Ghani, R., Mitchell, T., Riloff, E.: Active learning for information extraction with multiple view feature sets. ECML-03 Workshop on Adaptive Text Extraction and Mining, Cavtat-Dubrovnik (2003)
32. Kaiser, K., Miksch, S.: Information extraction – a survey. Tech. Rep. Asgaard-TR-2005-6, Vienna University of Technology, Institute of Software Technology and Interactive Systems, Vienna (2005)
33. Lee, A., Passantino, M., Ji, H., Qi, G., Huang, T.S.: Enhancing multi-lingual information extraction via cross-media inference and fusion. In: COLING (Posters), Beijing, pp. 630–638 (2010)
34. Lehnert, W., Cardie, C., Fisher, D., McCarthy, J., Riloff, E., Soderland, S.: University of Massachusetts: MUC-4 test results and analysis. In: Proceedings of the 4th Message Understanding Conference. Morgan Kaufmann, McLean (1992)
35. Lehnert, W., Cardie, C., Fisher, D., Riloff, E., Williams, R.: University of Massachusetts: Description of the CIRCUS system as used for MUC-3. In: Proceedings of the 3rd Message Understanding Conference. Morgan Kaufmann, San Diego (1991)

36. Liao, S., Grishman, R.: Using document level cross-event inference to improve event extraction. In: Proceedings of ACL 2010, Uppsala, pp. 789–797. ACL (2010)
37. Liu, X., Zhang, S., Wei, F., Zhou, M.: Recognizing named entities in tweets. In: Proceedings of the 49th Annual Meeting of the Association for Computational Linguistics. Human Language Technologies, Vol. 1, pp. 359–367. Association for Computational Linguistics, Stroudsburg (2011)
38. Llytinen, S., Gershman., A.: ATRANS: automatic processing of money transfer messages. In: Proceedings of the 5th National Conference of the American Association for Artificial Intelligence. IEEE Computer Society Press (1986)
39. Locke, B., Martin, J.: Named entity recognition: adapting to microblogging. Senior Thesis, University of Colorado, Colorado (2009)
40. Makhoul, J., Kubala, F., Schwartz, R., Weischedel, R.: Performance measures for information extraction. In: Proceedings of DARPA Broadcast News Workshop, Herndon (1999)
41. Manning, C., Schütze, H.: Foundations of Statistical Natural Language Processing. MIT, Cambridge, MA (1999)
42. Maynard, D., Tablan, V., Cunningham, H., Ursu, C., Saggion, H., Bontcheva, K., Wilks, Y.: Architectural elements of language engineering robustness. J Nat. Lang. Engin. **8**(2/3), 257–274 (2002)
43. Mohri, M., Nederhof, M.: Regular approximation of context-free grammars through transformations. In: Junqua, J., van Noord, G. (eds.) Robustness in Language and Speech Technology, pp. 153–163. Kluwer, The Netherlands (2001)
44. Nadeau, D., Sekine, S.: A survey of named entity recognition and classification. Linguist. Investig. **30**(1), 3–26 (2007)
45. Neumann, G., Piskorski, J.: A shallow text processing core engine. Comput. Intell. **18**, 451–476 (2002)
46. Ng, V., Cardie, C.: Combining sample selection and error-driven pruning for machine learning of coreference rules. In: Proceedings of the 2002 Conference on Empirical Methods in Natural Language Processing, Philadelphia, pp. 55–62 (2002)
47. Patwardhan, S., Riloff, E.: Effective information extraction with semantic affinity patterns and relevant regions. In: Proceedings of the 2007 Joint Conference on Empirical Methods in Natural Language Processing and Computational Natural Language Learning, pp. 717–727 (2007)
48. Phillips, W., Riloff, E.: Exploiting strong syntactic heuristics and co-training to learn semantic lexicons. In: Proceedings of the 2002 Conference on Empirical Methods in Natural Language Processing (EMNLP 2002) (2002)
49. Piskorski, J.: ExPRESS – extraction pattern recognition engine and specification suite. In: Proceedings of FSMNLP 2007 (2007)
50. Piskorski, J., Belayeva, J., Atkinson, M.: On refining real-time multilingual news event extraction through deployment of cross-lingual information fusion techniques. In: EISIC, pp. 38–45. IEEE (2011)
51. Piskorski, J., Tanev, H., Atkinson, M., van der Goot, E., Zavarella, V.: Online news event extraction for global crisis surveillance. Trans. Comput. Collectiv. Intell. (5) (2011)
52. Piskorski, J., Wieloch, K., Sydow, M.: On knowledge-poor methods for person name matching and lemmatization for highly inflectional languages. Inf. Retr. **12**(3), 275–299 (2009)
53. Poibeau, T., Saggion, H. (eds.): In: Proceedings of the MMIES Workshop, RANLP: International Conference on Recent Advances in Natural Language Processing. Borovets, Bulgaria (2007)
54. Poibeau, T., Saggion, H., Yangarber, R. (eds.): In: Proceedings of the MMIES Workshop, COLING: International Conference on Computational Linguistics. Manchester (2008)
55. Recasens, M., Marquez, L., Sapena, E., Martï£¡, A., Taule, M., Hoste, V., Poesio, M., Versley, Y.: Proceedings of the 5th International Workshop on Semantic Evaluation (SemEval-2010), ACL 2010. Uppsala, Sweden. In: SemEval-2010 Task 1: Coreference Resolution in Multiple Languages, pp. 1–8 (2010)
56. Riloff, E.: Automatically constructing a dictionary for information extraction tasks. In: Proceedings of Eleventh National Conference on Artificial Intelligence (AAAI-93), Washington, DC, pp. 811–816. AAAI/MIT (1993)

57. Riloff, E.: Automatically generating extraction patterns from untagged text. In: Proceedings of Thirteenth National Conference on Artificial Intelligence (AAAI-96), Portland, pp. 1044–1049. AAAI/MIT (1996)
58. Ritter, A., Clark, S., Mausam, Etzioni, O.: Named entity recognition in tweets: an experimental study. In: Proceedings of the 2011 Conference on Empirical Methods in Natural Language Processing (EMNLP 2011), pp. 1524–1534. Association for Computational Linguistics, Edinburgh/Scotland (2011)
59. Sakaki, T., Okazaki, M., Matsuo, Y.: Earthquake shakes twitter users: real-time event detection by social sensors. In: Proceedings of WWW 2010, Raleigh, pp. 851–860. ACM (2010)
60. Settles, B.: Biomedical named entity recognition using conditional random fields and rich feature sets. In: In Proceedings of the International Joint Workshop on Natural Language Processing in Biomedicine and its Applications, pp. 104–107. NLPBA (2004)
61. Shinyama, Y., Sekine, S.: Preemptive information extraction using unrestricted relation discovery. In: Proceedings of the Main Conference on Human Language Technology Conference of the North American Chapter of the Association of Computational Linguistics, HLT-NAACL '06, pp. 304–311. Association for Computational Linguistics, Stroudsburg (2006)
62. Sidner, C.L., Schultz, T., Stone, M., Zhai, C. (eds.): Multi-Document Relationship Fusion via Constraints on Probabilistic Databases. The Association for Computational Linguistics (2007)
63. Stevenson, M., Greenwood, M.A.: A semantic approach to IE pattern induction. In: Knight, K., Ng, H.T., Oflazer, K. (eds.) ACL. The Association for Computer Linguistics (2005)
64. Sudo, K., Sekine, S., Grishman, R.: Cross-lingual information extraction system evaluation. In: Proceedings of the 20th International Conference on Computational Linguistics, COLING '04. Association for Computational Linguistics, Stroudsburg (2004)
65. Tanev, H., Piskorski, J., Atkinson, M.: Real-time news event extraction for global crisis monitoring. In: Proceedings of NLDB 2008, London, pp. 207–218 (2008)
66. Thelen, M., Riloff, E.: A bootstrapping method for learning semantic lexicons using extraction pattern contexts. In: Proceedings of the 2002 Conference on Empirical Methods in Natural Language Processing (EMNLP 2002) (2002)
67. Verma, S., Vieweg, S., Corvey, W., Palen, L., Martin, J., Palmer, M., Schram, A., Anderson, K.: Natural language processing to the rescue? extracting "situational awareness" tweets during mass emergency. In: Proceedings of the 5th International AAAI Conference on Weblogs and Social Media (ICWSM 2011), Barcelona, pp. 385–392. AAAI (2011)
68. Vieweg, S., Hughes, A., Starbird, K., Palen, L.: Microblogging during two natural hazards events: what twitter may contribute to situational awareness. In: Proceedings of the 28th International Conference on Human Factors in Computing Systems, pp. 1079–1088. ACM, New York (2010)
69. Wagner, E.J., Liu, J., Birnbaum, L., Forbus, K.D., Baker, J.: Using explicit semantic models to track situations across news articles. In: Proceedings of the 2006 AAAI Workshop on Event Extraction and Synthesis, pp. 42–47 (2006)
70. Witten, I.H., Frank, E.: Data mining: practical machine learning tools and techniques, 2nd edn. The Morgan Kaufmann Series in Data Management Systems. Morgan Kaufmann, San Francisco (2005)
71. Yangarber, R.: Counter-training in discovery of semantic patterns. In: Proceedings of the 41st Annual Meeting of the Association for Computational Linguistics, Sapporo (2003)
72. Yangarber, R.: Verification of facts across document boundaries. In: Proceedings IIIA-2006: International Workshop on Intelligent Information Access, IIIA-2006 (2006)
73. Yangarber, R., Jokipii, L.: Redundancy-based correction of automatically extracted facts. In: Proceedings of the conference on Human Language Technology and Empirical Methods in Natural Language Processing, HLT '05, pp. 57–64. Association for Computational Linguistics, Stroudsburg (2005)
74. Yangarber, R., Lin, W., Grishman, R.: Unsupervised learning of generalized names. In: Proceedings of COLING: the 19th International Conference on Computational Linguistics, Taipei (2002)

75. Yates, A., Banko, M., Broadhead, M., Cafarella, M.J., Etzioni, O., Soderland, S.: Textrunner: Open information extraction on the web. In: HLT-NAACL (Demonstrations), Rochester, pp. 25–26 (2007)
76. Zavarella, V., Tanev, H., Piskorski, J.: Event extraction for Italian using a cascade of finite-state grammars. In: Proceedings of FSMNLP 2008, Ispra (2008)

Part II
Named Entity in a Multilingual Context

Chapter 3
Learning to Match Names Across Languages

Inderjeet Mani, Alex Yeh, and Sherri Condon

Abstract We report on research on matching names in different scripts across languages. We explore two trainable approaches based on comparing pronunciations. The first, a cross-lingual approach, uses an automatic name-matching program that exploits rules based on phonological comparisons of the two languages carried out by humans. The second, monolingual approach relies only on automatic comparison of the phonological representations of each pair. Alignments produced by each approach are fed to a machine learning algorithm. Results show that the monolingual approach results in machine-learning based comparison of person-names in English and Chinese at an accuracy of over 97.0 F-measure.

3.1 Introduction

The problem of matching pairs of names which may have different spellings or segmentation arises in a variety of common settings, including integration or linking database records, mapping from text to structured data (e.g., phonebooks, gazetteers, and biological databases), and text to text comparison (for information retrieval, clustering, summarization, coreference, etc.). For named entity recognition, a name from a gazetteer or dictionary may be matched against text input; even within monolingual applications, the forms of these names might differ. In multi-document summarization, a name may have different forms across different sources. Systems that address this problem must be able to handle variant spellings, as well as

---

I. Mani (✉) · A. Yeh
The MITRE Corporation, 202 Burlington Road, Bedford, MA, 01730, USA
e-mail: imani@mitre.org; inderjeet.mani@gmail.com; asy@mitre.org

S. Condon
The MITRE Corporation, 7515 Colshire Drive, McLean, VA, 22102, USA
e-mail: scondon@mitre.org

T. Poibeau et al. (eds.), *Multi-source, Multilingual Information Extraction
and Summarization 11*, Theory and Applications of Natural Language Processing,
DOI 10.1007/978-3-642-28569-1__3, © Springer-Verlag Berlin Heidelberg 2013

Table 3.1 Mapping of names to Mandarin

Roman script	IPA for Roman	Mandarin	Pinyin	IPA for Mandarin
Stewart	/s t u w ə r t/	斯图尔特	si tu er te	/s i tʰ u a ɻ tʰ e/
Elizabeth	/ I l I z ə b ɛ θ/	伊丽莎白	yi li sha bai	/I l I ş ɑ p aI/

abbreviations, missing or additional name parts, and different orderings of name parts.

In multilingual settings, where the names being compared can occur in different scripts in different languages, the problem becomes relevant to additional practical applications, including both multilingual information retrieval and machine translation. Commercial applications of identity management address the need to accurately identify clients, prospects, suppliers, taxpayers, or criminal suspects, and one data integration corporation claims that "most large identity databases now contain identity data from multiple languages, countries and cultures".[1] Businesses track identities across languages and cultures to investigate credit or fraud and to comply with regulations such as sanctions and embargoes enforced by the United States Office of Foreign Assets Control.

When matching names from different languages, special challenges are posed by the fact that there usually are not one-to-one correspondences between sounds across languages. Table 3.1 shows the mapping of names in Roman script to Mandarin and the corresponding Pinyin.

Further, in a given writing system, there may not be a one-to-one correspondence between orthography and sound, a well-known case in point being English. In addition, there may be a variety of variant forms, including dialectical variants, (e.g., *Bourguiba* can map to *Abu Ruqayba*), orthographic conventions (e.g., Anglophone *Wasim* can map to Francophone *Ouassime*), and differences in name segmentation (*Abd Al Rahman* can map to *Abdurrahman*). Given the high degree of variation and noise in the data, approaches based on machine learning are needed.

The considerable differences in possible spellings of a name also call for approaches which can compare names based on pronunciation. Recent work has developed pronunciation-based models for name comparison, e.g., [28, 29]. This paper explores trainable pronunciation-based models further.

Consider the problem of matching Chinese script names against their English (Pinyin) Romanizations. Chinese script has nearly 50,000 characters in all, with around 5,000 characters in use by the well-educated. However, there are only about 1,600 Pinyin syllables when tones are counted, and as few as 400 when they are not. This results in multiple Chinese script representations for a given Roman form name and many Chinese characters that map to the same Pinyin forms. In addition, one can find multiple Roman forms for many names in Chinese script, and multiple Pinyin representations for a Chinese script representation.

---

[1] www.informatica.com/solutions/identity_resolution_solution/Pages/index.aspx.

In developing a multilingual approach that can match names from any pair of languages, we compare an approach that relies strictly on **monolingual** knowledge for each language, specifically, grapheme-to-phoneme rules for each language, with a method that relies on **cross-lingual** rules which in effect map between graphemic and/or phonemic representations for the specific pair of languages.

The monolingual approach requires finding data on the phonemic representations of a name in a given language, which (as we describe in Sects. 3.3 and 3.4) may be harder than finding more graphemic representations. But once the phonemic representation is found for names in a given language, then as one adds more languages to a system, no more work needs to be done in that given language. In contrast, with the cross-lingual approach, whenever a new language is added, one needs to go over all the existing languages already in the system and compare each of them with the new language to develop cross-lingual rules for each such language pair. The engineering of such rules requires bilingual expertise, and knowledge of differences between language pairs. The cross-lingual approach is thus more expensive to develop, especially for applications which require coverage of a large number of languages.

Our paper investigates whether we can address the name-matching problem without requiring such a knowledge-rich approach, by carrying out a comparison of the performance of the two approaches. We present results of large-scale machine-learning for matching personal names in Chinese and English using large training sets to establish proof of the concept. We also present some preliminary results for English and Urdu based on very small training sets.

3.2 Related Work

3.2.1 Name Matching

Early research on name-matching focused on string comparison methods for searching and indexing records using person names. The Russell Soundex method, copyrighted in 1918, groups similar English names using numerical representations of character strings based on grapheme-phoneme correspondences [36]. Although Soundex does not provide a similarity score, it has been extended to other languages and inspired many variants, but studies have demonstrated that it performs poorly compared to other string similarity measures [17, 36]. These other measures include scores based on character n-grams [31] and a family of measures developed for the United States Census Bureau based on the number of common characters in the string and incorporating weights depending on the source files of the names [35].

Levenshtein edit-distance (or Damerau-Levenshtein distance) has been used extensively as a string comparison measure, and dynamic programming can be used to find optimal alignments that minimize the cost of edits [4, 18]. Much of the research that compares names using string similarity/distance measures has been conducted in the context of record linkage or deduplication of records, in which the name is only one of the strings that is compared in the record. While there is a substantial literature employing learning techniques for record linkage based on the theory developed by Fellegi and Sunter [5], researchers have only recently developed applications that focus on name strings and that employ methods which do not require features to be independent [3]. Ristad and Yianilos [25] have developed a generative model for learning string edit-distance that learns the cost of different edit operations during string alignment. Bilenko and Mooney [2] extend the approach of [25] to include gap penalties (where the gaps are contiguous sequences of mis-matched characters) and compare this generative approach with a vector similarity approach that does not carry out alignment.

McCallum et al. use Conditional Random Fields (CRFs) to learn edit costs, arguing in favor of discriminative training approaches and against generative approaches, based in part on the fact that the latter approaches "cannot benefit from negative evidence from pairs of strings that (while partially overlapping) should be considered dissimilar" [21]. Such CRFs model the conditional probability of a label sequence (an alignment of two strings) given a sequence of observations (the strings).

3.2.2 Automatic Transliteration

A related thread of research is work on automatic transliteration. Here the goal is not to match names, but to translate a name in one language (and orthography) into a phonemically similar and acceptable name in another language (and orthography). Some of the approaches to English-Chinese transliteration incorporate handcrafted rules and mappings. The mapping system developed by Wan and Verspoor is entirely handcrafted and rule-based, and they adopt the strategy of inserting vowels into the sequence of English sounds to produce syllables that can be mapped to Pinyin forms [33]. Meng et al. also use handcrafted rules to normalize sequences of English phonemes, which are mapped to sequences of Chinese phonemes using rules derived from transformation-based error driven learning [22]. The Chinese phonemes are mapped to Pinyin using a bigram language model. Finally, Jung, Hong, and Paek's tool for transliterating English to Korean begins with a handcrafted table that maps English phonemes to Korean phonemes [14].

The fully data-driven methods often employ weighted finite state transducers (WFST) to generate transliterations, as proposed by Knight and Graehl for Japanese to English backward transliteration [16]. In their system, cascading WFSTs map between graphemes and phonemes in each language and from English to Japanese phonemes. Expectation maximization (EM) is used to align the latter for training. Gao, Wong and Lam also use EM to align training samples and obtain probabilities for a WFST that maps English phonemes into Pinyin characters [8]. In this system, dummy symbols, usually corresponding to epenthetic vowels, are inserted by pre-processing operations that combine sequences of characters in both English and Chinese strings before alignment.

Working with English and Arabic, Al-Onaizan and Knight obtained improvements in performance when they combined the phonetic approach in [16] with one based on spelling alone [1]. Li, Zhang, and Su experimented with a joint source-channel model based only on spelling using EM to learn n-gram alignments between Roman and Pinyin characters [20]. To find English transliterations for named entities in Chinese, Huang, Vogel, and Waibel compute a transliteration probability for a pair of English and Chinese named entities as the product of their character transliteration probabilities, and the probabilities are learned by aligning names using the Levenshtein edit-distance algorithm with edit costs reduced for phonetically similar substitutions [10]. They also incorporate a similarity score based on semantic context vectors, which incorporate content words that surround the names and weights based on the context words POS tags, locations, and frequency of co-occurrence with the names.

Virga and Khudanpur treat the transliteration problem as a translation problem (over the "vocabulary" of phonemes) and use the IBM source-channel model [32]. In this approach the epenthetic vowels needed to make English words conform to Chinese syllable structure are treated as zero fertility words. Ji et al. have also experimented with statistical machine translation techniques to translate the sequences of characters in Arabic and Chinese names to sequences of characters in English names [11]. They compare this method to one in which a structured perceptron was trained to emit character edit operations on a name string to generate an English version of the name. They obtain comparable performance from the two approaches and report that combining the two approaches increases performance [7].

Jiampojamarn et al. also adopt a discriminative training framework to learn which pairs have the most heavily weighted feature vectors, using EM to align name pairs in the training data [12]. Their features include n-grams in the source string that surround a source substring as it generates a target substring, as well as bigram context in the target substring to constrain well-formedness of the output.

3.2.3 Exploiting Phonological Information

The hypothesis that phonological information can improve matching performance has been explored by a number of researchers cited above. Jiampojamarn et al. not only compare orthographic representations, but they also boost performance using Kondrak's ALINE tool, which is described in more detail below, to carry out alignment of the names based on phonemic representations of Roman script English and Pinyin Chinese [12].

Among the studies cited above, [12] and [14] exploit the context of phonemes within the name when computing alignment probabilities. In [14] Markov chains are used to approximate conditional probabilities for an "extended Markov window", which takes into account the phonemes immediately preceding the current Korean and English phonemes as well as the phoneme that immediately follows the

English phoneme. Oh, Choi, and Isahara experiment with use of context and several other variables for transliteration of English to Korean and Japanese [24]. They systematically compare (1) models based on functions that map graphemes directly to graphemes such as [20], (2) models based on functions that map source graphemes to source phonemes plus functions that map source phonemes to target graphemes such as [14, 16, 22], and (3) hybrid models that are linear interpolations of the previous two models with varying interpolation parameters such as [1].

The comparison includes a fourth type of model that Oh, Choi and Isahara [24] call correspondence-based, based on functions that map source graphemes to source phonemes plus functions that map source grapheme-phoneme correspondences to target graphemes. Unlike hybrid models, correspondence-based models take into account the dependence between the source grapheme and source phoneme in the mapping process. The implementations are designed to compare results from models that incorporate contexts extending from 1 to 5 units to the left and right of each grapheme or phoneme. The best performances for both languages are achieved by the correspondence-based approach implemented using a maximum entropy model with the context extending 3 units left and right of each grapheme or phoneme. Like the results reported in [14], these results are based on transliteration of general vocabulary, rather than personal names.

In other phonologically oriented research, Sproat, Tao, and Zhai have compared names from comparable and contemporaneous English and Chinese texts, scoring matches by training a learning algorithm to compare the phonemic representations of the names in the pair, in addition to taking into account the frequency distribution of the pair over time [28]. Tao et al. obtain similar results using frequency and a similarity score based on a phonetic cost matrix [29].

3.3   Basic Approaches

3.3.1   *Cross-Lingual Approach*

Our cross-lingual approach (called MLEV) is based on earlier work at MITRE by Freeman et al., who use a modified Levenshtein string edit-distance algorithm to match Arabic script person names against their corresponding English versions [6]. The Levenshtein edit-distance algorithm counts the minimum number of insertions, deletions or substitutions required to make a pair of strings match. Freeman et al. [6] use (1) insights about phonological differences between the two languages to create rules for equivalence classes of characters that are treated as identical in the computation of edit-distance and (2) the use of normalization rules applied to the English and transliterated Arabic names based on mappings between characters in the respective writing systems.

For example, Table 3.2 shows a rule which normalizes low and high diphthongs to Arabic. As [6] point out, these insights are not easy to come by: "These rules are

Table 3.2 An Arabic-English normalization rule

English diphthongs	Normalization	Arabic script
low	w	و
high	y	ي

Table 3.3 Matching *Ashburton* and Pinyin for Ashburton

Method	Roman	Chinese (Pinyin)	Alignment										Score
LEV	ashburton	ashenbodu	❘ a s h b u r t o n ❘										0.67
			❘ a s h e n b o d u ❘										
MLEV	ashburton	ashenbodu	❘ a s h - - b u r t o n ❘										0.72
			❘ a s h e n b o - d u - ❘										
MALINE	asVburton	aseCnpotu	❘ a sV - b < u r t o ❘ n										0.48
			❘ a s eC n p o - t u ❘ -										

based on first author Dr. Andrew Freeman's experience with reading and translating Arabic language texts for more than 16 years" (ibid., p. 474).

Table 3.3 shows the representation and comparison of a Roman-Chinese name pair obtained from the Chinese-English name pairs corpus from the Linguistic Data Consortium (LDC2005T34).[2] This corpus provides name part pairs, the first element in English (Roman characters) and the second in Chinese characters, created by the LDC from the Xinhua Newswire proper name and who's who databases. The name part can be a first, middle or last name. We compare the English form of the name with a Pinyin Romanization of the Chinese. Note that many Chinese characters have more than one possible Pinyin representation. For instance, there are actually 16 different Chinese pinyinizations of the Chinese transliteration of *Ashburton*, according to the tables our system uses to map a Chinese character to its Pinyin. For the experiments reported here, the tool which maps Chinese characters to Pinyin arbitrarily selects one representation, rather than trying all combinations. Consequently, it does not always find the best mapping. In this case, the alternative representation "a shi bo dun" provides a pronunciation that is closer to *Ashburton*. Also, since the Chinese is being compared with English, which is toneless, the tone part of Pinyin is being ignored throughout this paper.

For this study, the Levenshtein edit-distance score (where a perfect match scores zero) is normalized to a similarity score as in [6], where the score ranges from 0 to 1, with 1 being a perfect match. This similarity score is shown in the LEV row of Table 3.3.

The MLEV row of Table 3.3, under the Chinese column, shows an English-oriented normalization of the Pinyin for *Ashburton*. Certain characters or character sequences in Pinyin are pronounced differently than in English. We therefore apply certain transforms to the Pinyin; for example, the following substitutions are applied at the start of a Pinyin syllable, which makes it easier for an English speaker to see

---

[2] www.ldc.upenn.edu/Catalog/CatalogEntry.jsp?catalogId=LDC2005T34.

how to pronounce it and renders the Pinyin more similar to English orthography: "u:"(umlaut"u") ⇒ "u", "zh" ⇒ "j", "c" ⇒ "ts", and "q" ⇒ "ch" (so the Pinyin "Qian" is more or less pronounced as if it were spelled as *Chian*, etc.). The MLEV algorithm uses equivalence classes that allow "o" and "u" to match, which results in a higher score than the generic score using the LEV method.

3.3.2 Monolingual Approach

Instead of relying on rules that require extensive knowledge of differences between a language pair, the monolingual approach first builds phonemic representations for each name, and then aligns them. Earlier research by Kondrak [15] uses dynamic programming to align strings of phonemes, representing the phonemes as vectors of phonological features, which are associated with scores to produce similarity values. His program ALINE[3] includes a 'skip' function in the alignment operations that can be exploited for handling epenthetic segments, and in addition to 1:1 alignments, it also handles 1:2 and 2:1 alignments. In this research, we made extensive modifications to ALINE to add the phonological features for languages like Chinese and Arabic and to normalize the similarity scores, producing a system called MALINE.

In Table 3.3, the MALINE row[4] shows that the English name has a palato-alveolar modification on the "s" (expressed as "sV", i.e., voiceless postalveolar fricative ʃ), so that we get the sound corresponding to "sh". The Pinyin name inserts a centered "e" vowel (expressed as "C", i.e., close-mid central unrounded vowel ɘ), and devoices the bilabial plosive /b/ to /p/. The MALINE score reflects the phonetic similarity of the fricatives and plosives in the two representations.

3.4 Experimental Setup

3.4.1 Machine Learning Framework

Neither of the two basic approaches described so far uses machine learning. Our machine learning framework is based on learning from alignments produced by

---

[3] webdocs.cs.ualberta.ca/~kondrak/aline1.1.zip.

[4] For the MALINE row in Table 3.3, the ALINE documentation explains the notation as follows: "every phonetic symbol is represented by a single lowercase letter followed by zero or more uppercase letters. The initial lowercase letter is the base letter most similar to the sound represented by the phonetic symbol. The remaining uppercase letters stand for the feature modifiers which alter the sound defined by the base letter. By default, the output contains the alignments together with overall similarity scores. The aligned subsequences are delimited by 'l' signs. The '<' sign signifies that the previous phonetic segment has been aligned with two segments in the other sequence, a case of compression/expansion. The '−' sign denotes a "skip", a case of insertion/deletion."

either approach. To view the learning problem as one amenable to a statistical classifier, we need to generate labeled feature vectors so that each feature vector includes an additional class feature that can have the value "true" or "false". Given a set of such labeled feature vectors as training data, the classifier builds a model which is then used to classify unlabeled feature vectors with the right labels.

A given set of attested name pairs constitutes a set of positive examples. To create negative pairs, we have found that randomly selecting elements that have not been paired will create negative examples in which the pairs of elements being compared are so different that they can be trivially separated from the positive examples. The experiments reported here used the MLEV score as a threshold to select negatives, so that examples below the threshold are excluded. As the threshold is raised, the negative examples should become harder to discriminate from positives (with the harder problems mirroring some of the "confusable name" characteristics of the real-world name-matching problems this technology is aimed at). Positive examples below the threshold are also eliminated. Other criteria, including a MALINE score, could be used, but the MLEV scores seemed adequate for these preliminary experiments.

Raising the threshold reduces the number of negative examples. It is highly desirable to balance the number of positive and negative examples in training, to avoid the learning being biased by a skewed distribution. However, when one starts with a balanced distribution of positive and negatives, and then excludes a number of negative examples below the threshold, a corresponding number of positive examples must also be removed to preserve the balance. Thus, raising the threshold reduces the size of the training data. Machine learning algorithms, however, can benefit from more training data. Therefore, in the experiments below, thresholds which provided woefully inadequate training set sizes were eliminated.

One can think of both the machine learning method and the basic name comparison methods (MLEV and MALINE) as taking each pair of names with a known label and returning a system-assigned class for that pair. Precision, Recall, and F-Measure can be defined in an identical manner for both machine learning and basic name comparison methods. In such a scheme, a threshold on the similarity score is used to determine whether the basic comparison match is a positive match or not. Learning the best threshold for a dataset can be determined by searching over different values for the threshold.

In short, the methodology employed for this study involves two types of thresholds: the MLEV threshold used to identify negative examples and the threshold that is applied to the basic comparison methods, MLEV and MALINE, to identify matches. To avoid confusion, the term *negative threshold* refers to the former, while the term *positive threshold* is used for the latter.

The basic comparison methods were used as baselines in this research. To be able to provide a fair basic comparison score at each negative threshold, we ran each basic comparison matcher at 20 different positive thresholds on the same training set used by the learner. For each negative threshold, we picked the positive threshold that gave the best performance on the training data, and used that to score the matcher on the same test data as used by the learner.

3.4.2 Feature Extraction

We used two types of features: edit features and context features. The edit features, called *match unigrams*, represent the particular insertion, deletion, or substitution operation between a source character and the target character it generates. Consider the MLEV alignment in Table 3.3. It can be seen that the first three characters are matched identically across both strings; after that, we get an "e" inserted, an "n" inserted, a "b" matched identically, a "u" matched to an "o", an "r" deleted, a "t" matched to a "d", an "o" matched to a "u", and an "n" deleted. The match unigrams are thus *[a:a], [s:s], [h:h], [-:e], [-:n], [b:b], [u:o], [r:-], [t:d], [o:u],* and *[n:-]*.

In addition to edit features, we also used a limited representation of context in terms of edit operations carried out in the immediately prior context. These first-order Markov features, called *match bigrams*, were generated by considering any insertion, deletion, and (non-identical) substitution unigram, and noting the unigram, if any, to its left, prepending that left unigram to it (delimited here by a comma). Thus, the match bigrams in the above example include*[h:h,-:e], [-:e,-:n], [b:b,u:o], [u:o,r:-],[r:-,t:d], [t:d,o:u], [o:u,n:-]*.

The above match unigram and match bigram features are generated from just a single MLEV match. The composite feature set is the union of all the match unigram and bigram features across the matches. Given the composite feature set, each match pair is turned into a feature vector consisting of the following features: string1, string2, the match score according to each of the basic comparison matchers (MLEV and MALINE), and the Boolean value of each feature in the composite feature set.

3.4.3 Data Set

Our data is a (roughly 470,000 attested-pair) subset of the Chinese-English personal name pairs in LDC2005T34. About 150,000 of the pairs had more than one way to pronounce the English and/or Chinese. For these, to keep the size of the experiments manageable from the point of view of training the learners, one pronunciation was randomly chosen as the one to use. (Even with this restriction, a minimum negative threshold results in over half a million examples consisting of positive and negative pairs.) Chinese characters were mapped into Hanyu Pinyin representations, which are used for MLEV alignment and string comparisons. Since the input to MALINE uses a phonemic representation that encodes phonemic features in one or more letters, both Pinyin and English forms were mapped into the MALINE notation.

There are a number of slightly varying ways to map Pinyin into an international pronunciation system like IPA. For example, Wikipedia [34] and Safalra [26] have mappings that differ from each other and also each of these two sources have changed its mapping over time. We used a version of Safalra [26] (but we ignored the ejectives).

For English, the CMU pronouncing dictionary [30] provided phonemic representations that were then mapped into the MALINE notation. The dictionary had entries for only 12% of our data set. For the names not in the CMU dictionary, a simple grapheme to phoneme script provided an approximate phonemic form. We did not use a monolingual mapping of Chinese characters (Mandarin pronunciation) into IPA because we did not find any.

Note that we could insist that all pairs in our dataset be distinct, requiring that there be exactly one match for each Roman name and exactly one match for each Pinyin name. This in our view is unrealistic, since large corpora will be skewed towards names which tend to occur frequently (e.g., international figures in news) and occur with multiple translations. We included attested match pairs in our test corpora, regardless of the number of matches that were associated with a member of the pair.

3.5 Results

A variety of machine learning algorithms were tested. Results are reported, unless otherwise indicated, using SVM Lite [13], a Support Vector Machine (SVM) classifier that scales well to large data sets. We used a linear kernel function in our SVM experiments; using polynomial or radial basis kernels did not improve performance.

Testing with SVM Lite was done with a 90/10 train-test split. Further testing was carried out with the weka SMO SVM classifier[9], which used built-in cross-validation. Although the latter classifier did not scale to the larger data sets we used, it did show that cross-validation did not change the basic results for the data sets it was tried on.

3.5.1 Machine Learning with Different Feature Sets

Table 3.4 shows the F-measure for learning for monolingual features (M, based on MALINE), cross-lingual features (X, based on MLEV), and a combined feature set (C) of both types of features at different negative thresholds (NT). Baseline scores are shown in square brackets.

Fig. 3.1 displays the same F-measure scores graphically, at different negative thresholds (shown on the horizontal axis). Baselines are shown with the suffix B, e.g., the basic MALINE without learning is MB.

The X curve is more or less under the C curve. When using both monolingual and cross-lingual features (C), the baseline (CB) is set to a system response of 'true' only when both the MALINE and MLEV baseline systems by themselves respond 'true'.

Table 3.4 F-measure with different feature sets (Baselines in square brackets)

NT	Monolingual (M)	Cross-lingual (X)	Combined (C)
0	95.20 [90.79]	96.82 [91.66]	97.38 [90.89]
0.1	95.75 [85.20]	98.15 [89.54]	98.51 [85.81]
0.2	96.22 [85.29]	98.15 [89.33]	98.53 [85.68]
0.3	96.35 [82.88]	98.48 [88.44]	98.75 [82.64]
0.4	97.48 [79.61]	98.74 [88.91]	99.27 [78.95]
0.5	96.58 [73.14]	99.71 [90.63]	99.67 [72.77]
0.6	97.40 [71.13]	98.32 [88.83]	99.68 [66.68]
0.7	97.56 [69.10]	97.29 [78.9]	97.29 [64.13]

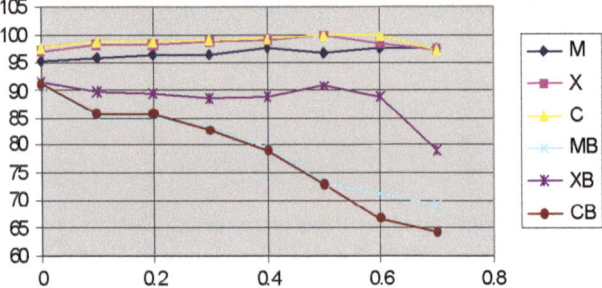

Fig. 3.1 F-measure (graphical summary) with different feature sets

Table 3.5 Precision and recall with different feature sets (Baseline scores in square brackets)

NT	Examples	Monolingual (M) P	R	Cross-lingual (X) P	R	Combined (C) P	R
0	538,621	94.69 [90.6]	95.73 [91.0]	96.5 [90.0]	97.15 [93.4]	97.13 [90.8]	97.65 [91.0]
0.1	307,066	95.28 [87.1]	96.23 [83.4]	98.06 [89.2]	98.25 [89.9]	98.4 [87.6]	98.64 [84.1]
0.2	282,214	95.82 [86.2]	96.63 [84.4]	97.91 [88.4]	98.41 [90.3]	98.26 [86.7]	98.82 [84.7]
0.3	183,188	95.79 [80.6]	96.92 [85.3]	98.18 [86.3]	98.8 [90.7]	98.24 [80.6]	99.27 [84.8]
0.4	72,176	96.31 [77.1]	98.69 [82.3]	97.89 [91.8]	99.61 [86.2]	98.91 [77.1]	99.64 [80.9]
0.5	17,914	94.62 [64.6]	98.63 [84.3]	99.44 [89.4]	100.0 [91.9]	99.46 [63.8]	99.89 [84.7]
0.6	2,954	94.94 [66.1]	100 [77.0]	98.0 [85.2]	98.66 [92.8]	99.37 [61.3]	100.0 [73.1]
0.7	362	95.24 [52.8]	100 [100.0]	94.74 [78.9]	100.0 [78.9]	100.0 [47.2]	94.74 [100.0]

Table 3.5 shows the number of examples at each negative threshold and the Precision and Recall for these methods, along with baselines using the basic methods shown in square brackets.

These results show that (i) the learning method outperforms the baselines (basic methods), and (ii) the gap between learning and basic comparison widens as the problem becomes harder (i.e., as the threshold is raised). The latter result demonstrates how strongly evaluation of similarity-based matching algorithms can depend on the composition of the test set. Matchers may achieve high scores when the test sets contain large proportions of very dissimilar non-matching pairs, but

Table 3.6 Accuracy with different feature sets

NT	Examples	Monolingual (M)		Cross-lingual (X)		Combined (C)	
		Learning	Baseline	Learning	Baseline	Learning	Baseline
0	538,621	95.2	90.8	96.8	91.5	97.4	90.9
0.1	307,066	95.7	85.5	98.2	89.5	98.5	86.1
0.2	282,214	96.2	85.4	98.2	89.2	98.5	85.9
0.3	183,188	96.3	82.4	98.5	88.2	98.7	82.2
0.4	72,176	97.5	78.9	98.7	89.3	99.3	78.4
0.5	17,914	96.5	69.1	99.7	90.5	99.7	68.3
0.6	2,954	97.3	68.8	98.3	88.3	99.7	63.5
0.7	362	97.5	55.3	97.2	78.9	97.4	44.1

they can experience significant decreases in performance when test sets consist of more similar non-matching pairs. For example, when the negative pairs in the test set were maximally similar according to the MLEV matcher, the F-measure for the MALINE matcher dropped from 90.79 to 69.10 (from Table 3.4). Consequently, conventionally used string comparison methods such as Levenshtein edit-distance may perform reasonably well in some contexts, but their weaknesses are revealed when equally small differences are significant for some name pairs, but not for others. In contrast, our approach and similar methods provide an opportunity for the matcher to learn which differences are more likely to reflect different names.

Statistical significance between these F-measures is not directly computable since the overall F-measure is not an average of the F-measures of the data samples. Instead, we checked the statistical significance of the increase in accuracy due to learning over the baseline. The statistical significance test was done by assuming that the accuracy scores were binomials that were approximately Gaussian. When the Gaussian approximation assumption failed (due to the binomial being too skewed), a looser, more general bound was used (Chebyshev's inequality, which applies to all probability distributions with a finite variance). Accordingly, in Table 3.6, we provide along with the number of examples at each negative threshold, the Predictive Accuracy for these methods.[5]

For separate monolingual and cross-lingual learning, the increase in accuracy of the learning over the baseline (non-learning) results was statistically significant at all negative thresholds except 0.6 and 0.7. For learning with combined monolingual and cross-lingual features (C), the increase over the baseline (non-learning) combined results was statistically significant at all negative thresholds except for 0.7. All statistically significant differences are at the 1% level (2-sided).

In comparing the mono-lingual and cross-lingual learning approaches, however, the only statistically significant differences were that the cross-lingual features were more accurate than the monolingual features at the 0–0.4 negative thresholds. This suggests that (iii) the mono-lingual learning approach is as viable as the cross-lingual one as the problem of confusable names becomes harder. However, using the

---

[5]The Predictive Accuracy was computed with exactly half the test examples being positive.

combined learning approach (C) is better than using either one at lower negative thresholds. Learning accuracy with both monolingual and cross-lingual features is statistically significantly better than learning with monolingual features at the 0.0–0.4 negative thresholds, and better than learning with cross-lingual features at the 0.0–0.2, and 0.4 negative thresholds. Nevertheless, although the learning accuracy of the combined approach equals or exceeds the learning accuracy based on cross-lingual features alone at each negative threshold, the highest learning accuracy is the same for both methods.

3.5.2 Feature Set Analyses

The unigram features reflect common correspondences between Chinese and English pronunciation. For example, [28] note that Chinese /l/ is often associated with English /r/, and the feature *[l:r]* is among the most frequent unigram mappings in both the MLEV and MALINE alignments. At a frequency of 103,361, it is the most frequent unigram feature in the MLEV mappings, and it is the third most frequent unigram feature in the MALINE alignments (56,780). Systematic correspondences among plosives are also captured in the MALINE unigram mappings. The unaspirated voiceless Chinese plosives /p, t, k/ contrast with aspirated plosives /p^h, t^h, k^h/, whereas the English voiceless plosives (which are aspirated in predictable environments) contrast with voiced plosives /b, d, g/. As a result, English /b, d, g/ phonemes are usually transliterated using Chinese characters that are pronounced /p, t, k/, while English /p, t, k/ phonemes usually correspond to Chinese /p^h, t^h, k^h/. The examples of *Stewart* and *Elizabeth* in Sect. 3.1 illustrate the correspondence of English /t/ and Chinese / t^h/ and of English /b/ with Chinese /p/ respectively. All six of the unigram features that result from these correspondences occur among the 20 most frequent in the MALINE alignments, ranging in frequency from 23,602 to 53,535.

3.5.3 Comparison with Other Learners

To compare with other machine learning tools, we used the WEKA toolkit [9]. Table 3.7 shows the comparisons on the MLEV data for a fixed size at one positive threshold (PT). Except for SVM Light, the results are based on ten-fold cross validation. The other classifiers appear to perform relatively worse at that setting for the MLEV data, but the differences in accuracy are not statistically significant even at the 5% level. A large contributor to the lack of significance is the small test set size of 66 pairs (10% of 660 examples) used in the SVM Light test.

Table 3.7 Comparison of different classifiers

PT	Examples	Method	P	R	F	Accuracy
0.65	660	SVM light	90.62	87.88	89.22	89.39
0.65	660	WEKA SMO	80.6	83.3	81.92	81.66
0.65	660	AdaBoost M1	84.9	78.5	81.57	82.27

Table 3.8 Urdu-Roman name-matching results with random negatives

Method	PT	Examples	P	R	F
WEKA SMO	0.55 (MALINE)	206 (MALINE)	84.8 [81.5]	86.4 [93.3]	85.6 [87.0]
WEKA SMO	0.85 (MLEV)	584 (MLEV)	89.9 [93.2]	94.7 [91.2]	92.3 [92.2]

3.5.4 Other Language Pairs

Some earlier experiments for Arabic-Roman comparisons were carried out using a Conditional Random Field learner (CRF), using the Carafe toolkit.[6] The method computes its own Levenshtein edit-distance scores, and learns edit-distance costs from that. The scores obtained, on average, had only a 0.6 correlation with the basic comparison Levenshtein scores. However, these experiments did not return accuracy results, as ground-truth data was not specified for this task.

Several preliminary machine learning experiments were also carried out on Urdu-Roman comparisons. The data used were Urdu data extracted from a parallel corpus recently produced by the Linguistic Data Consortium (LCTL_Urdu 20060408.[7]) The results are shown in Table 3.8.

Here a 0.55 MALINE score and a 0.85 MLEV score were used for selecting positive examples by basic comparison, i.e., a positive threshold (PT), and negative examples were selected at random. The MALINE method (row 1) using the weka SMO SVM made use of a threshold based on a MALINE score. In these earlier experiments, machine learning does not really improve the system performance (F-measure decreases with learning on one test and only increases by 0.1% on the other test). However, since these earlier experiments did not benefit from the use of different negative thresholds, there was no control over problem difficulty.

3.6 Our Approach in Perspective

The Named Entity Workshop at ACL 2009 (NEWS-2009) [23] focused on a shared transliteration task, which involved taking as input a test set of source language names and generating a ranked list of transliteration candidates in the

---

[6] sourceforge.net/projects/carafe.

[7] projects.ldc.upenn.edu/LCTL/.

target language for each of those names. These candidates were then scored against a reference set of target language names for those source names. The dataset comprised 38K matched English-Chinese name pairs, along with up to 10K matched name pairs for six other language pairs, all with English as the source language. The best accuracy obtained (measuring correctness of the first transliteration candidate in the candidate list) was for English to Chinese, and was 73.1 [19].

In contrast, our name-matching task is not a generative one; it involves taking pairs of source and target names as input and producing as output a score for how similar one is to the other. Our scores for English-Chinese are not directly comparable with transliteration scores due to differences in the task formulation and the size and type of the training data. However, the transliteration research underscores a reason why cross-script name-matching is not typically performed by transliterating one of the names: the transliteration procedure may introduce unpredictable errors into the process, which is likely to reduce performance.

There has been some follow-on work at MITRE related to the experiments described in this chapter. Samuel et al. applied a different phonological approach to the matching of English Roman script and Chinese names [27]. They leveraged the insights of Wan and Verspoor, who observe that an individual Chinese character corresponds not to a single phoneme as in the case of (English) Roman script but a sequence of one to three phonemes that has a single consonant, or else a vowel preceded by an optional consonant and/or succeeded by an optional nasal phoneme [33]. Such sequences are called 'subsyllable units'. The system of Samuel et al. [27] takes each Roman-script name and creates a phonemic representation which is then, following [33], split into subsyllable units. The subsyllable units are then matched to the Chinese characters in the corresponding Chinese name, with alignment costs between subsyllables and Chinese characters learned from the training data. The results, on the same dataset LDC2005T34 as ours, showed an F-measure of 96.0, outperforming our learning using MALINE. Although their results used randomly generated negative examples, their research suggests the possibility that performance of our system might be further improved using such phonological representations.

Previous approaches have all developed special-purpose machine-learning archi-tectures to address the matching of string sequences. They take pairs of strings that have not been aligned, and learn costs or mappings from them, and once trained, search for the best match given the learned representation. Our approach, by contrast, takes pairs of strings along with an alignment, and using features derived from the alignments, trains a learner to derive the best match given the features. This offers the advantage of modularity, in that any type of alignment model can be combined with SVMs or other classifiers (we have preferred SVMs since they offer discriminative training). Our approach can thus leverage any existing alignments, which can lead to starting the learning from a higher baseline and less training data to get to the same level of performance. Since the learner itself does not compute the alignments, the disadvantage of our approach is the need to engineer features that communicate important aspects of the alignment to the learner.

In addition, our approach, as with McCallum et al. [21], allows one to take advantage of both positive and negative training examples, rather than positive ones alone. Our data generation strategy has the advantage of generating negative examples in order to vary the difficulty of the problem, allowing for more fine-grained performance measures. Metrics based on such a control are likely to be useful in understanding how well a name-matching system will work in particular applications, especially those involving confusable names.

Finally, systems which use machine learning are dependent on feature engineering. It can be seen, from Sect. 3.5.2, that our features are transparent enough to capture many of the linguistic intuitions that one has when comparing writing systems and pronunciation. This, we believe, is one of the important considerations in using machine learning to perform natural language processing tasks. For improved feature engineering and enhanced linguistic understanding, it is not sufficient to achieve high accuracy in a task; at the same time we need to be able to explain how the system results compare with our linguistic intuitions.

3.7 Conclusion

The work presented here has established a framework for application of machine learning techniques to multilingual name-matching. The results show that machine learning dramatically outperforms basic comparison methods, with F-measures as high as 97.0 on the most difficult problems.

Our approach is being embedded in a larger system that matches full names using a vetted database of full-name matches for evaluation. This larger system can take into account the matching of first-names and other name components. Here Chap. 4 by P. Driscoll in this volume is relevant, as it discusses a name-nickname matching method where evidence from morphology, web-discovered name-nickname pairs and hand-written rules is combined together within a machine learning framework. Since nicknames are not only shorter but often quite dissimilar in both orthography and pronunciation, Driscoll's method could be leveraged as an additional component in our larger system.

So far, we have confined ourselves to minimal feature engineering. Future work will investigate a more abstract set of phonemic features. We also hope to leverage ongoing work on harvesting name pairs from web resources, in addition applying them to less commonly taught languages, as and when appropriate resources for them become available. Experiments with smaller training sets are needed in order to assess whether the approach will be practical for languages with fewer resources. Additional experiments should design training and test sets so that the balancing factor can be viewed as another parameter and should include test data from other contexts.

Acknowledgements This research has been funded by the MITRE Innovation Program (Public Release Case Number 07–0752). We are also grateful to the reviewers for their insightful comments.

References

1. Al-Onaizan, Y., Knight, K.: Machine transliteration of names in Arabic text. In: Proceedings of the ACL Workshop on Computational Approaches to Semitic Languages, Philadelphia, pp. 1–13. Association for Computational Linguistics, Stroudsburg (2002)
2. Bilenko, M., Mooney, R.J.: Adaptive duplicate detection using learnable string similarity measures. In: Proceedings of Ninth ACM SIGKDD International Conference on Knowledge Discovery and Data Mining, Washington, DC, pp. 39–48. ACM, New York (2003)
3. Cohen, W.W., Richman, J.: Learning to match and cluster large high-dimensional data sets for data integration. In: Proceedings of The Eighth ACM SIGKDD International Conference on Knowledge Discovery and Data Mining, Edmonton, pp. 475–480. ACM, New York (2002)
4. Damerau, F.J.A.: Technique for computer detection and correction of spelling errors. Commun. ACM **7**(3), 171176 (1964)
5. Fellegi, I., Sunter, A.: A theory for record linkage. J. Am. Stat. Soc. **64**, 1183–1210 (1969)
6. Freeman, A., Condon, S., Ackermann, C.: Cross linguistic name matching in English and Arabic. In: Proceedings of the Human Language Technology Conference, New York, pp. 471–478. Association for Computational Linguistics, Stroudsburg (2006)
7. Freitag, D., Khadivi, S.: A sequence alignment model based on the averaged perceptron. In: Proceedings of EMNLP-CONLL, Prague (2007)
8. Gao, W., Wong, K., Lam, W.: Phoneme-based transliteration of foreign names for OOV problem. In: Proceedings of First International Joint Conference on Natural Language Processing (IJCNLP), Hainan Island, China, pp. 374–381 (2004)
9. Hall, M., Frank, E., Holmes, G., Pfahringer, B., Reutemann, P., Witten, I.H.: The WEKA data mining software: an update. SIGKDD Explor. **11**(1) (2009). www.cs.waikato.ac.nz/%ml/weka/
10. Huang, F., Vogel, S., Waibel, A.: Improving named entity translation combining phonetic and semantic similarities. In: Proceedings of HLT-NAACL, Boston (2004)
11. Ji, H., Grishman, R., Freitag, D., Blume, M., Wang, J., Khadivi, S., Zens R., Ney, H.: Name extraction and translation for distillation. In: Olive, J., Christianson, C., McCary, J. (eds.) Handbook of Natural Language Processing and Machine Translation: DARPA Global Autonomous Language Exploitation, Springer (2011). DOI: 10.1007/978-1-4419-7713-7_3
12. Jiampojamarn, S., Bhargava, A., Dou, Q., Dwyer, K., Kondrak, G.: DIRECTL: a language-independent approach to transliteration. In: Proceedings of the 2009 Named Entities Workshop, ACL-IJCNLP, Singapore, pp. 28–31 (2009)
13. Joachims, T.: Making large-Scale SVM Learning Practical. In: Scholkopf, B., Burges, C., Smola, A. (eds.) Advances in Kernel Methods – Support Vector Learning. MIT Press, Cambridge, MA (1999). svmlight.joachims.org/
14. Jung, S., Hong, S., Paek, E.: An English to Korean transliteration model of extended Markov window. In: Proceedings of the 18th Conference on Computational Linguistics (COLING), Saarbrücken, Germany, vol. 1, pp. 383–389. Association for Computational Linguistics, Stroudsburg (2000)
15. Kondrak, G.: A new algorithm for the alignment of phonetic sequences. In: Proceedings of the First Meeting of the North American Chapter of the Association for Computational Linguistics, Seattle, WA, pp. 288–295. Association for Computational Linguistics, Stroudsburg (2000)
16. Knight, K., Graehl, J.: Machine transliteration. Comput. Linguist. **27**(4), 599–612 (1998)
17. Lait, A., Randell, B.: An assessment of name matching algorithms. Technical Report, Department of Computer Science, University of Newcastle upon Tyne, UK (1996)

18. Levenshtein, V.I.: Binary codes capable of correcting deletions, insertions and reversals. Sov. Phys. Dokl. **10**(8), 707–710 (1966)
19. Li, H., Kumaran, A., Pervouchine, V., Zhang, M.: Report of NEWS 2009 machine transliteration shared task. In: Proceedings of the 2009 Named Entities Workshop, ACL-IJCNLP, Singapore (2009)
20. Li, H., Zhang, M., Su, J.: A joint source-channel model for machine transliteration. In: Proceedings of Conference of the Association for Computation Linguistics, Barcelona, Spain, pp. 159–166. Association for Computational Linguistics, Stroudsburg (2004)
21. McCallum, A., Bellare, K., Pereira, F.: A conditional random field for discriminatively-trained finite-state string edit distance. In: Proceedings of the Conference on Uncertainty in AI, Edinburgh, Scotland, pp. 388–395 (2005)
22. Meng, H., Lo, W., Chen B., Tang, T.: Generating phonetic cognates to handle named entities in English-Chinese cross-language spoken document retrieval. In: Proceedings of the IEEE Automatic Speech Recognition and Understanding Workshop, Madonna di Campiglio, Italy (2001)
23. (NEWS-2009) 2009 named entities workshop: shared task on transliteration. In: Proceedings of the 2009 Named Entities Workshop, ACL-IJCNLP, Singapore (2009)
24. Oh, J., Choi, K., Isahara, H.: A comparison of different machine transliteration models. J. Artif. Intell. Res. **27**, 119–151 (2006)
25. Ristad, E.S., Yianilos, P.N.: Learning string edit distance. In: IEEE Transactions on Pattern Recognition and Machine Intelligence, pp. 522–532. IEEE Computer Society, Washington, DC (1998)
26. Safalra: www.safalra.com/science/linguistics/pinyin-pronunciation/ (2006)
27. Samuel, K., Rubenstein, A., Condon, S., Yeh, A.: Name matching between Chinese and Roman scripts: machine complements human. In: Proceedings of the 2009 Named Entities Workshop, Singapore, pp. 152–160. ACL-IJCNLP, Stroudsburg (2009)
28. Sproat, R., Tao, T., Zhai, C.: Named entity transliteration with comparable corpora. In: Proceedings of the Conference of the Association for Computational Linguistics, Sydney, Australia, pp. 73–80. Association for Computational Linguistics, Stroudsburg (2006)
29. Tao, T., Yoon, S., Fister, A., Sproat, R., Zhai, C.: Unsupervised named entity transliteration using temporal and phonetic correlation. In: Proceedings of the Empirical Methods in Natural Language Processing Conference, Sydney, Australia, pp. 250–257. Association for Computational Linguistics, Stroudsburg (2006)
30. The CMU Pronouncing Dictionary: ftp://ftp.cs.cmu.edu/project/speech/dict/ (2008)
31. Ukkonnen, E.: Approximate string-matching with Q-grams and maximal matches. Theor. Comput. Sci. **92**, 191–211 (1992)
32. Virga, P., Khudanpur, S.: Transliteration of proper names in cross-lingual information retrieval. In: Proceedings of the ACL 2003 Workshop on Multilingual and Mixed-language Named Entity Recognition, Sapporo, Japan. Association for Computational Linguistics, Stroudsburg (2003)
33. Wan, S., Verspoor, C.M.: Automatic English-Chinese name transliteration for development of multilingual resources. In: Proceedings of the 36th Annual Meeting of the Association for Computational Linguistics, Montreal, Quebec, pp. 1352–1356. Association for Computational Linguistics, Stroudsburg (1998)
34. Wikipedia: Pinyin. en.wikipedia.org/wiki/Pinyin (2006)
35. Winkler, W., Thibaudeau, Y.: An application of the fellegi-sunter model of record linkage to the 1990 U.S. decennial census. Technical Report RR91/09, Energy Information Administration, Washington, DC (1991)
36. Zobel, J., Dart, P.: Finding approximate matches in large lexicons. Softw. Pract. Exp. **25**(3), 331–345 (1995)

Chapter 4
Computational Methods for Name Normalization Using Hypocoristic Personal Name Variants

Patricia Driscoll

Abstract A growing body of research addresses name normalization as part of coreference and entity resolution systems, but the problem of hypocoristics has not been systematically addressed as a component to such systems. In many languages, these name variants are governed by morphological and morphophonological constraints, providing a dataset rich in features that may be used to train and run matching systems. This paper gives a full treatment to the phenomenon of hypocoristics and presents a supervised learning method that takes advantage of their properties to untangle the relationships between hypocoristic name variants and corresponding full form names.

4.1 Introduction

As social communication moves increasingly to the Internet, the study of entity disambiguation, which aims to cluster large numbers of name mentions according to entity referents, has become more popular (See Chaps. 5 and 6). As a crucial part of this process, personal name normalization aims to create linguistically-motivated links between personal names using information about their structure and morphology that can be mined using a variety of sources. Nicknames are an often-overlooked source of ambiguity in personal name reference. As informal textual data sources such as social networking sites become more popular, informal name variants like nicknames have increasing utility. In this paper, we do a thorough analysis of the nickname formation process and identify hypocoristics (internally derived, linguistically related nicknames) as having regular construction properties. We take advantage of these regular properties to devise a computational methodology for

P. Driscoll (✉)
Columbia University, New York, NY, USA
e-mail: pdriscoll@gmail.com

T. Poibeau et al. (eds.), *Multi-source, Multilingual Information Extraction and Summarization 11*, Theory and Applications of Natural Language Processing, DOI 10.1007/978-3-642-28569-1__4, © Springer-Verlag Berlin Heidelberg 2013

linking full form names with their hypocoristic variants.[1] Experiments over a set
of representative data validate this methodology for English nicknames and may be
easily extended to other languages.

4.2   Related Work

Personal name normalization has been studied both for its place in the entity disambiguation process, where it can aid tasks like co-reference resolution [6, 22, 30], and
for the production of stand-alone tools and entity disambiguation resources, such as
proper name ontologies [18, 20], onomastica [32], and fuzzy name-matching tools
[23] that accept candidate pairs as input.

The specific case of entity disambiguation on the Internet has assumed particularly intense investigation. The formation of the web people search workshop
(WePS) [3–5] run since 2008 has engendered a significant amount of work. In that
evaluation, the input is a set of web pages with a shared common name (e.g. 'John
Smith'), and the goal is to cluster the pages so that each cluster represents a unique
real referent. Typically, no effort is made in the workshop to resolve name variants
that would complicate the problem by expanding the confusion set. One purpose
of this task is clear, as an aid to internet searches, and a number of companies are
capitalizing on this technology (e.g. intellius.com).

A related area of current research, name normalization systems have the capability to take into account occupation-related titles, honorifics, and variation in
capitalization and punctuation [37]. Problems similar to personal name normalization, including name transliteration and cognate matching for common nouns,
have been studied in the context of machine translation [17, 21, 33]. The process of
name variant untangling may additionally be useful for cross-language information
retrieval [36] and speech recognition [28] where infrequent name variants may be
able to be more effectively handled by relating them to a smaller set of known
canonical names.

Standardized personal name variants like hypocoristic nicknames of personal
names have been little-studied elements of the name normalization problem. Those
systems that include standardized nicknames as equivalent to their corresponding
full forms typically do so by inclusion of a pre-packaged dataset such as a
nickname pair list [16], or by simple string-matching methods that do not take
into consideration the morphological relationship between nicknames and the names
generating them, leaving the systems susceptible to error [23]. No research to date
has addressed dynamic matching of names and nicknames, which would have the
potential to overcome problems presented by language change, incomplete datasets,
and languages for which few standard data sources are available.

In addition to personal name normalization problems, there are many settings
where there is a need for normalizing other kinds of noun variants. For example,

---

[1] An earlier version of this work appeared as Driscoll and Yarowsky 2007 [10].

in bioinformatics, gene name normalization has been significantly addressed [15]. Disease names (e.g. those found in MeSH) have also been a target for disambiguation efforts [2], as has species resolution [38]. All of these tasks are directed at an open set of nouns that may have complicated variants.

4.3 Full Form Names and Hypocoristics

Nicknames are used for personal address in a large number of the world's languages [25]. However, few of the extant references present an adequate analysis of the phenomenon from a computational perspective. In this section, we explore the phenomenon and lay the groundwork for the computational methods presented in Sect. 4.4.

Nicknames for personal names can be divided into two classes: externally derived, in which the formation relates to qualities of the recipient, or internally derived, formed based on linguistic characteristics of the base name [1]. In computational contexts, understanding externally derived nicknames is potentially useful in entity resolution for a single individual, such as Aristocles → Plato, a name attributed variously to the philosopher's robust figure or broad style [26]. Internally derived nicknames are formed based on linguistic features characteristic to a particular language, and thus can provide more general information that is extensible to the population of language speakers. Our focus is this latter type of nickname.

Internally derived nicknames are members of a class of linguistically-based modifications called hypocoristics. Hypocoristics may be used to indicate familiarity as well as endearment, affection, diminutive nature of the referent [25], playfulness, warmth, teasing, or disrespect [27]. Hypocoristics are not restricted to personal names, and can be formed for other proper nouns, such as day or place names (see Table 4.1 or [25, 27, 34]), as well as common nouns (see Table 4.2 or [12, 13]). Hypocoristics have been studied in a large number of the world's languages; see references in [25].

Table 4.1 Australian place names [34]

Full name	Hypocoristic
St. Ives	Snives
Utopia	Utopes
Macquarie university	Macker

Table 4.2 Italian common nouns (called Accorciamenti) [35]

Full name	Hypocoristic
Bbiblioteca	Biblio
Pomeriggio	Pome
Frigorifero	Frigo

Table 4.3 Spanish hypocoristics can be predicted for nonsense names by extending patterns [19]

Full name	Hypocoristic
Federico	Quico
Enriqu	Quique
Josefa	Fefa
Rodolfo	Fofo
Nonsense name	Hypocoristic
*Justoneco	Queco
*Panurdo	Dudo

Table 4.4 Foreign names in Bernese German [12]

Full name	Hypocoristic
Karim	Käru
Sandro	Sändu
Giuséppe	Tschüsu
Jean-Pierre	Schämpu

4.3.1 Productive Hypocoristic Formation

Typically subject to a set of morphological and morphophonological processes (see [7, 9, 12, 19, 25, 27]), hypocoristic formation can be understood in terms of general processes like truncation and reduplication, and specific constraints that vary by language and can often be complex. Optimality Theory (OT), a model initially described by Prince and Smolensky [29] to describe observed language, is often used to analyze the linguistic relationship between the base name and the nickname (see [7, 13, 24]). OT proposes a mechanism for generating a list of candidate outputs, a list of ordered constraints, and an evaluation metric to choose between the candidates based on the constraints. The candidate that violates the fewest highest-ranking constraints is generally chosen. Constraints described in nickname formation include anchoring constraints, which determine the edge of a base name that is retained when forming a nickname. Left- or right- anchoring may be the default in a particular language, but anchoring constraints may be violated in cases where they conflict with higher-ranking constraints, like onset constraints that require the nickname to begin with a consonant.

Although the set of processes and constraints described for hypocoristic formation in any given language is large and often quite complex, the rules are useful because the formation is generally productive, meaning novel names can be given nicknames using such a system (see Table 4.3). This is notable for our task because it means in any given language, the set of name-nickname pairs is an open set, and static lists by definition would not be complete. Nicknames for novel names, like names imported from other languages (see Table 4.4), would always be in the process of being added to the lexicon. That hypocoristic formation is subject to constraints, however complex, bodes well for the potential utility of computational methods to assist in matching or normalization tasks where exhaustive lists do not exist.

Table 4.5 Examples of truncation, reduplication, affixation

Czech truncation & affixation [7]		French truncation & reduplication [24]	
Vladimir	Vlàdá	Nicola	NiNi
Stanislav	Standa	Thomas	ToTo
Antonín	Tonouš	Michelle	MiMi

Table 4.6 Hausa names and their hypocoristic variants [25]

Type	Full name	Hypocoristic	Full name	Hypocoristic
A	Yaarò (m)	Yàarooro (m)	Bintù (f)	Bìntuutu (f)
B	Gařbà (m)	Gàřbaatii (m)	Yàlwa (m/f)	Yàlwaatii (f)
C	Sandà (m)	Sandaloo (m)	Habiibà (f)	Habiibaloo (f)
D	YaaÕù (m)	YaaÕulle (m)	Hajìaa (f)	Hajíyalle (f)
E	Baawà (m)	Bàabàndi (m)	Uwa (f)	Uwàale (f)
F	Bàaba (m)	Bàabàndi (m)		
G	Kàbîř (m)	Kàbiiřùwaa (m)	Hànnatù (f)	Hànnatùwaa (f)

While hypocoristic formation is subject to complex constraints, a number of basic processes can be described that tend to be present in many cases in a large number of languages.

- Truncation, often described as the primary process in creating short-form nicknames, refers to shortening by eliminating one or more parts of the word, typically syllables. Constraints governing truncation often include anchoring constraints to determine which part of the base name is truncated (see [12, 13]).
- In reduplication, part of the base name is repeated to create the hypocoristic.
- Affixation is the addition of new morpheme, like a prefix or suffix, to the base name to form the hypocoristic.

While these process sometimes operate independently, in many cases more than one is at work, complicating the determination of the resulting hypocoristic (see Table 4.5).

4.3.2 Example Derivations in Hausa and Bernese Swiss German

The range of basic templates for hypocoristic formation in a single language may be seen in Table 4.6. These examples from Hausa, a Chadic language spoken in West and Central Africa, are drawn from the work of Newman and Ahmad [25]. The seven hypocoristic types described are suffixal reduplication, suffix (aa)tii, suffix aloo (-alaa), suffix lle, suffix (a)le, suffix ndi, suffix uwaa. Which type is used depends both on intended meaning and on characteristics of the base name. While hypocoristics in Hausa follow these general patterns, constraints also exist to determine the exact hypocoristic form in any specific case. One example of

Table 4.7 Bernese
u-formation [12]

a.	Lukas	Lüku
	Simon	Simu
	David	Dävu
	Moritz	Möru
	Gregor	Gregu
	Thomas	Tömu
	Gabriel	Gäbu
	Daniel	Dänu
	Michael	Michu
b.	Eugen	Genu
	Adolf	Dölfu
	August	Güsto
c.	Erich	Richu, Eru
	Oliver	Livu, Ölu
d.	Markus	Küsu, Märku
	Marcel	Selu, Märsu
e.	Konrad	Könu
	Oswald	Ösu
	Silvan	Silu
	Siegfried	Sigu
f.	Hans	Hänsu
	Viktor	Viktu
	Walter	Wältu
g.	Norbert	Nörbu, Nöbu
	Bernhard	Bernu, Benu
	Jörg	Jörgu, Jöggu

reduplication given by the authors that does not fit the most basic template is Bàlaa → Bàleele, rather than Bàlaa → *Bàlaala.

To illustrate the more complex kinds of constraints involved in end-to-end process of hypocoristic formation, the theory of Bernese Swiss German u-formation may be useful [12]. As seen in Table 4.7, while left-anchoring is preferred, it is not always present (e.g. names in sections b, c, and d). The choice of syllable from the base is seen as resulting from a competition between correspondence and position-sensitive markedness constraints. Correspondence constraints include left anchoring and contiguity, that the portion retained represents a contiguous string. Markedness constraints include a constraint that all syllables have a consonantal onset and one that penalizes a vocalic /r/ in the nucleus (the central part of the syllable). Though there are several additional features not represented by this explanation, like an acquired Umlaut or the reduction of complex onset clusters, correspondence and markedness constraints can be seen as adequately modeling the majority of the u-formation data.

Although hypocoristics are widely used, resources to make use of the information they contain are not generally available. In rare circumstances, baby name lists (such as the one described in Sect. 4.5) can be mined to extract pairs and produce a static dictionary of limited use. In many low-density languages, this data is not available at all. Yet in all the languages described in the literature, there are a number of productive processes for forming hypocoristics. As described above, these processes are typically regular but complex, making them ideal candidates for computational approaches. In light of this, we present a methodology that is language-independent and can be used where pre-existing resources are limited or unavailable.

4.4 Computational Methods

Leveraging hypocoristics may augment the process of personal name normalization for entity disambiguation in multiple ways. Given a novel name variant, full form name resources like onomastica might be used in conjunction with a system to draw a number of potential links. A large name-nickname resource, like a dictionary, might also be constructed using a list of full form names and a body of text, to which new pairs are added as new nicknames are encountered. Both of these may be achieved by a system that scores arbitrary candidate pairs, going through a resource of full form names with each new nickname and retaining the pairs that score above some threshold. The full form name resource might vary, from a large dictionary to a single document in which multiple entities are being discussed.

In this section, we draw on the insights discussed previously to develop a computational methodology that is sufficiently robust to handle novel names in a variety of languages, including those where pre-existing resources may be scarce. We address three tiers of complexity in hypocoristic formation: (1) the small set of pairs that may be derived via regular truncation as described in Sect. 4.3; (2) the large set of constrained but complex morphologically and morphophonologically-derived pairs, which make use of one or more constraints often described using OT; and (3) highly irregular pairs for which the pattern of derivation is rare or difficult to pinpoint. For each stage, we assume limited resources, very little time and linguistic knowledge, and in some cases allow small amount of seed data. Components can be combined to improve results.

To link standardized personal name variants with corresponding full form names, several methods were chosen for strengths in accuracy, flexibility, and requirements for human annotation, supervision, and data. Used both individually and in combination, the methods chosen are likely to work well in a variety of settings, including quick ramp-up for languages with little available data.

4.4.1 Web-Based Extraction

Although there are few languages for which entire name variant dictionaries are available for download from the Internet, the Internet can nonetheless be a reliable

Table 4.8 Web extraction
component finds nicknames
in a variety of contexts

Seed nickname	Candidate full form
Katia	Katarina
Lynn	Madeline
Debbie	Phyllis
Lee	My

tool to use in the creation of such resources. Personal web pages, increasingly available in many languages, are rich in name information which can fill in gaps left by other systems, providing a subset of hard-to-reach pairs to supplement methods with wider recall. In particular, web extraction often covers name pairs that are not related by simple morphological rules, and thus difficult to access using other methods. Findings of the web extraction method may also be used to give a boost to correct pairs which are found in the results of other methods, but are ranked below erroneous pairs.

There has been much recent work on web-based extraction, in which systems typically start with a hand-picked seed phrase [11] or seed instances [8]. For the web extraction component, we started from the seed phrase "My name is *full form* ... friends call me *hypocoristic*", issued this query to Internet search engine yahoo.com, and collected the first page of results for each seed nickname. This weakly supervised method ensures quick and easy extension to other languages. More strongly supervised methods with added alternate seed phrases and use of results beyond the one-page range would likely improve performance of this component in English, both by eliminating erroneous hits and by expanding coverage.

Many of the recovered full form/variant pairs retrieved are complex but correct (Table 4.8). Because of the potential unreliability of this extraction method, candidate full forms that did not appear in the census data (e.g. Lee ← My) were discarded. In an end-to-end system, in the absence of census data this filtering would easily be done using a nametagger. Another source of error was the fact that the my name is match immediately preceding the seed nickname on a particular web page was not always relevant: in some cases, long lists of online personal ads included phrases like "My name is" and "friends call me" as interchangeable (Debbie ← Phyllis).

4.4.2 Morphological Analyzer

To exploit the feature-rich, highly constrained morphological derivation process involved in nickname formation, we used a toolkit as described in [39]. The WordFrame model is a supervised, noise-robust morphological analyzer originally designed for modeling verb inflections across a diverse set of languages. Its use here to model nominal derivations is novel.

Table 4.9 The morphological analyzer learns the common morphological nickname inflections from supervised training data

Morphological rule	Nickname	Full form
IE → A	Elsie	Elsa
IE → ERTA	Albie	Alberta
EE → INA	Rosalee	Rosalina
EE → ENE	Charlee	Charlene

The WordFrame model is an extension to the end-of-string model (EOS) presented in [40]. EOS breaks all morphological changes into three pieces: the stem, a word-final stem change, and a suffix addition:

$$\text{EOS}(w_{1..n}) = \text{stem}_{1..j} + \text{word-final-stem-change}_{j..k} + \text{suffix}_{k..n}.$$

The WordFrame model extends this model by allowing prefixation, suffixation, and one internal vowel change.

$$\text{WordFrame}(w_{1..n}) = \text{prefix}_{1..f} + \text{word-initial-stem-change}_{f..g}$$
$$+ \text{secondary stem}_{g..h} + \text{vowel change}_{h..i}$$
$$+ \text{primary stem}_{i..j} + \text{word-final stem change}_{j..k}$$
$$+ \text{suffix}_{k..n}$$

Prefixes and suffixes can be induced by the WordFrame model, or the model can take advantage of prefix and suffix lists if they are available. Internally, the WordFrame model builds a smoothed tries to represent the stem changes necessary for transformation.

The WordFrame model was developed to induce morphological rules for arbitrary input languages that exhibit prefixation, suffixation, and internal vowel shifts, and has been shown successful with infixation, agglutination, and partial reduplication as well. In [39], the model was applied to regular and irregular verb inflections in over 30 languages with a median accuracy of 97.5%. Because the model is able to induce morphological rules without knowledge of language specifics, it has the potential to be useful in name variant analysis for arbitrary input languages, such as low-density languages for which resources are unlikely to be available. The model is noise-robust, so can be particularly useful when high-accuracy training set is not available.

Because the WordFrame model is designed to model verb inflections, it requires training data consisting of inflection-root pairs. For this task, we substituted hypocoristic-full form pairs, where the nickname was treated as an inflection of the full form name. Pre-specified lists of prefixes and suffixes are optional, and we included a small list of suffixes (IE, Y, EE, I, ITA, K, A, INA, EY, E) obtained by inspecting the training data. WordFrame also requires a list of vowels in the language being used, which was provided for English. Training data consisted of 1,000 hypocoristic-full form pairs. Table 4.9 gives examples of some of the rules learned by the morphological analyzer for English nickname formation.

Table 4.10 The handwritten truncation rules give four levels of matches

1. Exact nickname matches to the beginning or the end of full form.
2. Exact nickname matches anywhere in full form.
3. Lemmatized nickname matches at the beginning or the end of full form.
4. Lemmatized nickname matches anywhere in the full form.

4.4.3 Handwritten Truncation Rules

In addition, a small set of simple handwritten truncation rules was used to supplement results from the weakly supervised components. The rules were developed using basic knowledge of left and right truncation with vowel-only augmentation, so as to exploit just those features requiring only limited knowledge and limited implementation time. Because truncation is a hypocoristic formation mechanism found in many languages, similar rules might be easily written for other languages. As described in Sect. 4.3, nickname formation is a complex phenomenon and attempts to provide hardcoded rules for all languages are too limited to capture all of the requisite variation.

These rules (Table 4.10) provide a rough cut at matching, and were used both to form a baseline and as a supplement to the system, since they provided information on the most likely guesses when highly regular matches existed.

Using the rules above, nicknames using basic truncation such as Elizabeth → Liz, Lizzie, or Beth would be recognized, while more complex forms (Elizabeth → Betsy) would not. The handwritten rules did little to constrain the truncation by pruning out unlikely examples, so Elizabeth would also be matched with candidate nicknames like Eli, Zoë, and Bea. Simple phonological changes, although highly regular, were not modeled to limit size of rule set and allow for extensions to other languages.

4.4.4 Classifier Combination

To generate a composite result, a standard linear regression model was applied over the scores from the individual components and scored each proposed full form name for a given hypocoristic. Weights were trained using 145 pairs with names appearing in the top 10% of census data. Since both the web extraction and morphological analysis components were generative and thus likely to be relatively sparse, we also included a binary feature indicating whether the candidate appeared on each of these lists, regardless of score. Based on data indicating that number of nicknames varies according to full form prior (see Fig. 4.1), these prior probabilities were also included as a candidate feature.

$$y = \alpha_1 \, \texttt{morph score} + \alpha_2 \, \texttt{morph list} + \alpha_3 \, \texttt{web score} + \alpha_4 \, \texttt{web list}$$

$$+ \alpha_5 \, \texttt{truncation rule} + \alpha_6 \, \texttt{full form prior}$$

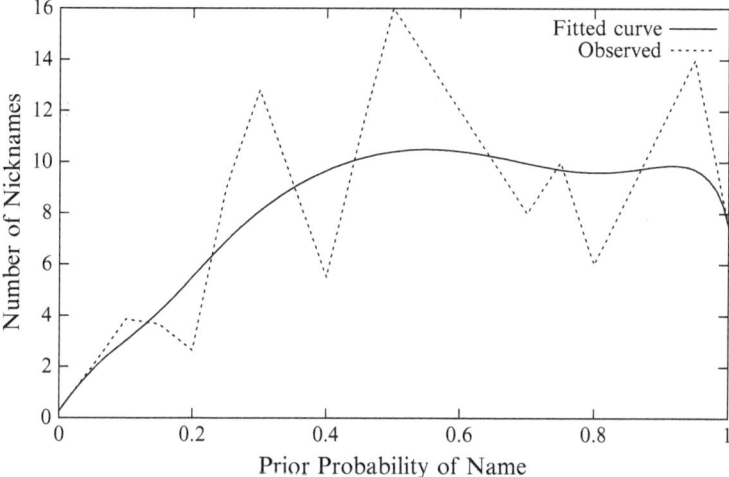

Fig. 4.1 More common names have more nicknames

Table 4.11 Final weights for system combination

Name	Feature	Weight
α_4	Appears in the web extraction list	0.286
α_5	Handwritten truncation rule score	0.214
α_1	Morphological analyzer score	0.143
α_3	Web extraction ranking	0.143
α_6	Full form prior probability	0.143
α_2	Appears in morphological analyzer list	0.071

Because Levenshtein distance, with a score cutoff of 4, gave no improved performance, it was not included in the final feature set. The final weights are shown in Table 4.11.

4.5 Experiments

To obtain training and evaluation data for English, first names taken from 1990 U.S. Census data were used to query a nickname database at www.oxygen.com/babynamer, which yielded a set of 2,543 name-nickname pairs using 907 of the census names. 1,837 nicknames were represented in the data, which often included multiple nicknames for particular first names (Jennifer → Jen, Jenny) as well as nicknames which were associates with multiple first names (Robert, Roberto → Bob). The resource, although it contained many name variants, was not exhaustive,

Table 4.12 Levenshtein
distance typically judges
rhyming name or unrelated
names to be closer than full
form names

Nickname	Confusors (distance)	True full form (distance)
Pammie	Mammie (1) Tommie (2)	Pamela (3)
Jess	Bess (1) Jose (2)	Jessica (3)
Lenny	Benny (1) Wendy (2)	Leonard (6)

thus presenting challenges for evaluation. One example of the lack of coverage occurred with spelling variants of full form names, in which nicknames would be linked to one form but not another.

An interesting property of the data was that more common first names (using probabilities given by census.gov) had more nicknames than their less common counterparts (Fig. 4.1).

Our initial data set of 2,543 name-nickname pairs was split into three portions: test data, development data, and training for the morphological analyzer. Names from the top 10% of the census data were used to create pairs for the test and development sets. Test data using 278 pairs of these commonly occurring names and their nicknames was then chosen at random, and was expanded to include all correct names and nicknames seen in testing. Development data using 145 pairs was chosen at random from the remains of this set, and training data of 1,000 pairs was selected using only nicknames not seen in test or development.

Baselines were a simple substring match ranked by proportion nickname comprised of full form, Levenshtein distance with a cutoff of 4, and the handwritten truncation rules described above. Table 4.12 gives examples that demonstrate why Levenshtein distance is an unsuitable approximation to the complex morphological processes involved in hypocoristict formation.[2]

The dataset, although relatively formal and thorough, did not escape some of the issues inherent in the use of static datasets for standardized name variant matching. One particularly challenging obstacle was a seeming lack of full recall in the test set. A type of legitimate match that was often not given credit in the test set was seen with spelling variants of other full form names for which more extensive pair lists existed (e.g. Deborah → Debbie was in the test data, but Debra → Debbie was not). The inclusion of relatively obscure matches also made the problem of scoring a difficult one: if systems are expected to include matches such as (Daisy ← Marguerite) before being granted full credit, recall scores will be unrealistically low. To address these issues, components were built to produce ranked lists and scored by means of precision/recall curves. Thus was the most straightforward approach

---

[2]Alternative noncontextual learned edit distance measures such as [31] would suffer from problems similar to those seen in traditional Levenshtein distance. Instead we look to the morphological analyzer presented in Sect. 4.4.2 as a linguistically motivated modeling tool.

Table 4.13 Each of the different components captures an important aspect to the nickname process: the simple truncation component allows high recall, the morphological analysis component allows more flexible matches than simple truncation rules, and the web extraction component is a high precision matcher

Nickname	Full form	Top component choices		
		Rule-based	Morphological analyzer	Web extraction
Steve	**Stephen**	1. Stevie	1. Steve	**1. Stephen**
	Steven	2. **Steven**	2. Stevie	**2. Steven**
Vikie	**Victoria**	1. Vikki	**1. Victoria**	*none*
		2. Viki	2. Viki	
Chas	**Charles**	1. Chastity	*none*	**Charles**
	Chasity			

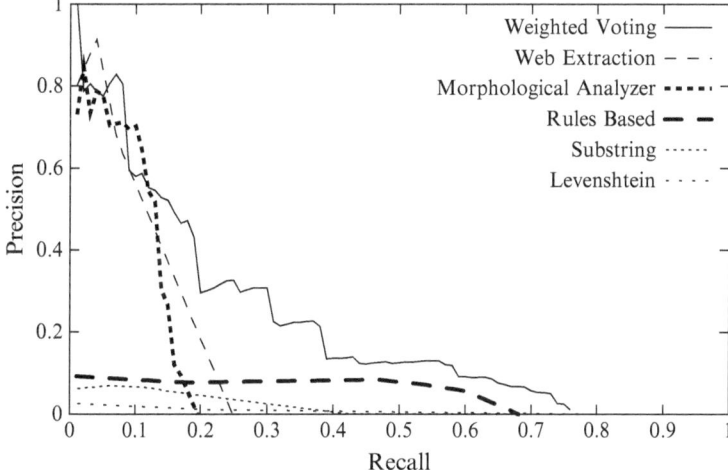

Fig. 4.2 The combined system is able to take advantage of the benefits of all of the component systems: the high precision of the web extractor and the morphological analyzer, and the high recall of the rule-based system

for integrating and evaluating information from a variety of heterogeneous sources, as some components produced reliable ranked lists, while others were more useful as isolated features into the system.

Table 4.13 shows examples of the kind of matches discovered by each of the system components. Figure 4.2 shows the performance of the components in isolation as well as the combined system performance. The combined system performance yields significantly superior performance to the baseline systems. It is able to combine the high precision/low recall performance of the web extraction component and the morphological analyzer with the high recall/low precision performance of the simple truncation rule component.

The weighted voting method, using features described above, outperformed all other methods as recall climbed above 15%. Web extraction and morphological analysis components were competitive at lower recall, indicating that these components were able to find certain pairs with good confidence. Truncation rules, substring, and Levenshtein methods were not strong on precision, although truncation rules remained steady as recall increased, indicating its strength as a component which selects many candidate matches among noisy pair results.

Exploratory Analysis of Spanish Nicknames

Previous approaches to hypocoristics have relied on knowledge by a language expert, and the output is a set of rules similar to the hand written truncation rules presented above as a baseline method. One of the great strengths of our method is that it doesn't rely on a language expert to train the system, but instead relies on general computational methods, and a small amount of training data that can be obtained by a non-expert. In particular, the morphological analyzer requires only seed pairs (nickname and full form) and the web-based extractor relies on a single language pattern (which can be obtained via public machine translation systems).

In this section, we demonstrate that the morphological analyzer and the web extraction can be applied to Spanish nicknames by a non-speaker of the language (namely, the author). For Spanish, unlike English, there were no large online repositories of nickname-full form pairs, and so to compensate, data was collected from multiple sources online and combined into a full set of 784 pairs composed of 300 unique names and 471 unique nicknames.[3] While this resource is sufficient to train and validate the sub-methods, it isn't enough to evaluate a complete system (it is less than 1/3 of the size of the English resource and noisy).

Morphological Analysis

Table 4.14 presents a selected set of rules learned by the morphological analyzer given the training data. As with English, these rules were induced from nickname-full form pairs, but in Spanish, no suffix list was provided. Even with this smaller training set, suitable morphological rules were found that modeled both long and short transformations. The morphological system performs in a manner similar to the performance on English. This is to be expected since the analysis of

---

[3]Data was collected from the following websites:

http://forum.wordreference.com/showthread.php?t=247679.

http://foro.enfemenino.com/forum/prenoms/_f4180_prenoms-Nombres-y-sus-di minutivos.html.

http://en.wiktionary.org/wiki/Appendix:Spanish_diminutives_of_given_names.

http://www.transparent.com/spanish/diminutivo-de-los-nombres-propios/.

Table 4.14 Spanish morphlogical rules: common morphological nickname inflections learned from supervised training data

Morph. rule	Nickname	Full form	Morph. rule	Nickname	Full form
I → ANDA	Yoli	Yolanda	CHE → SE	Joche	Jose
ε → LA	Pame	Pamela	LI → CIA	Luli	Lucia
I → A	Ani	Ana	LU → A	Marilu	Maria
ε → CIO	Igna	Ignacio	E → A	Julie	Julia
I → ANA	Susi	Susana	UJA → IA	Maruja	Maria
ITA → A	Sarita	Sara	MA → -MANUEL	Juanma	Juan-Manuel
MA → -MARIA	Josema	Jose-Maria	Y → IRA	May	Maira
ε → EL	Rafa	Rafael	U → ICIA	Letu	Leticia
ε → ANA	Lili	Liliana	ε → NIA	Silvi	Silvinia

hypocoristics given in Sect. 4.3 shows that regular morphological processes give rise to hypocoristics and the morphological analyzer has sufficient degrees of freedom to learn the relevant rules, no matter what the language may be.

Web-Based Extraction

For web-based extraction, it is less clear that the same method that works in English will transfer to Spanish, because it relies largely on the Internet behavior of individuals in different populations. To test performance in Spanish, the author used an automatic translation system to find a lexical pattern which was then seeded with hypocoristics to find hypocoristic-full form pairs, following the same strategy as before. Table 4.15 shows some of the pairs retrieved by the pattern search. As with English, many of the proposed results are legitimate, but, just as with English, some of the pairs proposed are erroneous. Since the web-based extraction system is by design a low-precision method that can be used to boost scores of irregular pairs, this is an acceptable result, demonstrating that conversation about one's own or another's nickname is part of the Internet discourse in Spanish as well as English.

4.6 Discussion

The use of nicknames is ubiquitous, especially in the colloquial, informal speech that constitutes the expanding internet (such as found on social sites like Twitter or Facebook). However, systems that attempt entity-coreference typically ignore this phenomenon and do not propose merges across name variants. When systems do incorporate nickname resolution, they use static nickname-full form lists or handwritten rules. Since hypocoristic nickname formation is a regular, productive and complex process (see Sect. 4.3), neither approach works very well.

Table 4.15 Selected web
extraction results

Seed nickname	Candidate full form
Ale	Alejandra
Ali	Alicia
Andy	Andrea
Ange	Ismael
Carito	Carolina
Carlitos	Carlos
Carol	Carolina
Ceci	Cecilia
Charly	Carlos
Evi	Evelyn
Jandro	Alejandro
Jano	Juan
Jani	Javier
Jorgito	George
Juli	Julian
Kike	Esteban
Tito	Roberto
Tina	Cristina
Tom	Lady
Tony	Antonio
Pancho	Francisco
Pepe	Jose
Perico	Pedro
Pili	Pilar

In this paper, we introduce a set of computational methods for hypocoristic name normalization. Unlike previous approaches, these methods do not rely on the existence of extensive onomastica or even competency in the target language. As demonstrated in the previous section, these computational methods can be built from small source dictionaries and web searches. Significantly, the composite system not only performs better than Levenshtein and substring baselines, but also outperforms handwritten rules, which are the current state-of-the-art for automatic hypocoristic name normalization.

The resulting system is not sufficiently high-precision to generate an exhaustive static resource. Still, these methods would be useful as part of a coreference resolution system where many sources of information can be taken into account to compare mentions (e.g. topic similarity). Without robust hypocoristic methods, there would be many mentions that could never be merged so even a noisy system is useful. Coreference systems have not yet been applied to the social internet where the use of hypocoristics are especially common, so no existing test sets exist. To complete an in situ test of the use of hypocoristics, a full coreference resolution system and a new test set which has hypocoristic mentions is required. We leave this evaluation to future work.

The proof of concept presented in Sect. 4.5 is just a first step toward a robust handling of hypocoristics cross-lingually. There are a number of clear directions for future work. One possibility is to improve classifier combination with the existing components via a machine learning approach. The web extraction component can be significantly extended by using a larger set of seed phrases and expanding to include search over social media.

Because the WordFrame morphological analyzer is designed to be robust to noise, it can be co-trained with a minimally supervised algorithm such as the web extraction component presented in Sect. 4.4. This is an especially attractive option in languages where data is scarce, virtually eliminating the need for seed data. Alternatively, the web and morphological analysis components might be done jointly to simultaneously learn a model of hypocoristic formation. Nickname lists produced by crowdsourcing [14] may also be a source for noisy nickname data which can then be used as input to the WordFrame model.

The methods presented in this paper might also be extended to other types of name variants. In particular, personal name spelling variants (e.g. Ashley:Ashleigh) might be well-suited to the methods we use here. Such work would likely have the creation of personal name equivalence classes as a goal, which could then extend nickname results to all members of target classes.

While important, work in languages other than English has been inhibited by the difficulty of test data collection, which speaks further for the need for building reliable systems in these languages. While nickname formation in languages other than English has seen a limited amount of attention in the theoretical linguistics community, the extension of the full set of methods applied in this paper to a larger set of languages is a promising area of further research.

References

1. Aceto, M.: Ethnic personal names and multiple identities in anglophone caribbean speech communities in latin america. Lang. Soc. **31**, 577–608 (2002)
2. Alexopoulou, D., Andreopoulos, B., Dietze, H., Doms, A., Gandon, F., Hakenberg, J., Khelif, K., Schroeder, M., Wächter, T.: Biomedical Word Sense Disambiguation with Ontologies and Metadata: Automation Meets Accuracy. BMC Bioinformatics, London, UK (2004)
3. Artiles, J., Gonzales, J., Sekine, S.: The semeval-2007 weps evaluation: establishing a benchmark for the web people search. In: SemEval. Association for Computational Linguistics, Stroudsburg (2007)
4. Artiles, J., Gonzalo, J., Sekine, S.: Weps 2 evaluation campaign: overview of the web people search clustering task. In: WWW, Madrid (2009)
5. Artiles, J., Sekine, S., Gonzalo, J.: Web people search: results of the first evaluation and the plan for the second. In: WWW, Beijing. ACM, New York (2008)
6. Bagga, A., Baldwin, B.: Entity-based cross-document coreferencing using the vector space model. In: Boitet, C., Whitelock. P. (eds.) Proceedings of the Thirty-Sixth Annual Meeting of the Association for Computational Linguistics and Seventeenth International Conference on Computational Linguistics, pp. 79–85. Morgan Kaufmann, San Francisco (1998). citeseer.nj. nec.com/226397.html

7. Bethin, C.: Metrical quantity in czech: evidence from hypocoristics. In: Formal Approaches to Slavic Linguistics 11. Michigan Slavic, Ann Arbor, MI (2003)
8. Brin, S.: Extracting patterns and relations from the world wide web. In: WebDB Workshop at 6th International Conference on Extending Database Technology, EDBT'98, Valencia, pp. 172–183 (1998). citeseer.nj.nec.com/brin98extracting.html
9. Davis, S., Zawaydeh, B.: Arabic hypocoristics and the status of the consonant root. Linguist. Inq. **32**(3), 512–520 (2001)
10. Driscoll, P., Yarowsky, D.: Disambiguation of standardized personal name variants. In: MMIES, Borouets, Bulgaria (2007)
11. Etzioni, O., Cafarella, M., Downey, D., Kok, S., Popescu, A.M., Shaked, T., Soderland, S., Weld, D., Yates, A.: Web-scale information extraction in knowitall. In: WWW, Manhattan ACM, New York (2004)
12. Grüter, T.: Hypocoristics: The case of u-formation in bernese swiss german. J. Ger. Linguist. **15**(1), 27-63 (2003)
13. Halicki, E.: Accorciamenti, hypocoristics, and foot structure: Against the ternary foot in italian. Indiana University Linguistics Club Working Papers Online, Bloomington, IN (2007)
14. Higgins, C., McGrath, E., Moretto, L.: Mturk crowdsourcing: A viable method for rapid discovery of arabic nicknames? In: Workshop on Creating Speech and Language Data with Amazons Mechanical Turk, Los Angeles, CA (2010)
15. Hirschman, L., Colosimo, M., Morgan, A., Yeh, A.: Overview of biocreative task 1b: normalized gene lists. BMC Bioinf. **6**(Suppl 1), S11 (2005)
16. Kazi, Z., Ravin, Y.: Who's who? identifying concepts and entities across multiple documents. In: International Conference on System Sciences. IEEE Computer Society, Washington, DC (2000)
17. Knight, K., Graehl, J.: Machine transliteration. Comput. Linguist. **24**(4), 599–612 (1998)
18. Krstev, C., Dusko, V., Maurel, D., Tran, M.: Multilingual ontology of proper names. In: Language and Technology Conference, Poznan, 21–23 April. Wydawnictwo Poznanskie Sp. z o.o, Poznan (2005)
19. Lipski, J.: Spanish hypocoristics: towards a unified prosodic analysis. Hispanic Linguist. **6/7**, 387–434 (1995)
20. Mann, G.S.: Building a proper noun ontology for question answering. In: Proceedings of SemaNet02: Building and Using Semantic Networks, pp. 16–22. Association for Computational Linguistics, Stroudsburg (2002)
21. Mann, G.S., Yarowsky, D.: Multipath translation lexicon induction via bridge languages. In: NAACL. Association for Computational Linguistics, Stroudsburg (2001)
22. Mann, G.S., Yarowsky, D.: Unsupervised personal name disambiguation. In: Proceedings of the Conference on Natural Language Learning, pp. 33–40. Association for Computational Linguistics, Stroudsburg (2003)
23. Navarro, G., Baeze-Yates, R., Arcoverde, J.: Matchsimile: a flexible approximate matching tool for searching proper names. JASIST **54**(1), 3–15 (2003)
24. Nelson, N.: Mixed anchoring in french hypocoristic formation. In: RuLing Papers 1, Seattle, WA (1998)
25. Newman, P., Ahmad, M.: Hypocoristics names in hausa. Anthropol. Linguist. **34**(1/4), 159–172 (1992)
26. Notopoulos, J.: The name of plato. Class. Philol. **34**(2), 135–45 (1939)
27. Obeng, S.: From morphophonology to sociolinguistics: the case of akan hypocoristic day-names. Multilingua **16**(1), 39–56 (1997)
28. Palmer, D., Ostendorf, M.: Improving out-of-vocabulary name resolution. Comput. Speech Lang. **19**(1), 107–128 (2005)
29. Prince, A., Smolensky, P.: Optimality Theory: Constraint Interaction in Generative Grammar. Rutgers University, New Brunswick, NJ (1993)
30. Ravin, Y., Kazi, Z.: Is hillary rodham clinton the president? disambiguating names across documents. In: Proceedings of the ACL '99 Workshop on Coreference and its Applications, College Park (1999)

31. Ristad, E., Yianilos, P.: Learning string edit distance. IEEE Trans. PAMI **20**(5), 522–532 (1998)
32. Sheremetyeva, S., Cowie, J., Nirenburg, S., Zajac, R.: Multilingual onomasticon as a multipurpose nlp resource. In: LREC, Grenada, Spain (1998)
33. Sherif, T., Kondrak, G.: Substring-based transliteration. In: ACL, Prague (2007)
34. Simpson, J.: Hypocoristics of place-names in australian english. In: Varieties of English: Australian English, Benjamins (2001)
35. Thornton, A.: On some phenomena of prosodic morphology in italian: Accorciamenti, hypocoristics and prosodic delimitation. Probus **8**(1), 81–112 (1996)
36. Virga, P., Khudanpur, S.: Transliteration of proper names in cross-lingual information retrieval. In: ACL. Association for Computational Linguistics, Stroudsburg (2003)
37. Wacholder, N., Ravin, Y., Choi, M.: Disambiguation of proper names in text. In: Proceedings of Fifth Conference on Applied Natural Language Processing, pp. 202–208. Association for Computational Linguistics, Stroudsburg (1997). citeseer.nj.nec.com/wacholder97disambiguation. html
38. Wang, X., Matthews, M.: Species disambiguation for biomedical term identification. In: BioNLP. Association for Computational Linguistics, Stroudsburg (2008)
39. Wicentowski, R.: Multilingual noise-robust supervised morphological analysis using the word-frame model. In: ACL SIGPHON. Association for Computational Linguistics, Stroudsburg (2004)
40. Yarowsky, D., Wicentowski, R.: Minimally supervised morphological analysis by multimodal alignment. In: ACL. Association for Computational Linguistics, Stroudsburg (2000)

Chapter 5
Entity Linking: Finding Extracted Entities in a Knowledge Base

Delip Rao, Paul McNamee, and Mark Dredze

Abstract In the menageric of tasks for information extraction, entity linking is a new beast that has drawn a lot of attention from NLP practitioners and researchers recently. Entity Linking, also referred to as record linkage or entity resolution, involves aligning a textual mention of a named-entity to an appropriate entry in a knowledge base, which may or may not contain the entity. This has manifold applications ranging from linking patient health records to maintaining personal credit files, prevention of identity crimes, and supporting law enforcement. We discuss the key challenges present in this task and we present a high-performing system that links entities using max-margin ranking. We also summarize recent work in this area and describe several open research problems.

D. Rao (✉)
Department of Computer Science, Johns Hopkins University, 3400 N. Charles St., Baltimore, MD-21218, USA
e-mail: delip@cs.jhu.edu

P. McNamee
Human Language Technology Center of Excellence, Johns Hopkins University, Baltimore, MD, USA
e-mail: paul.mcnamee@jhuapl.edu

M. Dredze
Department of Computer Science, Johns Hopkins University, 3400 N. Charles St., Baltimore, MD-21218, USA

Human Language Technology Center of Excellence, Johns Hopkins University, Baltimore, MD, USA
e-mail: mdredze@cs.jhu.edu

T. Poibeau et al. (eds.), *Multi-source, Multilingual Information Extraction and Summarization 11*, Theory and Applications of Natural Language Processing, DOI 10.1007/978-3-642-28569-1__5, © Springer-Verlag Berlin Heidelberg 2013

5.1   Introduction

Information extraction involves the processing of natural language text to produce structured knowledge, suitable for storage in a database for later retrieval or automated reasoning. An active area of research for over 20 years, the community has developed several core information extraction tasks that comprise an extraction pipeline.

Named Entity Recognition: Identify boundaries of named entities in text and classify the tokens into a predefined set of named entities, such as people, organizations and locations. See for example [1–5], and the review article by Nadeau and Sekine [6].

Coreference Resolution: Group two or more named entities and other anaphoras in a document or a set of documents that refer to the same real world entity. For example, "Bush", "Mr. President", "G. W. Bush", and "George Bush" occurring in a set of documents might refer to the same entity [7–10].

Relation Extraction: Given two named entities, identify relationships between the entities expressed in the text. For instance, given two person names in news documents about crime and violence, identify the victim and the perpetrator [11,12]. Most relation extraction methods can be classified as open- or closed-domain depending on the restrictions on extractable relations. Closed domain systems extract a fixed set of relations while in open-domain systems, the number and type of relations are unbounded.

This document-centric view of information extraction has received considerable attention. However, the end result, a groups of entities and relations, often are not the only structured knowledge product. In a development environment, new extractions must be merged with previously extracted information, often stored in a structured information database, a knowledge base (KB). This last step is critical for automatic knowledge base population, which requires linking mentions in text to entries in a KB, determining information duplication between the text and KB, exploiting existing knowledge in improving information extraction, and detecting when to create new entries in the knowledge base. These challenges are exacerbated by the scale of the data, often involving hundreds of thousands of documents and several million entities.

To the discerning human eye, the "Bush" in "Mr. Bush left for the Zurich environment summit in Air Force One." is clearly the US president. Further context may reveal him to be the 43rd president, George W. Bush, and not the 41st president, George H. W. Bush. The ability to disambiguate a polysemous entity mention or infer that two orthographically different mentions are the same entity is crucial in updating an entity's KB record. This task has been variously called entity disambiguation, record linkage, or entity linking. When performed without a KB, entity disambiguation reduces to the traditional document coreference resolution problem where entity mentions either within the same document or across multiple documents are clustered together, where each cluster corresponds to a single real world entity. The emergence of large scale publicly available KBs like Wikipedia and DBpedia has spurred an interest in linking textual entity references to their entries in these public KBs. Bunescu and Pasca [13] and Cucerzan [14] presented important pioneering work in this area, but suffer from several limitations including Wikipedia specific dependencies, scale, and the assumption of a KB entry for each entity.

In this chapter, we review some common approaches to entity disambiguation and discuss in detail an entity disambiguation system for linking entity mentions (also called an entity linking query) to an entry in a knowledge base or declare if no such entry exists. We adopt a supervised machine learning approach, where each of the possible entities contained in the KB are scored for a possible match to the query entity. Our system is designed for open domains, where a large percentage of entities will not be linkable since they do not appear in the knowledge base. For this scenario, our system learns when to withhold a link when an entity has no matching KB entry, a task that has largely been neglected in prior research in cross-document entity coreference. We also describe techniques to deal with large knowledge bases, such as Wikipedia and DBpedia, which contain millions of entries. Our system produces high quality predictions compared with recent work on this task.

5.2   Prior Art

Information extraction is concerned with both identifying structured information in text and disambiguating extracted information and entities. The ambiguity of entity names, especially on large corpora like the Web or citations in scholarly articles, has served to motivate research on entity resolution. To address ambiguity in personal name search, Mann and Yarowsky [15] disambiguates person names using biographic facts, like birth year, occupation and affiliation. When present in text, biographic facts extracted using regular expressions help disambiguation. More recently, the Web People Search Task, see [16] for example, clustered web pages for entity disambiguation.

The related task of cross document coreference resolution has been addressed by several researchers starting from Bagga and Baldwin [7]. Poesio et al. [17] built a cross document coreference system using features from encyclopedic sources like Wikipedia. This continues to be a popular task [18] that considers new data sets [19]. Entity linking has been scaled to consider hundreds of thousands of unique entities, whereas operating on this scale is a challenge for cross document coreference resolution. Recent approaches to scaling this task have included distributed graphical models over a compute cluster [20] and a streaming coreference algorithm [21]. Successful coreference resolution is insufficient for correct entity linking, as the coreference chain must still be correctly mapped to the proper KB entry.

A related task is within document coreference, or anaphora resolution, in which co-referent named entity, pronominal, and nominal mentions are linked together in an entity chain. This task has a long history in the NLP community [22] which still receives significant attention [23–25]. Interestingly, coreference systems have now been used as part of larger information extraction systems, such as relation extraction [26].

By comparison, entity linking is a recent task. The earliest work on the task by Bunescu and Pasca [13] and Cucerzan [14] aims to link entity mentions to their corresponding topic pages in Wikipedia. These authors do not use the term entity linking and they take different approaches. Cucerzan uses heuristic rules and Wikipedia disambiguation markup to derive mappings from surface forms of entities to their Wikipedia entries. For each entity in Wikipedia, a context vector is derived as a prototype for the entity and these vectors are compared (via dot-product) with the context vectors of unknown entity mentions. His work assumes that all entities have a corresponding Wikipedia entry, but this assumption fails for a significant number of entities in news articles and even more for other genres, like blogs. Bunescu and Pasca on the other hand suggest a simple method to handle entities not in Wikipedia by learning a threshold to decide if the entity is not in Wikipedia. Both works mentioned rely on Wikipedia-specific annotations, such as category hierarchies and disambiguation links.

The Entity Linking problem not only disambiguates entity mentions that occur in text but also link these mentions to entries in the knowledge base. This is the focus of this chapter. Since the the Text Analytics Conference on Knowledge Base Population (TAC-KBP) included the task of entity linking [27], the task has grown in popularity with many different approaches [28, 29]. Examples include the use of information retrieval techniques for retrieving the correct KB entry, such as query expansion [30], and generative clustering models for entities in text based on KB entries [31].

The work described in this paper was developed as one of the first entity linking systems. Subsequent systems have built on our approach [31–34].

5.3 Entity Linking

We now describe the details of building such a system and summarize other systems built for this task. We define *entity linking* as matching a textual entity mention, possibly identified by a named entity recognizer, to a KB entry, such as a Wikipedia page that is a canonical entry for that entity. An entity linking *query* is a request to link a textual entity mention in a given document to an entry in a KB. The system can either return a matching entry or NIL to indicate there is no matching entry. In this work we focus on linking organizations, geo-political entities and persons to a Wikipedia derived KB. While the problem is applicable for any language, in this paper we restriction our attention to matching English names to an English knowledge base.

5.3.1 Key Issues

There are three challenges to entity linking:

Name Variations

An entity often has multiple mention forms, including abbreviations (Boston Symphony Orchestra vs. BSO), shortened forms (Osama Bin Laden vs. Bin Laden), alternate spellings (Osama vs. Ussamah vs. Oussama), and aliases (Osama Bin Laden vs. Sheikh Al-Mujahid). Entity linking must find an entry despite changes in the mention string.

Entity Ambiguity

A single mention, like Springfield, can match multiple KB entries, as many entity names, like people and organizations, tend to be polysemous.

Absence

Processing large text collections virtually guarantees that many entities will not appear in the KB (NIL), even for large KBs.

    The combination of these challenges makes entity linking especially challenging. Consider an example of "William Clinton." Most readers will immediately think of the 42nd US president. However, the only two William Clintons in Wikipedia are "William de Clinton" the 1st Earl of Huntingdon, and "William Henry Clinton" the British general. The page for the 42nd US president is actually "Bill Clinton". An entity linking system must decide if either of the William Clintons are correct, even though neither are exact matches. If the system determines neither matches, should it return NIL or the variant "Bill Clinton"? If variants are acceptable, then perhaps "Clinton, Iowa" or "DeWitt Clinton" should be acceptable answers?

5.3.2 Contributions

We address these entity linking challenges.

Robust Candidate Selection

Our system is flexible enough to find name variants but sufficiently restrictive to produce a manageable candidate list despite a large-scale KB.

Ranking and Features for Entity Disambiguation

We developed a rich and extensible set of features based on the entity mention, the source document, and the KB entry. We use a machine learning ranker to score each candidate.

Learning NILs

We modify the ranker to learn NIL predictions, which obviates hand tuning and importantly, admits use of additional features that are indicative of NIL.

Our contributions differ from previous efforts [13, 14] in several important ways. First, previous efforts depend on Wikipedia markup for significant performance gains. We make no such assumptions, although we show that optional Wikipedia features lead to a slight improvement. Second, Cucerzan does not handle NILs while Bunescu and Pasca address them by learning a threshold. Our approach *learns* to predict NIL in a more general and direct way. Third, we develop a rich feature set for entity linking that can work with any KB. Finally, we apply a novel finite state machine method for *learning* name variations.[1]

The remaining sections describe the candidate selection stage, our ranking algorithm and features, and our novel approach to learning NILs.

5.4   Candidate Selection for Name Variants

The first system component addresses the challenge of name variants. As the KB contains a large number of entries (818,000 entities, of which 35% are PER, ORG or GPE), we require an efficient selection of the relevant candidates for a query.

Previous approaches used Wikipedia markup for filtering—only using the top-k page categories [13]—which is limited to Wikipedia and does not work for general KBs. We first consider a KB independent approach to selection that also allows for tuning candidate set size. This involves a linear pass over KB entry names (Wikipedia page titles): a naive implementation took 2 min per query.

5.4.1   Brute Force Candidate Selection

For a given query, the system selects KB entries using the following approach:

- Titles that are exact matches for the mention.
- Titles that are wholly contained in or contain the mention (e.g., *Nationwide* and *Nationwide Insurance*).

---

[1] http://www.clsp.jhu.edu/~markus/fstrain

- The first letters of the entity mention match the KB entry title (e.g., *OA* and *Olympic Airlines*).
- The title matches a known alias for the entity (aliases described in Sect. 5.5.2).
- The title has a strong string similarity score with the entity mention. We include several measures of string similarity, including: character Dice score > 0.9, skip bigram Dice score > 0.6, and Hamming distance $<= 2$.

We did not optimize the thresholds for string similarity, but these could obviously be tuned to minimize the candidate sets and maximize recall. For a comprehensive survey on string similarity metrics for duplicate names, we refer the reader to [35].

All of the above features are general for any KB. However, since our evaluation used a KB derived from Wikipedia, we included a few Wikipedia specific features. We added an entry if its Wikipedia page appeared in the top 20 Google results for a query.

On the training dataset (Sect. 5.7) the selection system attained a recall of 98.8% and produced candidate lists that were three to four orders of magnitude smaller than the KB. Some recall errors were due to inexact acronyms: ABC (Arab Banking; 'Corporation' is missing), ASG (Abu Sayyaf; 'Group' is missing), and PCF (French Communist Party; French reverses the order of the pre-nominal adjectives). We also missed International Police (Interpol) and Becks (David Beckham; Mr. Beckham and his wife are collectively referred to as 'Posh and Becks').

5.4.2   Sublinear Candidate Selection

Our previously described candidate selection relied on a linear pass over the KB, but we seek more efficient methods. We observed that many of the above string similarity filters, such as aliases and exact string matches, can be pre-computed and stored in an index, resulting in significant speedups. Additionally, the skip bigram Dice score can be computed using an index of skip bigrams to KB titles, removing from consideration the vast of titles which have no skip bigram overlap with the query. Other string similarity scores were omitted without significantly hurting the recall of the filtering stage. These changes collectively enable us to avoid a linear pass over the KB. Finally we obtained speedups by serving the KB concurrently using four processes, each of which execute queries against a portion of the KB. This allows parallelization can be extended for larger KBs. The results from each process are collected for the second ranking stage. We implemented this approach in Python and our system achieved up to an 80 times speedup compared to naive implementation, serving each query in under 2 s on average. Recall was nearly identical to the full system described above: only two more queries failed. Additionally, more than 95% of the processing time was consumed by Dice score computation, which was only required to correctly retrieve less than 4% of the training queries. Omitting the Dice computation yielded results in a few milliseconds on average. A related approach is that of canopies for scaling clustering for large amounts of bibliographic citations [36]. In contrast, our setting focuses on alignment vs. clustering mentions, for which overlapping partitioning approaches like canopies are applicable.

5.5 Entity Linking as Ranking

We consider a supervised machine learning approach to entity linking. Given a query represented by a D dimensional vector $\mathbf{x}$, where $\mathbf{x} \in \mathbb{R}^D$, and we aim to select a single KB entry y, where $y \in \mathcal{Y}$, a set of possible KB entries for this query produced by the selection system above, which ensures that $\mathcal{Y}$ is small. The ith query is given by the pair $\{\mathbf{x}_i, y_i\}$, where we assume at most one correct KB entry. Using these training examples, we can learn a system that produces the correct y for each query.

To evaluate each candidate KB entry in $\mathcal{Y}$ we create feature functions of the form $f(\mathbf{x}, y)$, dependent on both the example $\mathbf{x}$ (document and entity mention) and the KB entry y. The features address name variants and entity disambiguation. We categorize the features as atomic features and combination features. Atomic features are derived directly from the named entity in question and its context while combination features are logical expressions of atomic features in conjunctive normal form (CNF).

One natural approach to learning would be classification, in which each possible $y \in \mathcal{Y}$ is classified as being either correct or incorrect. However, such an approach enforces strong constraints: we not only require the correct KB entry to be classified positively, but all other answers to be classified negatively. Additionally, we can expect very unbalanced training, in which the vast majority of possible answers are incorrect. Furthermore, it is unclear how to select a correct answer at test time when multiple KB entries can be classified as correct.

Instead, we select a single correct candidate for a query using a supervised machine learning ranker. A ranker will create an ordering over a set of answers $\mathcal{Y}$ given a query. Typically, the resulting order over all items is important, such as ranking results for web search queries. In our setting, we assume only a single correct answer and therefore impose a looser requirement, that the correct answer be ranked highest. This formulation addresses several of the challenges of binary classification. We require only that relative scores be ordered correctly, not that each entry be given a label of correct/incorrect. Training is balanced as we have a single ranking example for each query. And finally, we simply select the highest ranked entry as correct, no matter its score.

We take a maximum margin approach to learning: the correct KB entry y should receive a higher score than all other possible KB entries $\hat{y} \in \mathcal{Y}, \hat{y} \neq y$ plus some margin γ. This learning constraint is equivalent to the ranking SVM algorithm of Joachims [37], where we define an ordered pair constraint for each of the incorrect KB entries $\hat{y}$ and the correct entry y. Since we have a preference only for the relative ordering of a single entry compared to all others, we introduce a linear number of constraints for learning. Only the position of the correct entry is important. Training sets parameters such that $\text{score}(y) \geq \text{score}(\hat{y}) + \gamma$. We used the library SVM^{rank} to solve this optimization problem.[2] We used a linear kernel, set the slack parameter C as 0.01 times the number of training examples, and take the loss function as the total number of swapped pairs summed over all training examples. While previous work used a custom kernel, we found a linear kernel just as effective with our features. This has the advantage of efficiency in both training and prediction [3]—important considerations in a system meant to scale to millions of KB entries.

5.5.1 Features for Entity Disambiguation

200 atomic features represent **x** based on each candidate query/KB pair. Since we used a linear kernel, we explicitly combined certain features (e.g., acroynym-match AND known-alias) to model correlations. This included combining each feature with the predicted type of the entity, allowing the algorithm to learn prediction functions specific to each entity type. With feature combinations, the total number of features grew to 26,569. The next sections provide an overview; for a detailed list see [38].

5.5.2 Features for Name Variants

Variation in entity name has long been recognized as a bane for information extraction systems. Poor handling of entity name variants results in low recall. We describe several features ranging from simple string match to finite state transducer matching.

String Equality

If the query name and KB entry name are identical, this is a strong indication of a match, and in our KB entry names are distinct. However, similar or identical entry names that refer to distinct entities are often qualified with parenthetical expressions or short clauses. As an example, "London, Kentucky" is distinguished from "London, Ontario", "London, Arkansas", "London (novel)", and "London". Therefore, other string equality features were used, such as whether names are equivalent after some transformation. For example, "Baltimore" and "Baltimore

---

[2]www.cs.cornell.edu/people/tj/svm_light/svm_rank.html

[3]Bunescu and Pasca [13] report learning tens of thousands of support vectors with their "taxonomy" kernel while a linear kernel represents all support vectors with a single weight vector, enabling faster training and prediction.

City" are exact matches after removing a common GPE word like city; "University of Vermont" and "University of VT" match if VT is expanded.

Approximate String Matching

Many entity mentions will not match full names exactly. We added features for character Dice, skip bigram Dice, and left and right Hamming distance scores. Features were set based on quantized scores. These were useful for detecting minor spelling variations or mistakes. Features were also added if the query was wholly contained in the entry name, or vice-versa, which was useful for handling ellipsis (e.g., "United States Department of Agriculture" vs. "Department of Agriculture"). We also included the ratio of the recursive longest common subsequence [39] to the shorter of the mention or entry name, which is effective at handling some deletions or word reorderings (e.g., "Li Gong" and "Gong Li"). Finally, we checked whether all of the letters of the query are found in the same order in the entry name (e.g., "Univ Wisconsin" would match "University of Wisconsin").

Acronyms

Features for acronyms, using dictionaries and partial character matches, enable matches between "MIT" and "Madras Institute of Technology" or "Ministry of Industry and Trade."

Aliases

Many aliases or nicknames are non-trivial to guess. For example JAVA is the stock symbol for Sun Microsystems, and "Ginger Spice" is a stage name of Geri Halliwell. A reasonable way to do this is to employ a dictionary and alias lists that are commonly available for many domains.[4]

FST Name Matching

Another measure of surface similarity between a query and a candidate was computed by training finite-state transducers similar to those described in [40]. These transducers assign a score to any string pair by summing over all alignments and scoring all contained character n-grams; we used n-grams of length 3 and less.

---

[4] We used multiple lists, including class-specific lists (i.e., for PER, ORG, and GPE) lists extracted from Freebase [41] and Wikipedia redirects. PER, ORG, and GPE are the commonly used terms for entity types for people, organizations and geo-political regions respectively.

The scores are combined using a global log-linear model. Since different spellings of a name may vary considerably in length (e.g., *J Miller* vs. *Jennifer Miller*) we eliminated the limit on consecutive insertions used in previous applications.[5]

5.5.3 Wikipedia Features

Most of our features do not depend on Wikipedia markup, but it is reasonable to include features from KB properties. Our feature ablation study shows that dropping these features causes a small but statistically significant performance drop.

WikiGraph statistics

We added features derived from the Wikipedia graph structure for an entry, like indegree of a node, outdegree of a node, and Wikipedia page length in bytes. These statistics favor common entity mentions over rare ones.

Wikitology

KB entries can be indexed with human or machine generated metadata consisting of keywords or categories in a domain-appropriate taxonomy. Using a system called *Wikitology*, Syed et al. [42] investigated use of ontology terms obtained from the explicit category system in Wikipedia as well as relationships induced from the hyperlink graph between related Wikipedia pages. Following this approach we computed top-ranked categories for the query documents and used this information as features. If none of the candidate KB entries had corresponding highly-ranked Wikitology pages, we used this as a NIL feature (Sect. 5.6).

5.5.4 Popularity

Although it may be an unsafe bias to give preference to common entities, we find it helpful to provide estimates of entity popularity to our ranker as others have done [43]. Apart from the graph-theoretic features derived from the Wikipedia graph, we also used Google's PageRank by adding features indicating the rank of the KB entry's corresponding Wikipedia page in a Google query for the target entity mention.

---

[5]Without such a limit, the objective function may diverge for certain parameters of the model; we detect such cases and learn to avoid them during training.

5.5.5 Document Features

The mention document and text associated with a KB entry contain context for resolving ambiguity.

Entity Mentions

Some features were based on presence of names in the text: whether the query appeared in the KB text and the entry name in the document. Additionally, we used a named-entity tagger and relation finder, SERIF [44], which identified name and nominal mentions that were deemed co-referent with the entity mention in the document, and tested whether these nouns were present in the KB text. Without the NE analysis, accuracy on non-NIL entities dropped 4.5%.

KB Facts

KB nodes contain infobox attributes (or facts); we tested whether the fact text was present in the query document, both locally to a mention, or anywhere in the text. Although these facts were derived from Wikipedia infoboxes, they could be obtained from other sources as well.

Document Similarity

We measured similarity between the query document and the KB text in two ways: cosine similarity with TF/IDF weighting [45]; and using the Dice coefficient over bags of words. IDF values were approximated using counts from the Google 5-gram dataset as by Klein and Nelson [46].

Entity Types

Since the KB contained types for entries, we used these as features as well as the predicted NE type for the entity mention in the document text. Additionally, since only a small number of KB entries had PER, ORG, or GPE types, we also inferred types from Infobox class information to attain 87% coverage in the KB. This was helpful for discouraging selection of eponymous entries named after famous entities (e.g., the former U.S. president vs. "John F. Kennedy International Airport").

5.5.6 Feature Combinations

To take into account feature dependencies we created combination features by taking the cross-product of a small set of diverse features. The attributes used as combination features included entity type; a popularity based on Google's rankings; document comparison using TF/IDF; coverage of co-referential nouns in the KB node text; and name similarity. The combinations were cascaded to allow arbitrary feature conjunctions. Thus it is possible to end up with a feature *kbtype-is-ORG AND high-TFIDF-score AND low-name-similarity*. The combined features increased the number of features from roughly 200–26,000.

5.6   Predicting NIL Mentions

So far we have assumed that each example has a correct KB entry; however, when run over a large corpus, such as news articles, we expect a significant number of entities will not appear in the KB. Hence it will be useful to predict NILs.

We *learn* when to predict NIL using the SVM ranker by augmenting $\mathcal{Y}$ to include NIL, which then has a single feature unique to NIL answers. It can be shown that (modulo slack variables) this is equivalent to learning a single threshold τ for NIL predictions as in [13].

Incorporating NIL into the ranker has several advantages. First, the ranker can set the threshold optimally without hand tuning. Second, since the SVM scores are relative within a single example and cannot be compared across examples, setting a single threshold is difficult. Third, a threshold sets a uniform standard across all examples, whereas in practice we may have reasons to favor a NIL prediction in a given example. We design features for NIL prediction that cannot be captured in a single parameter.

Integrating NIL prediction into learning means we can define arbitrary features indicative of NIL predictions in the feature vector corresponding to NIL. For example, if many candidates have good name matches, it is likely that one of them is correct. Conversely, if no candidate has high entry-text/article similarity, or overlap between facts and the article text, it is likely that the entity is absent from the KB. We included several features, such as (a) the max, mean, and difference between max and mean for seven atomic features for all KB candidates considered, (b) whether any of the candidate entries have matching names (exact and fuzzy string matching), (c) whether any KB entry was a top Wikitology match, and (d) if the top Google match was not a candidate.

5.7   Evaluation

We evaluated our system on two datasets: the Text Analysis Conference (TAC) track on Knowledge Base Population (TAC-KBP) [27] and the newswire data used by Curcerzan in [14] (Microsoft News Data).

Table 5.1 Number of queries/entities by type and presence in the KB

Type	2009 queries			2010 queries		
	Total	KB	Missing	Total	KB	Missing
PER	627	255	372	751	213	538
ORG	2710	1013	1697	750	304	446
GPE	567	407	160	749	503	246
All	3904	1675	2229	2250	1020	1230

Since our approach relies on supervised learning, we begin by constructing our own training corpus.[6] We highlighted 1,496 named entity mentions in news documents (from the TAC-KBP document collection) and linked these to entries in a KB derived from Wikipedia infoboxes.[7] We added to this collection 119 sample queries from the TAC-KBP data. The total of 1,615 training examples included 539 (33.4%) PER, 618 (38.3%) ORG, and 458 (28.4%) GPE entity mentions. Of the training examples, 80.5% were found in the KB, matching 300 unique entities.

This set has a higher number of NIL entities than did Bunescu and Pasca [13] (10%) but lower than the TAC-KBP test set (43%).

All system development was done using a train (908 examples) and development (707 examples) split. The TAC-KBP and Microsoft News data sets were held out for final tests. A model trained on all 1,615 examples was used for experiments.

5.7.1 TAC-KBP 2009 Experiments

In 2009–2011, NIST conducted evaluations of entity linking technologies as part of the Text Analysis Conference (TAC). The Knowledge Base Population track (TAC-KBP) focused on two subtasks: linking mentions of entities to a standard KB, and gleaning novel attributes about and relationships between entities from a large corpus. In the entity linking subtask, each query consisted of a name string and a reference document that contained the name string and provided context to help determine which KB entity is being referred to. Each query was either a person (PER), organization (ORG), or geo-political entity (GPE; essentially an inhabited location) but the entity type was not provided in the query. A breakdown of queries by type is given in Table 5.1. In 2009 queries were not balanced by entity type or presence in the KB (e.g., organizations accounted for a majority of the of the queries—69%), but in 2010 a more uniform distribution was created. Persons and organizations were more likely to be absent than GPEs, which have broad coverage in the KB, as they do in Wikipedia. As of this work's publication, the 2011 task is underway.

---

[6]Data available from www.dredze.com

[7]http://en.wikipedia.org/wiki/Help:Infobox

Table 5.2 Sample queries for the entity linking task. Only a small excerpt from the provided document is shown

Query	Name	DOCID	Entity / KBID
EL2025	Michael Kennedy	NYT_ENG_20010122.0439.LDC2007T07	NIL

Van Brett Watkins, who confessed to the shooting, testified during the trial that Carruth planned to pay him $5,000 to kill Adams so that Carruth would not have to pay child support. Carruth's co-defendants, **Michael Kennedy**, who drove the car Watkins was riding in when he shot Adams, and Stanley Drew Abraham are awaiting trial.

EL2029	Michael Kennedy	NYT_ENG_19990717.0169.LDC2007T07	E0499939

Michael Kennedy, another of Robert and Ethel's children, was killed on Dec. 31, 1997, in a bizarre skiing accident. The 39-year-old skied into a tree in Aspen, Colo., while playing a game of ski football.

EL2030	Michael Kennedy	NYT_ENG_20070430.0025.LDC2009T13	NIL

The Revolution were pinned back and failed to control the ball in counterattacking opportunities in the early going. But things changed quickly as Twellman ran through the halfway line and was taken down by Dax McCarty, who was cautioned by referee **Michael Kennedy**.

EL2042	Mike Kennedy	AFP_ENG_20070414.0006.LDC2009T13	NIL

But the Victorian AIDS Council said overseas arrivals accounted for only nine of the 334 new HIV notifications in the state last year and it was wrong to single out immigrants as a source of infection. "That number is incredibly low," council president **Mike Kennedy** told Melbourne's Age newspaper. "In Australia the bulk of the epidemic is gay men."

We evaluated our approach on the 2099 data and describe the data in detail. The KB is derived from English Wikipedia pages that contained an infobox. Entries contain basic descriptions (article text) and attributes. The TAC-KBP query set contains 3,904 entity mentions for 560 distinct entities; entity type was only provided for evaluation. The majority of queries were for organizations (69%). Most queries were missing from the KB (57%). seventy-seven percent of the distinct GPEs in the queries were present in the KB, but for PERs and ORGs these percentages were significantly lower, 19% and 30% respectively.

Fictional entities, which are well-covered in Wikipedia, and time-sensitive entities (e.g., ORGs with dynamic membership such as the US Olympic men's ice hockey team) were deliberately excluded as targets. Also prohibited were names that can refer to a group of entities (e.g., Blue Devils might refer to any of Duke University's athletic teams). Care was taken to avoid using documents where the target entity name was internally ambiguous. Additional details about the target selection process are described in Simpson et al. [47]. In Table 5.2 several of the 2009 entity linking queries are presented, along with KB node that was judged to be correct.

Table 5.3 shows results on TAC-KBP data using all of our features as well a subset of features based on feature selection experiments on development data. We include scores for both micro-averaged accuracy—averaged over all queries—and macro-averaged accuracy—averaged over each unique entity—as well as the best and median reported results for these data [27]. We obtained the best reported results for macro-averaged accuracy, as well as the best results for NIL detection with

Table 5.3 Micro and macro-averaged accuracy for TAC-KBP data compared to best and median reported performance. Results are shown for all features as well as removing a small number of features using feature selection on development data

	Micro-averaged				Macro-averaged			
	Best	Median	All feats	Best feats	Best	Median	All feats	Best feats
All	0.8217	0.7108	0.7984	0.7941	0.7704	0.6861	0.7695	**0.7704**
non-NIL	0.7725	0.6352	0.7063	0.6639	0.6696	0.5335	0.6097	0.5593
NIL	0.8919	0.7891	0.8677	**0.8919**	0.8789	0.7446	0.8464	0.8721

micro-averaged accuracy, which shows the advantage of our approach to learning NIL. See [38] for additional experiments.

The candidate selection phase obtained a recall of 98.6%, similar to that of development data. Missed candidates included *Iron Lady*, which refers metaphorically to Yulia Tymoshenko, *PCC*, the Spanish-origin acronym for the Cuban Communist Party, and *Queen City*, a former nickname for the city of Seattle, Washington. The system returned a mean of 76 candidates per query, but the median was 15 and the maximum 2,772 (*Texas*).

In about 10% of cases there were four or fewer candidates and in 10% of cases there were more than 100 candidate KB nodes. We observed that ORGs were more difficult, due to the greater variation and complexity in their naming, and that they can be named after persons or locations.

5.7.2 Feature Effectiveness

We performed two feature analyses on the TAC-KBP data: an additive study—starting from a small baseline feature set used in candidate selection we add feature groups and measure performance changes (omitting feature combinations), and an ablative study—starting from all features, remove a feature group and measure performance.

Table 5.4 shows the most significant features in the feature addition experiments. The baseline includes only features based on string similarity or aliases and is not effective at finding correct entries and strongly favors NIL predictions. Inclusion of features based on analysis of named-entities, popularity measures (e.g., Google rankings), and text comparisons provided the largest gains. Although the overall changes are fairly small the changes in non-NIL precision are much larger.

The ablation study showed considerable redundancy across feature groupings. In several cases, performance could have been slightly improved by removing features. Removing all feature combinations would have improved overall performance to 81.05% by gaining on non-NIL for a small decline on NIL detection.

Table 5.4 Additive analysis: micro-averaged accuracy

Class	All	non-NIL	NIL
Baseline	0.7264	0.4621	0.9251
Acronyms	0.7316	0.4860	0.9161
NE analysis	0.7661	0.7181	0.8022
Google	0.7597	0.7421	0.7730
Doc/KB text similarity	0.7313	0.6699	0.7775
Wikitology	0.7318	0.4549	0.9399
All	0.7984	0.7063	0.8677

5.7.3   Experiments on Microsoft News Data

We downloaded the evaluation data used in [14][8]: 20 news stories from MSNBC with 642 entity mentions manually linked to Wikipedia and another 113 mentions not having any corresponding link to Wikipedia.[9] A significant percentage of queries were not of type PER, ORG, or GPE (e.g., "Christmas"). SERIF assigned entity types and we removed 297 queries not recognized as entities (counts in Table 5.5).

We learned a new model on the training data above using a reduced feature set to increase speed.[10] Using our fast candidate selection system, we resolved each query in 1.98 s (median). Query processing time was proportional to the number of candidates considered. We selected a median of 13 candidates for PER, 12 for ORG and 102 for GPE. Accuracy results are in Table 5.5. The high results reported for this dataset over TAC-KBP is primarily because we perform very well in predicting popular and rare entries—both of which are common in newswire text.

One issue with our KB was that it was derived from infoboxes in Wikipedia's Oct 2008 version which has both new entities, [11] and is missing entities.[12] Therefore, we manually confirmed NIL answers and new answers for queries marked as NIL in the data. While an exact comparison is not possible (as described above), our results (94.7%) appear to be at least on par with Cucerzan's system (91.4% overall accuracy).With the strong results on TAC-KBP, we believe that this is strong confirmation of the effectiveness of our approach.

---

[8]http://research.microsoft.com/en-us/um/people/silviu/WebAssistant/TestData/

[9]One of the MSNBC news articles is no longer available so we used 759 total entities.

[10]We removed Google, FST and conjunction features which reduced system accuracy but increased performance.

[11]2008 vs. 2006 version used in [14] We could not get the 2006 version from the author or the Internet.

[12]Since our KB was derived from infoboxes, entities not having an infobox were left out.

Table 5.5 Micro-average results for Microsoft data

	Num. queries		All	Accuracy	
	Total	Nil		non-NIL	NIL
NIL	452	187	0.4137	0.0	1.0
GPE	132	20	0.9696	1.00	0.8000
ORG	115	45	0.8348	0.7286	1.00
PER	205	122	0.9951	0.9880	1.00
All	452	187	**0.9469**	0.9245	0.9786
	Cucerzan (2007)		0.914	–	–

5.8 The TAC-KBP Entity Linking Task

We summarize general approaches and results for the 2009 and 2010 TAC-KBP entity linking tasks. A detailed summary can be found in [28].

5.8.1 Challenging Queries

In their 2009 overview paper McNamee and Dang [27] describe several types of errors prevalent among the most challenging queries.

- Ambiguous acronyms: in query EL1213 ("DRC") the article refers to the Democratic Republic of Congo as both "DCR" and "DRC".
- Related organizations: in query EL3871 "Xinhua Finance" is referring to Xinhua Finance Media Ltd., not its parent company Xinhua Finance Ltd.
- Metaphorical names: queries EL1717 and EL1718 ("Iron Lady") referred to two different women and it wasn't clear that the nickname was commonly used for them.
- Metonymic references—query EL2599 "New Caledonia" is in a document about World Cup rankings, and it is debatable whether or the the name in the article refers to the country or its national soccer team.
- Assessment errors: queries EL3334 and EL3335 "The Health Department", are referring to the New York City Department of Health and the New York State Department of Health, respectively, but the United States Department of Health and Human Services was incorrectly judged to be the proper response.

5.8.2 Approaches

The majority of systems divided the task into three parts: (a) identifying a subset of KB entries that are reasonable candidates for a query entity, and (b) selection of the most likely non-NIL candidate; and (c) deciding whether absence from the KB (i.e., NIL) is the correct response.

Approaches to candidate identification were generally based on name comparisons between query entities and KB entries, often using precomputed dictionaries or inverted files to quickly identify potential KB nodes. A variety of non-exact matching techniques were used, including: alias lists; character n-grams; phonetic matching; acronyms; Wikipedia links or redirects; external Web search; and relationship similarity. Intra-document coreference resolution was exploited by several groups.

A variety of machine learning approaches were used in the selection process. Our submission (HLTCOE [38]) and the QUANTA team [48] used learning to rank frameworks with good effect. In fact, Li et al. [48] propose an approach that bears a number of similarities to ours; both systems create candidate sets and then rank possibilities using differing learning methods, but the principal difference is in our approach to NIL prediction. Where we simply consider absence (i.e. the NIL candidate) as another entry to rank, and select the top ranked option, they use a separate binary classifier to decide whether their top prediction is correct, or whether NIL should be output. We believe relying on features that are designed to inform whether absence is correct is the better alternative. Other approaches for ranking candidates included binary classification and vector comparisons. Ji and Grishman [28] also discuss the TAC-KBP entity linking task in depth and they give a detailed comparison of approaches of different systems in the 2010 evaluation.

5.9 Beyond Entity Linking

As currently formulated, evaluations of entity linking suffer from a number of limitations. For example, at TAC-KBP, only named mentions (vs. pronouns) have been the linking targets, some names are unresolvable even by humans, and it is challenging to develop query sets to rigorously exercise systems. For example, if a random sample of names is taken, then prominent (and more easily linkable) entities form the majority of the query set; however it is non-trivial to identify challenging queries that contain confusable names of roughly equal prominence.

At present, the only available test sets are in English, although there should be no impediment to developing multilingual and non-English test collections.[13] Another difficulty is the lack of releasable, large-scale knowledge bases. Wikipedia has been the subject of much study, due to its generous licensing, and broad coverage; however, Wikipedia has many unique characteristics (e.g., being indexed by major search engines) that make its use as a test KB susceptible to solutions that do not

---

[13]As this article went to press we became aware of the efforts by Mayfield et al. [49] to construct a cross-language entity linking test collection where the language of the knowledge base is English, but query names are in many languages.

generalize to other knowledge bases. Further work on Entity Linking in non-English languages will have to deal with case inflections and the similarity metrics which proved most effecive in English might not always be optimal. However, the design of our Entity Linking architecture allows most of these components to be pluggable.

What might the future hold for entity linking? In the near term we expect to see work in multilingual entity linking, increasing interest in linking entity mentions in social media (e.g., Twitter, Facebook), and efforts to increase the diversity and granularity of entity types (e.g., products and brands, events, books, films, and works of art). If Semantic Web technologies continue to gain wider acceptance, then we might expect to see a proliferation of large-scale KBs, which could motivate entity linking beyond its current focus on Wikipedia. Ultimately we expect to see entity linking being used as a component of complex NLP and knowledge discovery applications.

Finally, we should point out that research community has currently split the two problems of cross-document entity coreference and entity linking to a reference knowledge base. Clustering techniques have been dominant in solving the former, while supervised machine learning appears to be the leading approach for the later. It would be beneficial if the research community could better articulate for which real world applications each problem formulation is most applicable, and if possible, develop a unification of these two highly-related problems.

5.10 Conclusion

We presented a state of the art system to disambiguate entity mentions in text and link them to a knowledge base. Unlike previous approaches, our approach readily ports to KBs other than Wikipedia. We described several important challenges in the entity linking task including handling variations in entity names, ambiguity in entity mentions, and missing entities in the KB, and we showed how to each of these can be addressed. We described a comprehensive feature set to accomplish this task in a supervised setting. Importantly, our method discriminately learns when not to link with high accuracy.

References

1. Sang, E.T.K., Meulder, F.D.: Introduction to the conll-2003 shared task: language-independent named entity recognition. In: Conference on Natural Language Learning (CONLL), Edmonton. The Association for Computational Linguistics, Stroudsburg (2003)
2. Asahara, M., Matsumoto, Y.: Japanese named entity extraction with redundant morphological analysis. In: Proceedings of the 2003 Conference of the North American Chapter of the Association for Computational Linguistics on Human Language Technology - vol. 1, NAACL '03, Stroudsburg, pp. 8–15, Association for Computational Linguistics, Stroudsburg (2003)

3. McCallum, A., Li, W.: Early results for named entity recognition with conditional random fields, feature induction and web-enhanced lexicons. In: Proceedings of the seventh conference on Natural language learning at HLT-NAACL 2003 - vol. 4, CONLL '03, Stroudsburg, pp. 188–191. Association for Computational Linguistics, Stroudsburg (2003)
4. Collins, M., Singer, Y.: Unsupervised models for named entity classification. In: In Proceedings of the Joint SIGDAT Conference on Empirical Methods in Natural Language Processing and Very Large Corpora, College Park, pp. 100–110. Association for Computational Linguistics, New Brunswick (1999)
5. Cucerzan, S., Yarowsky, D.: Language independent ner using a unified model of internal and contextual evidence. In: Proceedings of the 6th Conference on Natural Language Learning - vol. 20, COLING-02, Stroudsburg, pp. 1–4. Association for Computational Linguistics, Stroudsburg (2002)
6. Nadeau, D., Sekine, S.: A survey of named entity recognition and classification. Linguisticae Investigationes, vol. 30, pp. 3–26. John Benjamins, Amsterdam (2007)
7. Bagga, A., Baldwin, B.: Entity-based cross-document coreferencing using the vector space model. In: Conference on Computational Linguistics (COLING), Montreal. Association for Computational Linguistics, Stroudsburg (1998)
8. van Deemter, K., Kibble, R.: On coreferring: coreference in muc and related annotation schemes. Comput. Linguist. **26**, 629–637 (2000)
9. Yang, X., Zhou, G., Su, J., Tan, C.L.: Coreference resolution using competition learning approach. In: Proceedings of the 41st Annual Meeting of the Association for Computational Linguistics, Sapporo, pp. 176–183 (2003)
10. Ng, V.: Supervised noun phrase coreference research: The first fifteen years. In: Proceedings of the ACL, Uppsala, pp. 1396–1411. Association for Computational Linguistics, Stroudsburg (2010)
11. Banko, M., Etzioni, O.: The tradeoffs between open and traditional relation extraction. In: Association for Computational Linguistics, Columbus (2008)
12. Sutton, C., Mccallum, A.: Introduction to conditional random fields for relational learning. In: Getoor, L., Taskar, B., (eds.) Introduction to Statistical Relational Learning. MIT, Cambridge (2006)
13. Bunescu, R.C., Pasca, M.: Using encyclopedic knowledge for named entity disambiguation. In: European Chapter of the Assocation for Computational Linguistics (EACL), Trento (2006)
14. Cucerzan, S.: Large-scale named entity disambiguation based on wikipedia data. In: Empirical Methods in Natural Language Processing (EMNLP), Prague. Association for Computational Linguistics, Stroudsburg (2007)
15. Mann, G.S., Yarowsky, D.: Unsupervised personal name disambiguation. In: Conference on Natural Language Learning (CONLL), Edmonton. Johns Hopkins University, Baltimore (2003)
16. Artiles, J., Sekine, S., Gonzalo, J.: Web people search: results of the first evaluation and the plan for the second. In: WWW, Beijing. ACM, New York (2008)
17. Poesio, M., Day, D., Artstein, R., Duncan, J., Eidelman, V., Giuliano, C., Hall, R., Hitzeman, J., Jern, A., Kabadjov, M., Yong, S., Keong, W., Mann, G., Moschitti, A., Ponzetto, S., Smith, J., Steinberger, J., Strube, M., Su, J., Versley, Y., Yang, X., Wick, M.: Exploiting lexical and encyclopedic resources for entity disambiguation: final report. Technical Report, JHU CLSP 2007 Summer Workshop, Johns Hopkins University, Baltimore (2008)
18. Popescu, O.: Dynamic parameters for cross document coreference. In: Conference on Computational Linguistics (COLING), Beijing (2010)
19. Huang, J., Treeratpituk, P., Taylor, S., Giles, C.L.: Enhancing cross document coreference of web documents with context similarity and very large scale text categorization. In: Conference on Computational Linguistics (COLING), Beijing. Association for Computational Linguistics, Stroudsburg (2010)
20. Singh, S., Subramanya, A., Pereira, F., McCallum, A.: Large-scale cross-document coreference using distributed inference and hierarchical models. In: Association for Computational Linguistics, Portland. Association for Computational Linguistics, Stroudsburg (2011)

21. Rao, D., McNamee, P., Dredze, M.: Streaming cross document entity coreference resolution. In: Conference on Computational Linguistics (COLING), Beijing. Association for Computational Linguistics, Stroudsburg (2010)
22. Ng, V.: Supervised noun phrase coreference research: The first fifteen years. In: Association for Computational Linguistics, Uppsala. Association for Computational Linguistics, Stroudsburg (2010)
23. Raghunathan, K., Lee, H., Rangarajan, S., Chambers, N., Surdeanu, M., Jurafsky, D., Manning, C.: A multi-pass sieve for coreference resolution. In: Empirical Methods in Natural Language Processing (EMNLP), Massachusetts. Association for Computational Linguistics, Stroudsburg (2010)
24. Elsner, M., Charniak, E.: The same-head heuristic for coreference. In: Association for Computational Linguistics, Uppsala. Association for Computational Linguistics, Stroudsburg (2010)
25. Stoyanov, V., Cardie, C., Gilbert, N., Riloff, E., Buttler, D., Hysom, D.: Reconcile: A coreference resolution research platform. In: Association for Computational Linguistics, Uppsala (2010)
26. Gabbard, R., Freedman, M., Weischedel, R.: Coreference for learning to extract relations: Yes virginia, coreference matters. In: Association for Computational Linguistics, Portland. Association for Computational Linguistics, Stroudsburg (2011)
27. McNamee, P., Dang, H.T.: Overview of the TAC 2009 knowledge base population track. In: Text Analysis Conference (TAC), Gaithersburg (2009)
28. Ji, H., Grishman, R.: Knowledge base population: successful approaches and challenges. In: Proceedings of the 49th Annual Meeting of the Association for Computational Linguistics (ACL-HLT), Portland. Association for Computational Linguistics, Stroudsburg (2011)
29. Zhang, W., Su, J., Tan, C.L.: Entity linking leveraging automatically generated annotation. In: Conference on Computational Linguistics (COLING), Beijing (2010)
30. Gottipati, S., Jiang, J.: Linking entities to a knowledge base with query expansion. In: Empirical Methods in Natural Language Processing (EMNLP), Edinburgh. Association for Computational Linguistics, Stroudsburg (2011)
31. Han, X., Sun, L.: A generative entity-mention model for linking entities with knowledge base. In: Association for Computational Linguistics, Portland. Association for Computational Linguistics, Stroudsburg (2011)
32. Lehmann, J., Monahan, S., Nezda, L., Jung, A., Shi, Y.: Lcc approaches to knowledge base population at tac 2010. In: Proceeding TAC 2010 Workshop, National Institute of Standards and Technology, Gaithersburg (2010)
33. Zhang, W., Sim, Y., Su, J., Tan, C.: Nus-i2r: Learning a combined system for entity linking. In: Proceeding TAC 2010 Workshop, National Institute of Standards and Technology, Gaithersburg (2010)
34. Zhang, W., Sim, Y.C., Su, J., Tan, C.L.: Entity linking with effective acronym expansion instance selection and topic modeling. In: International Joint Conference on Artificial Intelligence, Barcelona (2011)
35. Elmagarmid, A.K., Ipeirotis, P.G., Verykios, V.S.: Duplicate record detection: a survey. IEEE Trans. Knowl. Data Eng. **19**, 1–16 (2007)
36. McCallum, A., Nigam, K., Ungar, L.: Efficient clustering of high-dimensional data sets with application to reference matching. In: Knowledge Discovery and Data Mining (KDD), Boston. ACM, New York (2000)
37. Joachims, T.: Optimizing search engines using clickthrough data. In: Knowledge Discovery and Data Mining (KDD), Edmonton. ACM, New York (2002)
38. McNamee, P., Dredze, M., Gerber, A., Garera, N., Finin, T., Mayfield, J., Piatko, C., Rao, D., Yarowsky, D., Dreyer, M.: HLTCOE approaches to knowledge base population at TAC 2009. In: Text Analysis Conference (TAC), Gaithersburg (2009)
39. Christen, P.: A comparison of personal name matching: techniques and practical issues. Technical Report TR-CS-06-02, Australian National University, Australia (2006)

40. Dreyer, M., Smith, J., Eisner, J.: Latent-variable modeling of string transductions with finite-state methods. In: Empirical Methods in Natural Language Processing (EMNLP), Honolulu. Association for Computational Linguistics, Stroudsburg (2008)
41. Bollacker, K., Evans, C., Paritosh, P., Sturge, T., Taylor, J.: Freebase: a collaboratively created graph database for structuring human knowledge. In: SIGMOD Management of Data, Vancouver. ACM, New York (2008)
42. Syed, Z., Finin, T., Joshi, A.: Wikipedia as an ontology for describing documents. In: Proceedings of the Second International Conference on Weblogs and Social Media, Chicago. AAAI, Menlo Park (2008)
43. Fader, A., Soderland, S., Etzioni, O.: Scaling Wikipedia-based named entity disambiguation to arbitrary web text. In: WikiAI09 Workshop at IJCAI 2009, Pasadena (2009)
44. Boschee, E., Weischedel, R., Zamanian, A.: Automatic information extraction. In: Conference on Intelligence Analysis, Washington (2005)
45. Salton, G., McGill, M.: Introduction to Modern Information Retrieval. McGraw-Hill, New York (1983)
46. Klein, M., Nelson, M.L.: A comparison of techniques for estimating IDF values to generate lexical signatures for the web. In: Workshop on Web Information and Data Management (WIDM), Napa Valley. ACM, New York (2008)
47. Simpson, H., Parker, R., Strassel, S., Dang, H.T, McNamee, P.: Wikipedia and the web of confusable entities: experience from entity profile creation for tac knowledge base population. In: Proceedings of the Seventh International Language Resources and Evaluation Conference (LREC), Valletta. European Language Resources Association, Valletta (2010)
48. Li, F., Zhang, Z., Bu, F., Tang, Y., Zhu, X., Huang, M.: THU QUANTA at TAC 2009 KBP and RTE track. In: Text Analysis Conference (TAC), National Institute of Standards and Technology, Gaithersburg (2009)
49. Mayfield, J., Lawrie, D., McNamee, P., Oard, D.W.: Building a cross-language entity linking collection in twenty-one languages. In: Proceedings of the Cross Language Evaluate Forum (CLEF), Amsterdam (2011)

Chapter 6
A Study of the Effect of Document Representations in Clustering-Based Cross-Document Coreference Resolution

Horacio Saggion

Abstract Finding information about people on huge text collections or on-line repositories on the Web is a common activity. We describe experiments aiming at identifying the contribution of semantic information (e.g., named entities) and summarization (e.g., sentence extracts) in a cross-document coreference resolution system. Our system uses a clustering-based algorithm to group documents referring to the same entity. Clustering uses vector representations created by summarization and semantic tagging components. We investigate different clustering configurations and show that selection of the type of summary and the type of term to be used for vector representation is important to achieve good performance.

6.1 Introduction

A common activity of Internet users is to search for information about people (e.g., I would like documents about "John Smith") or other entities such as organizations or locations (e.g., I want documents about "London"), entities which can be denoted by a proper name. In ad-hoc retrieval of the type practised by Web search engines such as Yahoo! or Google a request for documents containing a particular name will typically return thousands of documents matching the name of the entity. The results will be ranked according to a particular criteria implemented in the search engine, however not all documents in the retuned list will refer to the same individual in the real word, for it is unlikely that there would be a one-to-one correspondence between a name and an individual. Names of entities are highly ambiguous, for example and according to the U.S. Census Bureau 90,000 different names are shared by

H. Saggion (✉)
Department of Information and Communication Technologies, Universitat Pompeu Fabra, Barcelona, Spain
e-mail: horacio.saggion@upf.edu

T. Poibeau et al. (eds.), *Multi-source, Multilingual Information Extraction and Summarization 11*, Theory and Applications of Natural Language Processing, DOI 10.1007/978-3-642-28569-1__6, © Springer-Verlag Berlin Heidelberg 2013

100 million people. In the Wikipedia on-line encyclopaedia (English version) there are 42 different famous "John Smith", there are of course many more non famous individuals with that name. There are also many different locations sharing identical names such as London in the United Kingdom and London in Ontario, Canada. For the problem of linking named entities in text to instances of individuals in a database or knowledge repository see work by Rao et al. Chap. 5 in this volume. Deciding if two mentions of the same name in two different documents refer to the same individual is a difficult problem known as cross-document coreference resolution. Automatic techniques for solving this problem are required not only for better access to information but also in natural language processing applications such as multi-document summarization, question answering, and information extraction. For example, one typical scenario in automatic text summarization is that of creating a profile of a person from a set of documents matching the person's name. This scenario was implemented in the Document Understanding Conference 2004 (an international evaluation of text summarization systems now replaced by the Text Analysis Conference) and input documents for that scenario were carefully selected by human analysts, so that the input set contained documents referring to a single individual. Having a coherent set of documents as input facilitates the task of the summarization system and may help avoid inconsistencies in the output.

One way of discovering what documents (or in some cases document fragments) refer to which individuals is by using document clustering. Clustering techniques applied to this problem were used in the field of cross-document coreference by Bagga and Baldwin [5]. In that work, documents for clustering were represented in the vector space model where each document was transformed into a vector of words and weights; they investigated two possible sources for extracting words: the full document or a summary of the document. A summary of the document was created as a set of sentences containing the name of the target person or any other expression coreferent with the target person. Only one person name was used in their experiments, and for that particular person name and set of documents, the use of summaries as document representation for clustering was deemed more effective than the use of full document representation. Bagga and Baldwin's work did not look into summaries other than "personal" summaries nor did they look into other vector representations such as semantic-based vectors of named entities. In this paper we investigate the role of various summarization techniques, full documents, and term representations (words and named entities) in clustering for cross-document coreference. As we will show, in a particular data set we have used, both summaries and full documents can achieve similar level of clustering accuracy provided that an appropriate term representation is used. We also show that the applied techniques achieve state of the art performance for this task. The work described here has been carried out in the context of the Web People Search Evaluation—an international evaluation program in the area of cross-source coreference which aims at measuring the performance of clustering for person name disambiguation, see Artiles et al. [3]. To the best of our knowledge this is the first work to study the effect of different summarization strategies for cross-source coreference resolution.

The rest of this contribution is organized as follows: In Sect. 6.2, we introduce approaches to cross-document coreference. Section 6.3 presents our clustering-based cross-document coreference algorithm and Sect. 6.4 explains how documents are analysed by extraction and summarization components. Sections 6.5 and 6.6 present experiments and results and Sect. 6.7 discusses our solution comparing with previous work. Finally, Sect. 6.8 closes the paper with conclusions and comments on further work.

6.2 Cross-Document Coreference Resolution

The problem of cross-document coreference has been studied for a number of years now. Bagga and Baldwin [5] used the vector space model together with summarization techniques to tackle the cross-document coreference problem. Their approach uses vector representations following a bag-of-words approach. Terms for vector representation are obtained from sentences where the target person appears. Another aspect of their work is the development of an evaluation framework called B-Cube algorithm which they argue is better fitted for evaluating this task than traditional information extraction precision and recall metrics. They have not presented an analysis of the impact of full document versus summary condition and their clustering algorithm is rather under-specified.

Mann and Yarowsky [15] used semantic information extracted from documents referring to the target person in an hierarchical agglomerative clustering algorithm. Semantic information here refers to factual information about a person such as the date of birth, professional career or education. Information is extracted using patterns some of them manually developed and others induced from examples. Phan et al. [17] follow Mann and Yarowsky in their use of a kind of biographical information about a person. They use a machine learning algorithm to classify sentences according to particular information types in order to automatically construct a person profile. Instead of comparing biographical information in the person profile altogether as in Mann and Yarowsky [15], they compare each type of information independently of each other, combining them only to make the final decision.

Bekkerman and McCallum [7] explore link analysis to find pages which relate to a set of people in a social network (e.g., researchers participating in a research project). This algorithm is suitable for people on the Web but not for finding documents for a given individual in text collections.

Aswani el al. [4] addressed the problem of cross-document coreference in the context of author name disambiguation. They mine information from the Web for authors including full name, personal page, and co-citation information to compute the similarity between two person names. Similarity is based on a formula which combines numeric features with appropriate weights experimentally obtained.

6.3 Document Clustering

Clustering is an important technique used in areas such as information retrieval, text mining, and data mining [9]. Clustering algorithms combine data points into groups such that: (i) data points in the same group are similar to each other; and (ii) data points in one group are "different" from data points in a different group or cluster. In information retrieval it is assumed that documents that are similar to each other are likely to be relevant for the same query, and therefore having the document collection organised in clusters can provide improved document access [28]. Different clustering techniques exist [29] the simplest one being the one-pass clustering algorithm [19].

We have implemented an agglomerative clustering algorithm which is relatively simple, has reasonable complexity, and as it will be shown gave us good results. Our algorithm operates in an exclusive way, meaning that a document belongs to one and only one cluster—while this is our working hypothesis, it might not be valid in some cases. The input to the algorithm is a set of document representations implemented as vectors of terms and weights ($term\_1 = weigth\_1, \ldots, term\_n = weigth\_n$). Initially, there are as many clusters as input documents; as the algorithm proceeds clusters are merged until a certain termination condition is reached. The algorithm computes the similarity between vector representations in order to decide whether or not to merge two clusters. The pseudo code of the algorithm is shown in Algorithm 1.

Algorithm 1 Clustering Algorithm

Given: LDOCS: a list of vector representations; THR: a similarity threshold
begin
for all $vector_i \in$ LDOCS **do**
    $cluster_i \leftarrow vector_i$
end for
compute similarity matrix for every pair of vectors ($sim_D(vector_i, vector_j)$)
$max\_sim \leftarrow 0; change \leftarrow$ true;
while $change$ **do**
    for all $active(cluster_i)$ **do**
        for all $active(cluster_j)$ $(i \neq j)$ **do**
            $current \leftarrow sim_C(cluster_i, cluster_j)$
            if $max\_sim \leq current$ **then**
                $max\_sim \leftarrow current$
                $cluster_1 \leftarrow cluster_i$
                $cluster_2 \leftarrow cluster_j$
            end if
        end for
    end for
    if $max\_sim > THR$ **then**
        /* Creates a new cluster with the vectors from $cluster_1$ and $cluster_2$ and marks $cluster_1$ and $cluster_2$ as inactive
        */
        $merge(cluster_1, cluster_2)$
    else
        $change \leftarrow false$
    end if
end while
return $clusters$
end

The similarity metric we use is the cosine of the angle between two vectors. This metric gives value one for identical vectors and zero for vectors which are orthogonal (no related). Various options have been implemented in order to measure how close two clusters are, but for the experiments reported here we have used the following approach: the similarity between two clusters (simC) is equivalent to the similarity (simD) between the two more similar documents in the two clusters—this is known as single linkage in the clustering literature; we take simD to be the cosine metric computed as follows:

$$\text{cosine}(d_1, d_2) = \frac{\sum_{i=1}^{n} w_{i,d_1} * w_{i,d_2}}{\sqrt{\sum_{i=1}^{n} (w_{i,d_1})^2} * \sqrt{\sum_{i=1}^{n} (w_{i,d_2})^2}}$$

Here d_1 and d_2 are document vectors and w_{i,d_k} is the weight of term i in document d_k.

If this similarity between two clusters is greater than a threshold—experimentally obtained—they are merged together. At each iteration in the algorithm, the most similar pair of clusters is merged. If this similarity is less than a certain threshold the algorithm stops. Merging two clusters consist of a simple step of set union, so there is no re-computation involved—such as computing a cluster centroid.

6.4   Document Analysis

In order to create representations for each document in the set to be disambiguated, we linguistically analyse the documents using information extraction and summarization techniques.

6.4.1   Information Extraction

Information extraction is the process of extracting from text specific facts in a given target domain [11]. For example, in extracting information about companies key elements to be extracted are the company address, contact phone, fax numbers, and e-mail address, products and services, members of the board of directors and so on. The information to be extracted is pre-specified and the system is tailored to extract those specific elements.

One essential component of any information extraction system is a named entity (NE) recognition component: the identification and extraction of key names from text. Today NE recognition is a mature technology which achieves performance levels of precision and recall above 90% for newswire texts where entities of interest are for example people, locations, times, organizations, etc.

Much research on NE recognition has been carried out in the context of the US sponsored Message Understanding Conferences (MUC) from 1987 until 1997 for research and development of information extraction systems. The ACE program was an extension of MUC but where the NE recognition task became more complex in the sense of being replaced by an entity detection and tracking task which involved, in addition to recognition, the identification of all mentions of a given entity. Two approaches are used to recognize entities in text: machine learning systems are given either an annotated corpus for training or a corpus of relevant and irrelevant documents together with only a few annotated examples of the extraction task, in this case some non-supervised techniques such as clustering can also be applied. Hand coded rule-based systems can be based on gazetteer lists—lists of keywords which can be used to identify known names (e.g. New York) or give contextual information for recognition of complex names (e.g. Corporation is a common postfix for a company name)—and cascades of finite state transducers which implement pattern matching algorithms over linguistic annotations (produced by various linguistic processors).

In the experiments reported here we make use of available named recognition technology provided by the GATE system [16]: a framework for the development and deployment of language processing technology. The text and linguistic processors used in our system are: document tokenisation to identify different kinds of words; sentence splitting to segment the text into sentence units; parts-of-speech tagging used for named entity recognition; named entity recognition using a gazetteer lookup module and regular expressions grammars; and name entity coreference module using a rule-based orthographic name matcher to identify name mentions considered equivalent (e.g., "John Smith" and "Mr. Smith"). Named entities of type Person, Organization, Address, Date, and Location are considered relevant document terms in our experiments. Coreference chains (e.g., sets of entities related by identity relations) are created in each input document for further use during summarization.

6.4.2 Text Summarization

The goal of automatic summarization is to take an information source, extract content from it, and present the most important content to the system's user in a condensed form which is sensitive to the user's task [13]. There are two main problems in text summarization: one is the problem of selecting the most relevant information from a source document or documents, the second problem is how to express that key information in the final summary. A compression parameter is usually specified as part of the process. This can be an absolute number of words to be produced or a percent of the original text. Approaches to the first problem have tried to come up with a list of relevant features that are believed to indicate the relevance of a sentence in a document or set of documents. These features can be used alone or in combination to produce sentence (or paragraph) relevance scores which in turn are used to rank sentences/paragraphs and select the top ranked ones (up to a certain compression) as the resulting summary content (e.g. an extract).

In this work we rely on available sentence extraction summarization technology provided by the SUMMA toolkit [22]: a set of components for the creation of summarization systems compatible with the GATE system.

SUMMA provides a set of algorithms to compute features (with numerical values) for each sentence in the input document which indicate how relevant the information in the sentence is for the feature. The computed values are combined— using a provided component—in a linear formula to obtain a score for each sentence which is used as the basis for sentence selection. Sentences are ranked based on their score and top ranked sentences selected to produce an extract. SUMMA has been used in experiments for image description (Aket et al. Chap. 14 in this volume) and also in multidocument summarization applications [23] among others.

In SUMMA, a corpus statistic module computes token statistics including term frequency—the number of times each term occurs in the document (tf). The vector space model [26] has been implemented and it is used to create vector representations of different text fragments. Each vector contains for each term occurring in the text fragment, the value tf*idf (term frequency * inverted document frequency). The inverted document frequency of a given term is the number of documents in a collection containing the term. Various modules in SUMMA use sentences represented as vectors for feature computation. Here we briefly describe some features used for our clustering experiments:

- Position feature: The sentence position module computes two features for each sentence: the absolute position of the sentence in the document and the relative position of the sentence in the paragraph. The absolute position of sentence i receives value i^{-1} while the paragraph feature receives a value which depends on the sentence being in the beginning, middle or end of paragraph—these values are parameters of the system and can be modified by the user.
- Cue feature: This feature is based on the idea that in a given application domain some expressions can be considered as important information carriers, and therefore their identification is useful to find information relevant for a summary [1]. Formulaic expressions such as "This paper presents," "Our results indicate," and "We conclude that" are used by paper authors to explicitly introduce key information. The computation of this feature is based on a pre-defined list of words and associated values. First, each sentence in the input document receives a cue feature which is the sum of the values of the words in the cue-list which are matched in the sentence. Then, the cue feature of each sentence is normalised to produce a value between 0 and 1.
- Query feature: In an information retrieval context, as the one we are investigating here, the user query can be used to boost the relative importance of a sentence [27]. Sentences sharing information (e.g. words) with the input query are considered more relevant than sentences not sharing words with the input query. In SUMMA this features is implemented in an information retrieval context, for each sentence (represented in the vector space) the cosine of the angle between the sentence and the query (also in the vector space) is computed. The value of the cosine (which is a number betwen 0 and 1) is the value of the query feature.

- Semantic feature: The presence of particular types of named entities has been used in text summarization. In SUMMA, a component counts the number of named entities (of types specified by the user) and computes a semantic feature which is the number of named entities present in the sentence. This semantic feature is normalized to yield values between 0 and 1.

Given these features various summarizers can be created. For example, a summary can be created using the semantic feature only, thus selecting sentences with high value of that particular feature (sentences containing many named entities). Another summarizer could use the position feature, therefore giving preference to sentences towards the beginning of the document (i.e., lead-based summary). Finally the various features could be combined using a linear formula or any other feature combination procedure.

Another type of summarization which we use in this work consist on extracting from each document a set of sentences containing the person name to be disambiguated (target person name) or a name which is coreferent with the target person name. This approach has been used in past approaches to cross-source coreference [5], thus we decided to investigate it in this work.

6.4.3   Document Representation

Documents are represented in the vector space module implemented by SUMMA. Using language resources creation modules from the SUMMA summarization tool, two frequency tables are created for each document set (or person) on-the-fly: (i) an inverted document frequency table for words (no normalisation is applied); and (ii) an inverted frequency table for named entities (the full entity string is used, no normalisation is applied). Term frequencies and inverted document frequencies over words and named entities in the corpus are computed. Each term (either a named entity or a simple word) received the weight tf*log(N/idf) where tf is the frequency of the term in the document, idf is the inverted frequency of the term in the corpus, and N is the number of documents in the collection.

In this work we explore two types of terms for document representation, as in [5] we use words (word condition) as terms as one possible representation and named entities (named entity condition) as a second possible representation. Where document condition is concerned we are interested in exploring the use of the full document (full document condition) as the source for extracting the terms and a document summary as a source to extract the terms.

Where the summaries are concerned, we have experimented with different types of summaries, one type of summary is a personal summary where the sentences for the summary are all those sentences containing a name coreferent with the target entity to be disambiguated (note that all sentences satisfying this condition are selected for the summary). Other summaries we have experimented with are:

- A lead based summary produced by ranking sentences using the position feature;
- A cue based summary produced by ranking sentences using the cue feature;

- A semantic based summary produced by ranking sentences using the semantic features;
- A query based summary produced by ranking sentences using the query based features where the input query is the name of the target entity to be disambiguated (first name and surname); and
- Two summaries produced by combining in a multiplicative formula two different features:

  - A semantic/query based features used both the query and the semantic feature, and
  - A cue/query based feature based on the use of query and cue features.

Note that for the lead, cue, semantic, query, and combined summaries unlike for the personal summaries, a compression rate has to be specified (what percentage of sentences to be selected from the document). We have experimented with compression rates of 10%, 20%, 30%, 40%, and 50% compression rate (where a 10% compression rate means to select 10% of the top ranked sentences of the document according to the scoring function). Note that our cue based summarizer uses a list of cue words induced from a set of biographies (a list of nouns, verbs, and adjectives was extracted from the set of biographies together with the number of times each word was observed in the corpus of biographies).

6.5   Experiments

Experiments have been carried out with different configurations of the clustering algorithm presented in Algorithm 1. Each experiment consists in the use of a particular threshold and document representation. After each experiment is conducted, the clusters returned by the algorithm are evaluated. The different configurations are then assessed by means of a statistical test to verify which configuration provides the best solution for the particular data-set used for experimentation.

6.5.1   Data Sources and Metrics

The People Web Search Evaluation has prepared data to investigate the cross-document coreference problem. The data consists of around 100 Web files per person name retrieved by a Web search engine (Yahoo!), which have been frozen and so, can be used as a static corpus. Each file in the corpus is associated with an integer number which indicates the rank at which the particular page was retrieved by the search engine. In addition to the files themselves, the following information was available: the page title, the URL, and the snippet.

In addition to the data itself, human assessments are provided which are used for evaluating the output of the automatic systems. The assessment for each person name is a file which contains a number of sets where each set is assumed to contain all (and only those) pages referring to one individual.

The development data is a selection of person names from different sources: (i) ten person names were selected from participants of the European Conference on Digital Libraries 2006; (ii) seven person names were selected from the on-line encyclopaedia Wikipedia; finally, (iii) 32 person names were made available from a corpus previously collected by Mann and Yarowsky [15]. The last source was transformed so that it is compatible with the structure described above. Note that there are other valuable resources for the study of the cross-document coreference problem, such as the corpus derived from the Automatic Content Extraction 2005 Corpus [10] that we intend to use in the continuation of this work.

The test data where our experiments are run consists of 30 person names from different sources: (i) ten names were selected from Wikipedia; (ii) ten names were selected from participants in the ACL 2006 conference; and finally, (iii) ten further names were selected from the US Census. One hundred documents were retrieved using the person name as a query using the search engine Yahoo!. The training and testing sets have very different characteristics, for example the average number of real entities in the training set is around ten (e.g. ten entities per person name) while the average number of real individuals per name in the testing set is around five. This is a real challenge for the clustering system, especially if the number of clusters to be produced is going to be optimised using the training data. Metrics used to measure the performance of automatic systems against the human output were borrowed from the clustering literature [12] and they are defined as follows:

$$\text{Precision}(A, B) = \frac{|A \cap B|}{|A|}$$

$$\text{Purity}(C, L) = \sum_{i=1}^{n} \frac{|C_i|}{n} max_j \text{Precision}(C_i, L_j)$$

$$\text{Inverse\_Purity}(C, L) = \sum_{i=1}^{n} \frac{|L_i|}{n} max_j \text{Precision}(L_i, C_j)$$

$$\text{F-Score}_\alpha(C, L) = \frac{\text{Purity}(C, L) * \text{Inverse\_Purity}(C, L)}{\alpha \text{Purity}(C, L) + (1 - \alpha)\text{Inverse\_Purity}(C, L)}$$

Where A and B are sets, $C = C\_1, \ldots C\_n$ is the set of n clusters produced by the system and $L = L\_1, \ldots L\_k$ is the set of k clusters produced by the human. Note that purity is a kind of precision metric which rewards a partition which has less noise. Inverse purity is a kind of recall metric. F-Score$_\alpha$ is the harmonic mean of purity and inverse purity, and is a weighting factor which has been set to $\alpha = 0.5$ for the Web People Search evaluation giving purity and inverse purity the same weight. F-Score is used to produce the ranking of participating systems, and thus establish which system has better performance (an average F-score of the whole data set tested is computed). Note that in the 2010 Web People Search evaluation campaign [2] the B-Cubed metric [6].

Table 6.1 Experiments Series I (X can be Location, Person, Organization, Date, Address)

Config.	Text condition	Term condition	Threshold
1	Full document	Words	0.10
2	Full document	All name entities	0.12
3	Personal summary	Words	0.10
4	Personal summary	All name entities	0.12
5	Full document	Name entity of type X	0.12
6	Personal summary	Name entity of type X	0.12

6.5.2 Initial Clustering Configuration

The configurations of the clustering algorithm with respect to document representation and threshold used can be seen in Table 6.1. First, we have carried out experiments comparing the use of full document condition versus summary condition and words versus the whole set of named entities types (experiments 1–4) and reported results in [20] in those experiments we identified that the whole set of name entities performed worst than using just words. These were unexpected results because in previous research [14] have shown that semantic information is important for the cross-document coreference tasks. We hypothesized that some named entities were contributing with noise because they do not have discriminative power. Therefore, in following experiments, we have examined the discriminative power of each entity types (Person, Organization, Location, Address, and Date) for name disambiguation and carried out "meta" experiments 5 and 6 (note that in configurations 5 and 6 vectors contain one and only one type of entity, so 5 and 6 are five different configurations each) and obtained a better picture of the effect of each type of name entity in the disambiguation problem [21].

6.5.3 Estimating the Parameters of the Clustering Algorithm

The clustering algorithm needs a stoping criteria to decide when it should stop merging clusters. This stoping criteria is a similarity threshold which indicates when the clusters are no longer similar and the algorithm should stop, returning the clusters constructed thus far. There are various ways to estimate this threshold (a real number bewteen 0 and 1): one critera would be to test thresholds such as 0.05, 0.10, 0.15, etc. on training data and decide which value gives the best overall performance; another way would be to average the optimal thresholds for each data point (in the test data) and use that average for clustering unseen data points. We have decided to estimate the threshold using this latter criteria: using the ECDL subset of the training data provided by the 2007 Web People Search Evaluation, we applied the clustering algorithm where the threshold was set to zero to each data point in the training set. At each iteration of the algorithm purity, inverse purity, and

Table 6.2 Word and all named entities experiments

Config.	Purity	I.Purity	F-score
1	0.68	0.85	0.74
2	0.62	0.85	0.68
3	0.84	0.70	0.74
4	0.65	0.75	0.64

Table 6.3 Full document and each named entity

Configuration	Purity	I.Purity	F-score
5 (X = Organization)	0.90	0.72	0.78
5 (X = Person)	0.81	0.72	0.75
5 (X = Address)	0.82	0.64	0.69
5 (X = Date)	0.58	0.85	0.67
5 (X = Location)	0.55	0.85	0.64

Table 6.4 Personal summary and each named entity

Configuration	Purity	I.Purity	F-score
6 (X = Person)	0.85	0.64	0.70
6 (X = Organization)	0.97	0.57	0.69
6 (X = Date)	0.82	0.60	0.68
6 (X = Location)	0.82	0.63	0.68
6 (X = Address)	0.93	0.54	0.65

F-score were computed, recording the similarity value of each newly created cluster. The similarity values for the best clustering results (best F-score) were recorded, and the maximum and minimum values discarded (considered outlines). The rest of the values were averaged to obtain an estimate of the optimal threshold to be applied to the test data. The thresholds obtained were 0.10 for word vectors and 0.12 for named entity vectors, thus the values seen in Table 6.1.

6.5.4 Initial Results

The test data of the 2007 People Web Search was used to run each of the experiments and compute F-score values comparing the automatic clusters with the human clusters. Results of the experiments in terms of Purity, Inverse Purity and F-score are reported in Tables 6.2–6.4.

Comparing the use of words and all named entities (and fixing the document condition) we can observe that a word based approach produces a better f-score than a name entities condition (0.74 versus 0.68 and 0.74 versus 0.64). A t-test was run to assess the difference in average F-score these F-scores are different at a 95% confidence level. Table 6.3 shows in detail, the value of each type of name entity in the disambiguation process. A clearer picture emerges from these configurations, we can observe that when disambiguation is carried out using as Organizations extracted from full documents in order to create the vectors for clustering, an F-score of 0.78 is obtained, which is better than an F-score of 0.74 obtained by a word based approach. A t-test was run to verify the differences. The algorithms are different at a 95% confidence level.

Where summary condition is concerned (Table 6.4), the use of Persons or Organizations extracted from personal summaries yields a better performance than using words extracted from personal summaries (0.70 or 0.69 versus 0.64).

6.6 Summarization-Based Experiments

The previous experiments showed the value of different types of entities in person name disambiguation, yet the use of personal summaries did not provide optimal performance. All results using personal summary condition were worst than those using full documents.

Our last set of experiments attempts to verify the value of different summarization strategies in person name disambiguation. In these experiments, we do not consider all named entities in bulk but as in the previous experiments test the value of each name entity type in turn. Table 6.5 describes the experiments; each row indicates what type of summary was used and what type of term was extracted from the summary. Note that each row represents many configurations since a compression and type of name entity has to be specified. Thus, meta-configuration 8 represents 25 different configurations (5 possible compressions and 5 possible named entity types).

6.6.1 Summary-Based Results

Here as well, the test data of the 2007 People Web Search was used to run each of the experiments and compute F-score values comparing the automatic clusters with the human clusters.

Because of the large number of experiments performed (over 150) we only report some of the most relevant results. Table 6.6 shows best and worst configurations: we report the type of summarization strategy (i.e., Config.), the percent compression of the summaries used (i.e., Compr.), the type of named entity used for document

Table 6.5 Summary-based experiments (multiple summaries generated at compressions 10, 20, 30, 40, and 50)

Config.	Summary	Term condition	Threshold
7	Semantic based	Entity X	0.12
8	Cue based	Entity X	0.12
9	Query based	Entity X	0.12
10	Semantic/query	Entity X	0.12
11	Cue/query	Entity X	0.12
12	Lead	Entity X	0.12

Table 6.6
Summarization-based
clustering with semantic
information (best and worst
results only)

Best results			
Config.	Compr.	Entity	F-score
7	40	Organization	0.79
7	50	Organization	0.78
11	50	Organization	0.78
7	30	Organization	0.78
11	40	Organization	0.77
12	50	Organization	0.77
Worst results			
Config.	Compr.	Entity	F-score
8	20	Address	0.60
8	10	Address	0.60
12	30	Address	0.59
12	20	Address	0.59
12	10	Address	0.58

representation (i.e., Entity), and the average F-score. In these summarization experiments the best F-score (0.79) is when the type of entity used is Organization and where the terms are extracted from a semantic based summary at 40% compression. The performance of this configuration goes down a bit when compression goes up to 50% but no statistical differences were found with the best configuration. The best lead-based summarization (configuration 12) result (i.e., F-score of 0.77) is obtained when compression rate is 50% and the extracted named entities are Organizations. There are statistical differences (at 95% confidence level) between this lead-based configuration and the best system. This appears to be an interesting result: lead-based summaries are usually very difficult to beat, however in this context summaries which are dense in named entities seem to perform better. Some summarization configurations proved not very useful, the worst configuration was one using a cue-based summarizer and Address as term type (F-score 0.58).

Where the semantic type of information used to create the vectors is concerned, we note that Organizations are better disambiguators than other types of entities in general. This has also been shown in our previous experiments. It appears that Address type of information performs worst in this task. This is not at all an unexpected result: Organizations appear to be less ambiguous than Persons or Addresses for example, therefore contributing to the disambiguation of person names. In fact the performance of the semantic-based summarization strategy goes down when the terms used for vector representation are different from Organization. Addresses and Dates seem to be ambiguous, therefore providing little help for the disambiguation task. It is important to point out that the extraction of semantic information was carried out automatically with a far from perfect tool and that other tools may lead to different results.

Where the compression parameter is concerned, it appears that compressions 30–50% provide the best results with very short summaries not providing enough context for disambiguation.

6.7 Discussion

The set of experiments presented in this paper aimed at a deeper understanding of the role of semantic information and summarization in person name disambiguation. We have shown that the use of summaries together with carefully selected semantic information in a clustering algorithm can lead to state-of-the-art performance for person name disambiguation. In Bagga and Baldwin [5] only word level information was used for clustering, and personal summaries were considered as better candidates for term extraction. Here, we have presented a clearer picture of the influence of summary vs. full document condition in the clustering process, and have shown that the type of summary used may influence the outcome of the clustering process. We have carried out more than 170 summarization experiments and assessed the value of the summarization strategies in the cross-source coreference task. The best 2007 Web People Search system (Chen and Martin [8]) achieved an F-score of 0.78 using techniques similar to ours: named entity recognition using off-the-shelf systems. However in addition to semantic information and full document condition they also explore the use of contextual information such as the url where the document comes from. They show that this information is of little help. Our best configuration achieves an F-score of 0.79 , thus similar to theirs.

To the best of our knowledge this is the first work to study and assess the contribution of different summarization strategies for cross-source coreference resolution. The reader may notice that we have used here only a few well known summarization strategies (e.g., lead summarization, cue-based summarization), there are however other summarization methods which could lead to improved performance. In a similar framework as in Saggion et al. [25] or Radev et al. [18] where information retrieval was used as a benchmark for summarization or in Saggion et al. [24] were summarization is tested in the rating-inference problem, here cross-source coreference could be used as a method to evaluate summarization and information extraction technology: *what combination of summarization and extraction strategies provides improved cross-source coreference resolution performance?* An interesting issue to investigate.

6.8 Conclusions and Future Work

Finding information about people on huge text collections or on-line repositories on the Web is a common activity. In ad-hoc Internet retrieval, a request for documents/pages referring to a person name may return thousand of pages which although containing the name, do not refer to the same individual. Cross-document coreference is the task of deciding if two entity mentions in two sources refer to the same individual. Because person names are highly ambiguous (i.e., names are shared by many individuals), deciding if two documents returned by a search engine such as Google or Yahoo! refer to the same individual is a difficult problem. We have presented a set of experiments on clustering-based cross-document coreference of person names in order to verify the value of various summarization and semantic tagging strategies in the clustering process.

The experiments presented here are an expansion of our initial experiments for the Web People Search Evaluation framework where the task consisted on disambiguating a set of 30 person names each named occurring in 100 different documents retrieved from the Web.

We have designed and implemented a solution which uses an in-house clustering algorithm and available extraction and summarization techniques to produce representations needed by the clustering algorithm.

In this paper different document representations have been tested and compared. In particular we have show that in the particular data set we have used (Web pages), the use of specific name type extracted from the full document or a semantic-based summary can achieve state of the art performance. However, and because of the characteristics of the input documents (Web pages), the effectiveness of the method in other data sets has to be evaluated.

Many avenues of improvement are expected. Where extraction technology is concerned, we have used an off-the-shelf system which is probably not the most appropriate for the type of data we are dealing with, and so adaptation is needed here. With respect to the clustering algorithm we plan to carry out further experiments to test the effect of different similarity metrics, different merging criteria including creation of cluster centroids, and cluster distances; with respect to the summarization techniques we intend to investigate how the extraction of sentences containing pronouns referring to the target entity affects performance, our current version only exploits name coreference. Our future work will also explore how (and if) the use of contextual information available on the Web can lead to better performance for example encyclopaedic knowledge provided by Wikipedia or other unstructured knowledge sources can inform the clustering algorithm with respect to known entities and lower limits for the number of clusters.

Acknowledgements We thank the reviewers for their comments and suggestions which helped improve the final version of this paper. Horacio Saggion is grateful to a fellowship from Programa Ramón y Cajal, Ministerio de Ciencia e Innovación, Spain. We acknowledge the support from the editors of this volume.

References

1. Abdalla R., Teufel, S.: A bootstrapping approach to unsupervised detection of cue phrase variants. In: Proceedings of COLING/ACL 2006, Sydney (2006)
2. Artiles, J., Borthwick, A., Gonzalo, J., Sekine, S., Amigó, E.: Weps-3 evaluation campaign: overview of the web people search clustering and attribute extraction tasks. In: CLEF - Notebook Papers/LABs/Workshops, Padova, Italy (2010)
3. Artiles, J., Gonzalo, J., Sekine, S.: The semEval-2007 wePS evaluation: establishing a benchmark for web people search task. In: Proceedings of Semeval 2007, Prague, Czech Republic. Association for Computational Linguistics, Stroudsburg (2007)

4. Aswani, N., Bontcheva, K., Cunningham, H.: Mining information for instance unification. In: 5th International Semantic Web Conference (ISWC2006), Athens. Springer, Berlin/Heidelberg (2006). http://gate.ac.uk/sale/iswc06/iswc06.pdf
5. Bagga, A., Baldwin, B.: Entity-based cross-document coreferencing using the vector space model. In: Proceedings of the 36th Annual Meeting of the Association for Computational Linguistics and the 17th International Conference on Computational Linguistics (COLING-ACL'98), Montreal, pp. 79–85. Association for Computational Linguistics, Stroudsburg (1998)
6. Bagga, A., Baldwin, B., Ramesh, G.: Methodology for cross-document coreference over degraded data sources. In: Angelova, G., Bontcheva, K., Mitkov, R., Nikolov, N., Nicolov, N. (eds.) Proceedings of Recent Advances in Natural Language Processing (RANLP'01), Tzigov Chark, Bulgaria, pp. 15–21 (2001)
7. Bekkerman, R., McCallum, A.: Disambiguating web appearances of people in a social network. In: Proceedings of WWW-05, the 14th International World Wide Web Conference, Chiba. ACM, New York (2005)
8. Chen, Y., Martin, J.: Cu-comsem: Exploring rich features for unsupervised web personal named disambiguation. In: Proceedings of SemEval 2007, Prague, pp. 125–128. Assocciation for Computational Linguistics, Stroudsburg (2007)
9. Cutting, D.R., Pedersen, J.O., Karger, D., Tukey, J.W.: Scatter/gather: A cluster-based approach to browsing large document collections. In: Proccedings of the Fifteenth Annual International ACM SIGIR Conference on Research and Development in Information Retrieval, Copenhagen, pp. 318–329 (1992)
10. Day, D., Hitzeman, J., Wick, J., Crouch, K., Poesio, M.: A corpus for cross-document co-reference. In: Proceedings of the Sixth International Language Resources and Evaluation (LREC'08), Marrakech, Morocco. European Language Resources Association, Paris, France (2008)
11. Grishman, R.: Information extraction: techniques and challenges. In: Pazienza, M.T. (ed.) Information Extraction: A Multidisciplinary Approach to an Emerging Information Technology, International Summer School (SCIE-97), Lecture Notes in Computer Science, vol. 1299, pp. 10–27. Springer, Frascati, Italy (1997)
12. Hotho, A., Staab, S., Stumme, G.: WordNet improves text document clustering. In: Proceeding of the SIGIR 2003 Semantic Web Workshop, Toronto (2003)
13. Mani, I.: Automatic Summarization. John Benjamins, Amsterdam/Philadelphia (2001)
14. Mann, G., Yarowsky, D.: Unsupervised personal name disambiguation. In: Proceedings of CoNLL, Edmonton. Association for Computational Linguistics, Stroudsburg (2003)
15. Mann, G.S., Yarowsky, D.: Unsupervised personal name disambiguation. In: Daelemans, W., Osborne, M. (eds.) Proceedings of the 7th Conference on Natural Language Learning (CoNLL-2003), Edmonton, pp. 33–40. Association for Computational Linguistics, Stroudsburg (2003)
16. Maynard, D., Tablan, V., Cunningham, H., Ursu, C., Saggion, H., Bontcheva, K., Wilks, Y.: Architectural elements of language engineering robustness. J. Nat. Lang. Eng. Spec. Issue Robust Methods Anal. Nat. Lang. Data 8(2/3), 257–274 (2002). http://www.gate.ac.uk/sale/robust/robust.pdf
17. Phan, X.H., Nguyen, L.M., Horiguchi, S.: Personal name resolution crossover documents by a semantics-based approach. IEICE Trans. Inf. Syst. 89, 825–836 (2006)
18. Radev, D.R., Teufel, S., Saggion, H., Lam, W., Blitzer, J., Qi, H., Çelebi, A., Liu, D., Drábek, E.: Evaluation challenges in large-scale document summarization. In: ACL, Sapporo, pp. 375–382 (2003)
19. Rasmussen, E., Willett, P.: Non-hierarchical document clustering using the icl distribution array processor. In: SIGIR '87: Proceedings of the 10th Annual International ACM SIGIR Conference on Research and Development in Information Retrieval, New Orleans, pp. 132–139. ACM Press, New York, NY, USA (1987)
20. Saggion, H.: Shef: Semantic tagging and summarization techniques applied to cross-document coreference. In: Proceedings of SemEval 2007, Prague, Czech Republic, pp. 292–295. Assocciation for Computational Linguistics, Stroudsburg, PA, USA (2007). http://gate.ac.uk/sale/semeval07/papers/shef-semeval07.pdf

21. Saggion, H.: Experiments on semantic-based clustering for cross-document coreference. In: Proceedings of the Third Joint International Conference on Natural Language Processing, AFNLP, Hyderabad, pp. 149–156 (2008)
22. Saggion, H.: SUMMA: a robust and adaptable summarization tool. Traitement Automatique des Langues **49**(2), 103–125 (2008)
23. Saggion, H., Gaizauskas, R.: Multi-document summarization by cluster/profile relevance and redundancy removal. In: Proceedings of the Document Understanding Conference 2004, Boston, USA. NIST, Gaithersburg, MD, USA (2004)
24. Saggion, H., Lloret, E., Palomar, M.: Using text summaries for predicting rating scales. In: Proceedings of the 1st Workshop on Computational Approaches to Subjectivity and Sentiment Analysis (WASSA), Lisbon, Portugal, pp. 44–51 (2010)
25. Saggion, H., Radev, D., Teufel, S., Wai, L., Strassel, S.: Developing infrastructure for the evaluation of single and multi-document summarization systems in a cross-lingual environment. In: 3rd International Conference on Language Resources and Evaluation (LREC 2002), Las Palmas, Gran Canaria, pp. 747–754 (2002)
26. Salton, G.: Automatic Text Processing. Addison-Wesley, Reading (1988)
27. Tombros, A., Sanderson, M., Gray, P.: Advantages of query biased summaries in information retrieval. In: Intelligent Text Summarization. Papers from the 1998 AAAI Spring Symposium. Technical Report SS-98-06, The AAAI Press, Standford, pp. 34–43 (1998)
28. van Rijsbergen, C.: Information Retrieval. Butterworths, London (1979)
29. Willett, P.: Recent trends in hierarchic document clustering: a critical review. Inf. Process. Manage. **24**(5), 577–597 (1988)

Part III
Information Extraction

Chapter 7
Interactive Topic Graph Extraction and Exploration of Web Content

Günter Neumann and Sven Schmeier

Abstract In the following, we present an approach using interactive topic graph extraction for the exploration of Web content. The initial information request, in the form of a query topic description, is issued online by a user to the system. The topic graph is then constructed from N Web snippets that are produced by a standard search engine. We consider the extraction of a topic graph to be a specific empirical collocation extraction task, where collocations are extracted between chunks. Our measure of association strength is based on the pointwise mutual information between chunk pairs which explicitly takes their distance into account. This topic graph can then be further analyzed by users so that they can request additional background information with the help of interesting nodes and pairs of nodes in the topic graph, e.g., explicit relationships extracted from Wikipedia or those automatically extracted from additional Web content as well as conceptual information of the topic in form of semantically oriented clusters of descriptive phrases. This information is presented to the users, who can investigate the identified information nuggets to refine their information search. An initial user evaluation shows that our approach is especially helpful for finding new interesting information on topics about which the user has only a vague idea or no idea, at all.

7.1 Introduction

Today's Web search is still dominated by a document-perspective: a user enters one or more keywords that represent the information of interest and receives a ranked list of documents. This technology has been shown to be very successful, because it

G. Neumann (✉) · S. Schmeier
German Research Center for Artificial Intelligence GmbH (DFKI), Stuhlsatzenhausweg 3, D-66123 Saarbrücken, Germany
e-mail: neumann@dfki.de; schmeier@dfki.de

T. Poibeau et al. (eds.), *Multi-source, Multilingual Information Extraction and Summarization 11*, Theory and Applications of Natural Language Processing, DOI 10.1007/978-3-642-28569-1_7, © Springer-Verlag Berlin Heidelberg 2013

very often delivers concrete documents or Web pages that contain the information the user is interested in. The following aspects are important in this context: (1) Users basically have to know what they are looking for. (2) The documents serve as answers to user queries. (3) Each document in the ranked list is considered independently.

If the user only has a vague idea of the information in question or just wants to explore the information space, the current search engine paradigm does not provide enough assistance for these kind of searches. The user has to read through the documents and then eventually reformulate the query in order to find new information. Seen in this context, current search engines seem to be best suited for "one-shot search" and do not support content-oriented interaction.

In order to overcome this restricted document perspective, and to provide more interactive searches to "find out about something", we want to help users with the Web content exploration process in two ways:

1. We consider a user query as a specification of a topic that the user wants to know and learn more about. Hence, the search result is basically a graphical structure of the topic and associated topics that are found.
2. The user can interactively explore this topic graph, in order to either learn more about the content of a topic or to interactively expand a topic with newly computed related topics.

In the first step, the topic graph is computed on the fly from a set of snippets that has been collected by a standard search engine using the initial user query. Rather than considering each snippet in isolation, all snippets are collected into one document from which the topic graph is computed. We consider each topic as an entity, and the edges between topics are considered as a kind of (hidden) relationship between the connected topics. The content of a topic are the set of snippets it has been extracted from, and the documents retrievable via the snippets' Web links.

The topic graph is then displayed either in a standard Web browser or on a mobile device (in our case an iPad). By just selecting a node, the user can either inspect the content of a topic (i.e., the snippets or Web pages) or activate the expansion of the graph through an on the fly computation of new related topics for the selected node.

In a second step, we provide additional background knowledge on the topic which consists of (1) explicit relationships that are either generated from an online Encyclopedia (in our case Wikipedia) or automatically extracted from the Web snippets through learned relation extraction rules, and (2) additional conceptual information of the topic in form of semantically oriented clusters of descriptive phrases automatically extracted from the Web.

The background knowledge approaches are based on and driven by specific patterns, e.g., the structure of infoboxes, in case of Wikipedia, predefined seed relations in the case of learned relation extraction rules or predefined patterns like "X is a Y" in the case of concept extraction. In contrast, the topic graph extraction process is much less pattern dependent. This means, that the topic graph extraction component is much more flexible in dealing with dynamic changes of Web content,

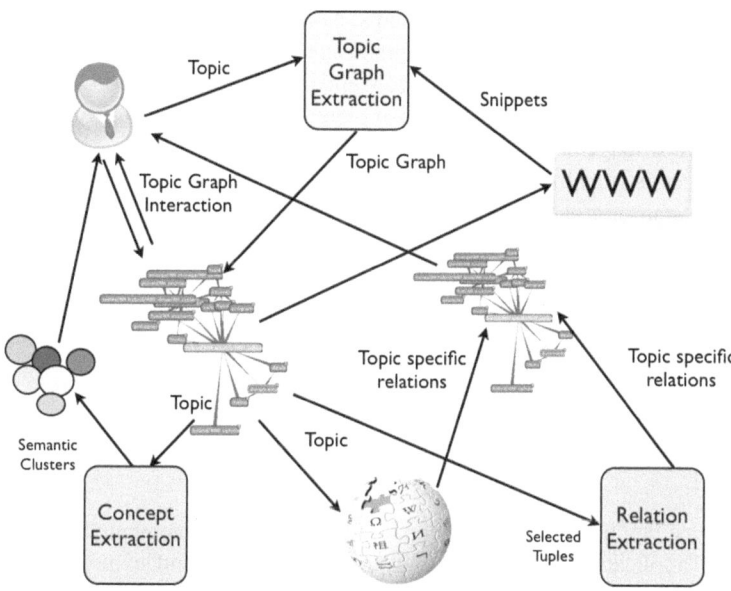

Fig. 7.1 Blueprint of the proposed system

whereas background knowledge components are much more fixed and stable with respect to possible changes.

This way the user can explore in a uniform way both new information nuggets and background information nuggets interactively. Figure 7.1 summarizes the main components and the information flow.

By selecting a node from a newly extracted topic graph, the user can request information from new topics on basis of previously extracted information. In such a dynamic information extraction situation, the user expects real–time performance from the underlying technology. The requested information cannot simply be pre–computed, therefore most of the relevant information has to be extracted online relative to the current user request. That is why we assume that the relevant information can be extracted from a search engine's *Web snippets* and hence avoid the costly retrieval and processing time for huge amounts of documents. Of course, direct processing of Web snippets also poses certain challenges for the Natural Language Processing (NLP) components.

Web snippets are usually small text summaries which are automatically created from parts of the source documents and are often only in part linguistically well–formed, cf. [17]. Thus the NLP components are required to possess a high degree of robustness and run–time behavior to process the Web snippets in real–time. Since, our approach should also be able to process Web snippets from different languages, the NLP components should be easily adaptable to many languages. Finally, no restrictions to the domain of the topic should be pre–supposed, i.e., the system should be able to accept topic queries from arbitrary domains. In order to fulfill all these requirements, we are favoring and exploring the use of shallow and highly data-oriented NLP components. Note that this is not a trivial or obvious design decision, since most of the current prominent information extraction methods advocate deeper NLP components for concept and relation extraction, e.g., syntactic and semantic dependency analysis of complete sentences and the integration of rich linguistic knowledge bases like Word Net.

The paper is organized as follows. In the next section, we illustrate the core concepts of our approach with an operating example. We then compare our work with that of others in Sect. 7.3. The major components of our approach are described in more detail in the Sects. 7.4–7.6. In Sect. 7.7, we present and discuss the results of an initial user evaluation, Sect. 7.8 concludes the paper and outlines some areas for future research.

7.2 Running Example

A prototype application of our approach has been implemented as a mobile touchable application for online topic graph extraction and exploration of Web content. The system has been implemented for operation on an iPad.

The following screenshots show some results for the search query "Justin Bieber" running on the current iPad demo–app. At the bottom of the iPad screen, the user can select whether to perform text exploration from the Web (via button labelled "i–GNSSMM") or via Wikipedia (touching button "i–MILREX"). The Figs. 7.2–7.5 show results for the "i–GNSSMM" mode, and Fig. 7.6 for the "i-MILREX" mode. General settings of the iPad demo-app can easily be changed. Current settings allow e.g., language selection (so far, English and German are supported)[1] or selection of the maximum number of snippets to be retrieved for each query. The other parameters mainly affect the display structure of the topic graph.

An entity can be ambiguous, i.e., two (or more) individuals can be referred to by the same name. For example, the name "Jim Clark" can refer to the well–known car racing driver or to the founder of Netscape but it can refer to even more individuals. Proper disambiguation of named entities is still a difficult problem that has not been resolved.[2]

So far, the extracted topic graph does not provide direct help for NE disambiguation, i.e., it is possible that the topic graph merges information from different individuals which cannot be easily distinguished. As an initial approximate solution towards NE disambiguation in the context of our approach, we provide an additional operator which we call the concept extractor (CE). It receives the label of a triggered (touched) node and searches the Web for descriptive phrases. All descriptive phrases

---

[1]Actually, both languages are only supported in the i–GNSSMM mode. In the case of the i–MILREX mode, we currently only support the English Wikipedia.

[2]Consult, for example, the Web page http://nlp.uned.es/weps/ for more information about the problem space.

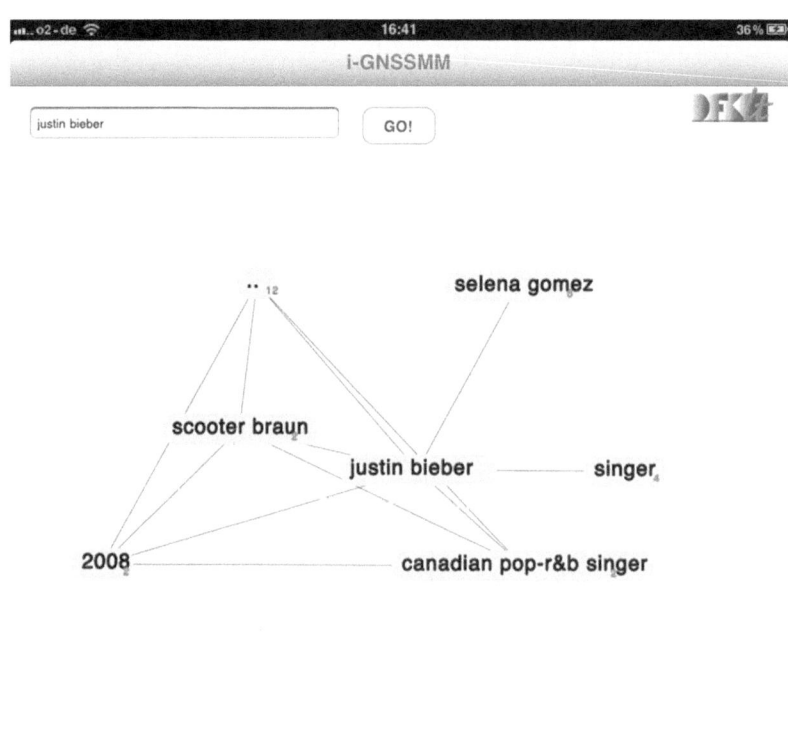

Fig. 7.2 The topic graph computed from the snippets for the query "Justin Bieber". The user can double touch on a node to display the associated snippets and Web pages. Since a topic graph can be very large, not all nodes are displayed. Nodes, which can be expanded are marked by the number of hidden immediate nodes. A single touch on such a node expands it, as shown in Fig. 7.3. A single touch on a node that cannot be expanded adds its label to the initial user query and triggers a new search with that expanded query

that are found are then clustered on the basis of latent semantic similarity. It is possible (although not in all cases) to induce from the resulting clusters whether a NE might refer to several individuals. Here is the output of this module for the name "Jim Clark" (we are using this example, because it nicely illustrates our approach; note that the clusters and their labels are computed completely automatically and unsupervised, see Sect. 7.5 for more details):

——————— Cluster [PRESIDENT]——————-
Jim Clark, who is the president of the Arizona Western Heritage Foundation, did a fantastic job of putting the whole event together.
——————— Cluster [MOTOR]——————-
1963: Scotland's Jim Clark became the youngest world motor racing champion
——————— Cluster [INDIANAPOLIS]——————-
1965 – Jim Clark becomes first foreigner in 49 years to win Indianapolis 500 car race

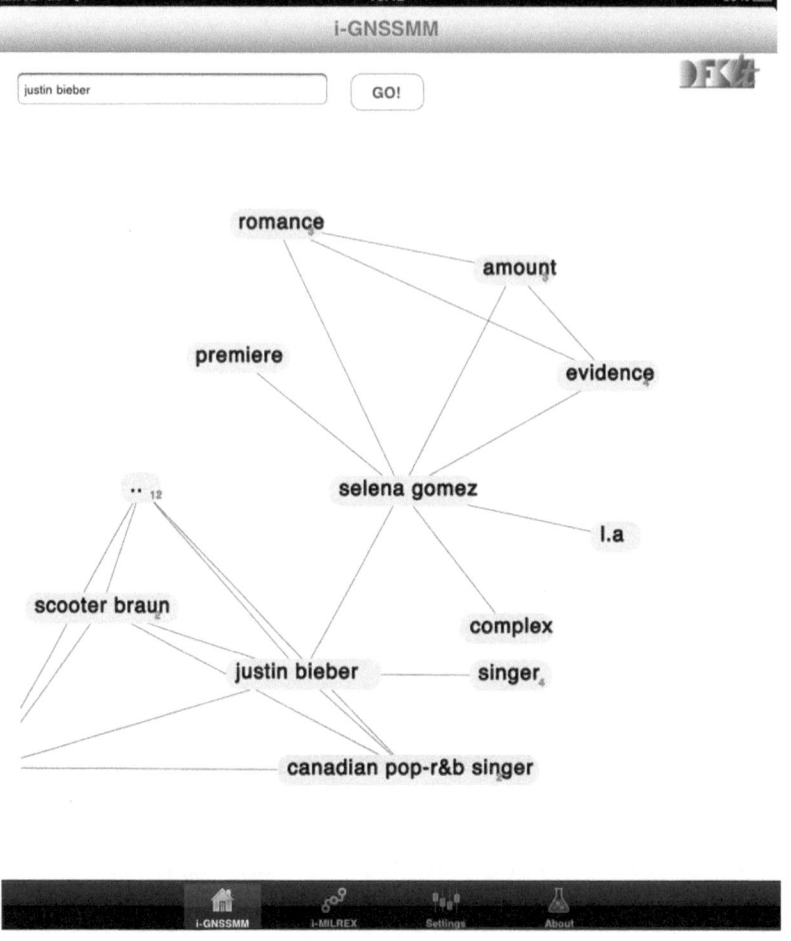

Fig. 7.3 The topic graph from Fig. 7.2 has been expanded by a single touch on the node labelled "selena gomez". Double touching on that node triggers the display of associated Web snippets (Fig. 7.4) and the Web pages (Fig. 7.5)

————— Cluster [GRAPHICS]—————

Andreessen partnered with Jim Clark, who a decade earlier had graduated from Stanford University in California to start up Silicon Graphics.

————— Cluster [GRAND]—————

Jim Clark became world champion grand prix driver in 1963 and 1965 and was the first non-American to win the Indianapolis 500 for nearly 50 years.

Jim Clark was a driver of such towering ability that he had few serious rivals during his 7 years in grand prix racing.

————— Cluster [FORMULA]—————

Jim Clark was a racing legend who racked up two Formula One World Championships, won some 25 Grand Prix races and even competed in NASCAR before his untimely death in 1968.

Justin Bieber and Selena Gomez take a stroll down Santa Monica Pier on a sunny Sunday afternoon (February 6) in California. The duo were seen smiling at each other ... 2011-02-09T17:46:00Zhttp://justjaredjr.buzznet.com/2011/02/06/justin-bieber-selena-gomez-santa-monica-sweeties/

Photos show Gómez and Bieber hugging and kissing, while enjoying the sun. 2011-02-09T10:39:00Zhttp://latino.foxnews.com/latino/entertainment/2011/02/08/selena-gomez-justin-bieber-embrace-intimate-moment-california-beach/

LOS ANGELES - The city got a bad case of Bieber fever on Tuesday. Hundreds of Justin Bieber fans - along with his famous friends Miley Cyrus, Will Smith, Usher and Selena Gomez - filled the L.A. Live complex for the premiere of the teen pop star's first ... Date: 2011-02-09T17:44:28Z URL:http://www.azcentral.com/thingstodo/movies/articles/2011/02/09/20110209justin-bieber-premiere-never-say-never.html

We apologize for continually breaking the hearts of Justin Bieber fans, but he and Selena Gomez really are an item. We've already posted a photo of this couple strolling along ... 2011-02-08T17:55:00Zhttp://www.thehollywoodgossip.com/2011/02/caught-on-tape-justin-bieber-and-selena-gomez/

LOS ANGELES (AP) ? The city got a bad case of Bieber fever on Tuesday, when hundreds of Justin Bieber fans ? along with his famous friends Miley Cyrus , Will Smith , Usher and Selena Gomez ? filled the L.A. Live complex for the premiere of the teen ... Date: 2011-02-09T07:36:01Z URL:http://www.orlandosentinel.com/entertainment/sns-ap-us-film-justin-bieber-premiere,0,4071818.story

Fig. 7.4 The snippets that are associated with the node label "selena gomez" of the topic graph from Fig. 7.3. In order to go back to the topic graph, the user simply touches the button labeled iGNSSMM on the *left upper corner* of the iPad screen

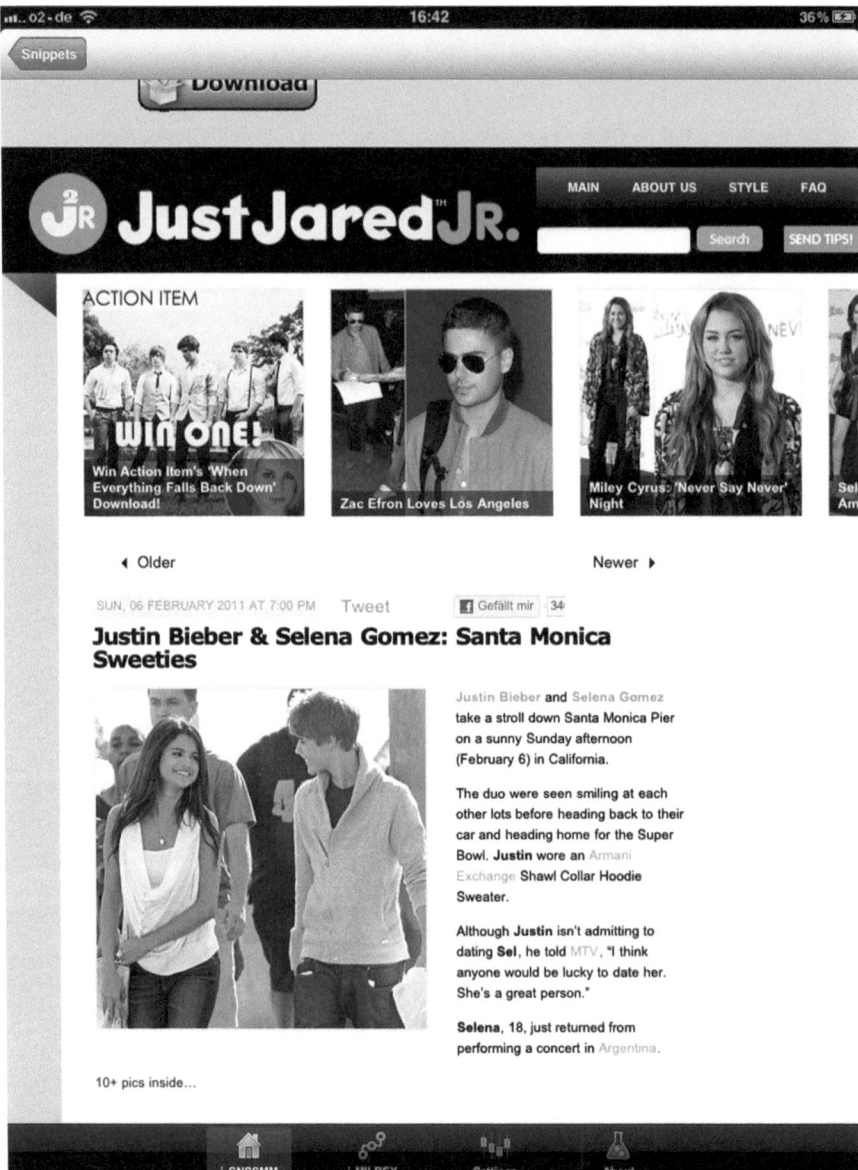

Fig. 7.5 The Web page associated with the first snippet of Fig. 7.4. A single touch on that snippet triggers a call to the iPad browser in order to display the corresponding Web page. The *left upper corner* button labeled "Snippets" has to be touched in order to go back to the snippets page

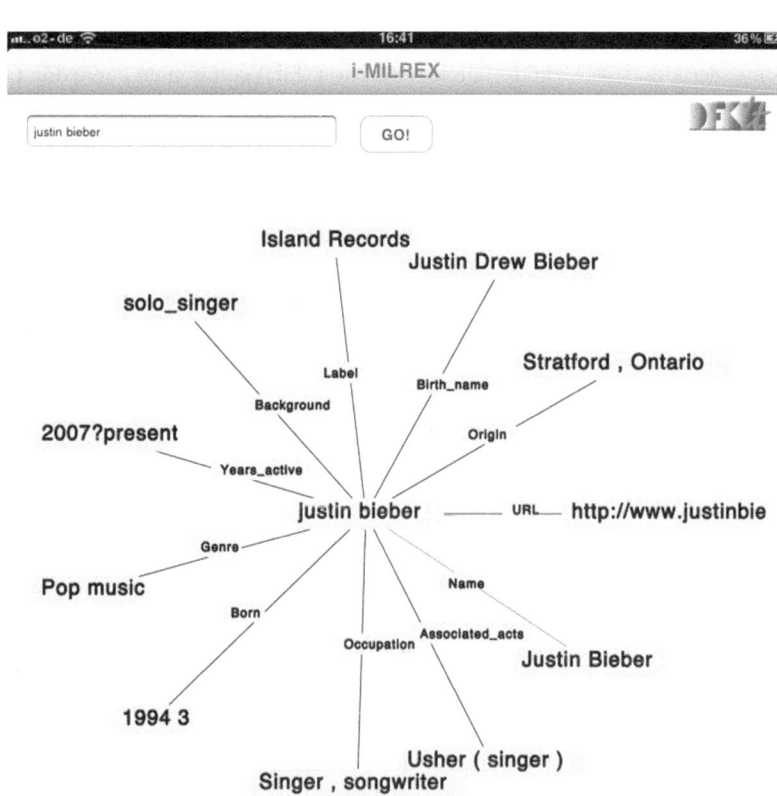

Fig. 7.6 If mode "i–MILREX" is chosen then text exploration is performed based on relations computed from the info–boxes extracted from Wikipedia. The *central node* corresponds to the query. The *outer nodes* represent the arguments and the inner nodes the predicate of a info–box relation. The center of the graph corresponds to the search query

──────── Cluster [DRIFT]────────
Jim Clark was a great exponent of the three wheel drift in his Lotus Cortina, campaigning the Lotus to win the British Touring Car Championship in 1964.

Although this initial clustering does not actually disambiguate named entities, it already provides hints that there are several different individuals named "Jim Clark", viz. the president of the Arizona Western Heritage Foundation, the Netscape founder (as a start up of Silicon Graphics) and the racing car driver. A major problem so far is that often "senses" are distributed across different clusters (e.g., in the case of the racing car driver), so that a clear identification of the possible senses is not currently possible.

One last step in our system is the discovery of hidden relations between concepts (see Fig. 7.6).[3] The Relation Extraction (RE) component can extract such information and add it to the topic graph by labeling its edges. Currently, it uses background knowledge from Wikipedia infoboxes. For missing relationships, i.e., if the desired relations are not contained in the infoboxes, we analyze the previously retrieved snippets based on relation extraction models that have already been learnt. Note that the component RE is the only module that requires some offline effort by the user, namely a small set of domain-independent relations and a small relation hierarchy, see Sect. 7.6. However, since we are using a minimally supervised machine learning component equipped with sophisticated inference mechanisms, the integration of new relations into the systems is simple and can be done by non-linguists.

7.3 Related Work

Our approach is unique in the sense that it combines interactive topic graph extraction and exploration with recently developed technology from text mining, information extraction and question answering methods. As such, it learns from and shares ideas with other search results. The most relevant ones are briefly discussed below.

Collocation Extraction We consider the extraction of a topic graph as a specific *empirical collocation extraction task*. However, instead of extracting collocations between words, which is still the dominating approach in collocation extraction research (e.g., [2]), we are extracting collocations between chunks, i.e., word sequences. Furthermore, our measure of association strength takes into account the distance between chunks and combines it with the PMI (pointwise mutual information) approach [23].

Concept Extraction During the last years, the problem of finding definitions for specific concepts (the *definiendum*) has been addressed by Question Answering Systems (QASs) in the context of the Text REtrieval Conference (TREC) and the Cross Language Evaluation Forum (CLEF). In TREC, QASs answer definition questions in English, such as *"What is a quasar?"*, by extracting as far as possible non-redundant descriptive information ('nuggets') about the *definiendum* from the ACQUAINT corpus.

In order to discover definition utterances, definition QASs usually align sentences with surface patterns in the target corpus at the word and/or the part-of-speech level [14]. Hence, the probability of matching sentences increases as the size of the target collection grows, and accordingly, performance improves substantially [15].

---

[3]The screenshots shows relations retrieved from Wikipedia infoboxes only. The component for detecting missing relationships is not yet integrated in the running system.

Along with surface patterns, definition QASs take advantage of wrappers around online resources, WordNet glossaries and Web snippets [5]. In addition, QASs, like Google, have also shown that definition Web–sites are a fertile source of descriptive information in English, concretely, in providing answers to 42 out of 50 TREC–2003 questions [5]. However, Web snippets have not yet been proven to be a valuable source of descriptive phrases.

QASs usually tackle redundancy by: (a) randomly removing one sentence from every pair that shares more than 60% of their terms [14], or (b) filtering out candidate sentences by ensuring that their cosine similarity to all previously selected utterances is below a certain threshold. It is also worth while to remark that definition QASs have not yet made effort to deal with disambiguation of the different senses of the *definiendum*.

Web Information Extraction Web Information Extraction (WIE) systems have recently been able to extract massive quantities of relational data from online text. The most advanced technologies are algorithmically based on Machine Learning methods taking into account different granularities of linguistic feature extraction, e.g., from PoS-tagging to full parsing. The underlying methods of the learning strategies for these new approaches can range anywhere from supervised or semi-supervised to unsupervised. Currently, there is a strong tendency towards semi-supervised and, more recently, unsupervised methods.

For example, [20] presents an approach for unrestricted relation discovery that is aimed at discovering all possible relations from texts and presents them as tables. Sekine [21] has further developed this approach to what he calls "on-demand information extraction". Major input to the system is topic based in form of keywords that are used to crawl relevant Web pages. The system then uses dependency parsing as a major tool for identifying possible relational patterns. The candidate relation instances are further processed by specialized clustering algorithms. A similar approach has been developed by [7] who further combines this approach with advanced user interaction.

Another approach of unsupervised IE has been developed by Oren Etzioni and colleagues, cf. [1, 8, 24]. They developed a range of systems (e.g., KnowItAll, Textrunner, Resolver) aimed at extracting large collections of facts (e.g., names of scientists or politicians) from the Web in an unsupervised, domain-independent, and scalable manner. In order to increase performance, specific Machine Learning based wrappers have been proposed for extracting subclasses, lists, and definitions.

Relation Extraction The bottleneck of Etzionis and his colleagues work is that they focus on the extraction of unary relations, although they claim these methods should also work in relations with greater arity. In a recent paper, [19] presented URES, an unsupervised Web relation extraction system, that is able to extract binary relations, on a large scale, e.g., CEO_of, InventorOf, MayorOf reporting precision values in the upper 80s. Furthermore, [6] presents a method that is able to handle sparse extractions.

Bunescu and Mooney [4] propose binary relation extraction based on Multiple Instance Learning (MIL). The process starts with some positive and negative instances of a relation, retrieves documents or snippets matching those instances and builds positive and negative training sets for a MIL algorithm. The generated model is then used to classify whether a text contains the relation or not. Our RE component (cf. Sect. 7.6) is based on a similar idea but with a MIL algorithm specialized on extracting relations using snippets and extensions to n-ary relations. Systems for extracting n-ary relations usually use parsers in combination with bootstrapping methods. See for example, the approaches presented by Greenwood and Stevenson [13], Sudo et al. [22], and McDonald et al. [18].

7.4 Topic Graph Extraction

The core idea of our topic graph extraction method is to compute a set of chunk–pair–distance elements for the N first Web snippets returned by a search engine for the topic Q, and to compute the topic graph from these elements.[4] In general for two chunks, a single chunk–pair–distance element stores the distance between the chunks by counting the number of chunks in–between them. We distinguish elements which have the same words in the same order, but have different distances. For example, (Peter, Mary, 3) is different from (Peter, Mary, 5) and (Mary, Peter, 3).

We begin by creating a document S from the N-first Web snippets so that each line of S contains a complete snippet. Each textline of S is then tagged with Part–of–Speech using the SVMTagger [12] and chunked in the next step. The chunker recognizes two types of word chains. Each chain consists of longest matching sequences of words with the same PoS class, namely noun chains or verb chains, where an element of a noun chain belongs to one of the extended noun tags,[5] and elements of a verb chain only contains verb tags. We finally apply a kind of "phrasal head test" on each identified chunk to guarantee that the right–most element only belongs to a proper noun or verb tag. For example, the chunk "a/DT british/NNP formula/NNP one/NN racing/VBG driver/NN from/IN scotland/NNP" would be accepted as proper NP chunk, where "compelling/VBG power/NN of/IN" is not.

Performing this sort of shallow chunking is based on the assumptions: (1) noun groups can represent the arguments of a relation, a verb group the relation itself, and (2) Web snippet chunking needs highly robust NL technologies. In general, chunking crucially depends on the quality of the embedded PoS tagger. However,

---

[4]For the remainder of the paper N = 1000. We are using Bing (http://www.bing.com/) for Web search.

[5]Concerning the English PoS tags, "word/PoS" expressions that match the following regular expression are considered as extended noun tag: "/(N(N|P))|/VB(N|G)|/IN|/DT". The English Verbs are those whose PoS tag start with VB. We are using the tag sets from the Penn treebank (English) and the Negra treebank (German).

it is known that PoS tagging performance of even the best taggers decreases substantially when applied on Web pages [11]. Web snippets are even harder to process because they are not necessary contiguous pieces of texts, and usually are not syntactically well-formed paragraphs due to some intentionally introduced breaks (e.g., denoted by ... betweens text fragments). On the other hand, we want to benefit from PoS tagging during chunk recognition in order to be able to identify, on the fly, a shallow phrase structure in Web snippets with minimal efforts. The assumption here is that we can tolerate errors caused by the PoS tagger because of the amount of redundancies inherent in the Web snippets.

The chunk–pair–distance model is computed from the list of noun group chunks.[6] This is done by traversing the chunks from left to right. For each chunk c_i, a set is computed by considering all remaining chunks and their distance to c_i, i.e., $(c_i, c_{i+1}, dist_{i(i+1)})$, $(c_i, c_{i+2}, dist_{i(i+2)})$, etc. We do this for each chunk list computed for each Web snippet. The distance $dist_{ij}$ of two chunks c_i and c_j is computed directly from the chunk list, i.e., we do not count the position of ignored words lying between two chunks.

The motivation for using chunk–pair–distance statistics is the assumption that the strength of hidden relationships between chunks can be covered by means of their collocation degree and the frequency of their relative Positions in sentences extracted from Web snippets; cf. [9] who demonstrated the effectiveness of this hypothesis for Web–based question answering. We are also making use of chunk–pair–distance statistics in the concept extractor, see Sect. 7.5.

Finally, we compute the frequencies of each chunk, each chunk pair, and each chunk pair distance. The set of all these frequencies establishes the chunk–pair–distance model CPD_M. It is used for constructing the topic graph in the final step. Formally, a topic graph $TG = (V, E, A)$ consists of a set V of nodes, a set E of edges, and a set A of node actions. Each node $v \in V$ represents a chunk and is labeled with the corresponding PoS tagged word group. Node actions are used to trigger additional processing, e.g., displaying the snippets, expanding the graph etc.

The nodes and edges are computed from the chunk–pair–distance elements. Since, the number of these elements is quite large (up to several thousands), the elements are ranked according to a weighting scheme which takes into account the frequency information of the chunks and their collocations. More precisely, the weight of a chunk–pair–distance element $cpd = (c_i, c_j, D_{ij})$, with $D_{ij} = \{(freq_1, dist_1), (freq_2, dist_2), \ldots, (freq_n, dist_n)\}$, is computed based on PMI as follows:

$$PMI(cpd) = log_2((p(c_i, c_j)/(p(c_i) * p(c_j))))$$
$$= log_2(p(c_i, c_j)) - log_2(p(c_i) * p(c_j))$$

---

[6]Currently, the main purpose of recognizing verb chunks is to improve proper recognition of noun groups. The verb chunks are ignored when building the topic graph.

where relative frequency is used for approximating the probabilities $p(c_i)$ and $p(c_j)$. For $log_2(p(c_i, c_j))$ we took the (unsigned) polynomials of the corresponding Taylor series[7] using $(freq_k, dist_k)$ in the k-th Taylor polynomial and adding them up:

$$PMI(cpd) = \left(\sum_{k=1}^{n} \frac{(x_k)^k}{k} \right) - log_2(p(c_i) * p(c_j)),$$

$$where \ x_k = \frac{freq_k}{\sum_{k=1}^{n} freq_k}$$

The visualized topic graph TG is then computed from a subset $CPD'_M \subset CPD_M$ using the m highest ranked cpd for fixed c_i. In other words, we restrict the complexity of a TG by restricting the number of edges connected to a node.

7.5 Concept Extraction

By Concept Extraction (CE for short), we mean the identification and clustering of descriptive sentences for the node of a topic graph that has been selected by the user. The particular approach that we are following is based on our own work on searching for definitional answers on the Web, using surface patterns, cf. [10]. We consider CE to be a member of the actions A_n associated to a node n with label w_n (a word or word group) such that CE is automatically evaluated with the input "define:w_n" which can be paraphrased as the definition question: "What is w_n?".

We will now summarize the major steps exploited by CE, see [10] for more details. CE aims at finding answers to definition questions from Web snippets. The major advantages are that it: (a) avoids downloading full-documents, (b) does not need specialized wrappers that extract definition utterances from definitional Web–sites, and (c) uses the redundancy provided by Web snippets to check whether the information is reliable or not. CE achieves these goals by rewriting the query in such a way that it markedly increases the probability of aligning well-known surface patterns with Web snippets. Matched sentences are therefore ranked according to three aspects: (a) the likelihood of words to belong to a description, (b) the likelihood of words to describe definition facets of the word being defined, and (c) the number of entities in each particular descriptive sentence. For this ranking purpose, CE takes advantage of a variation of Multi-Document Maximal Marginal Relevance and distinguishes descriptive words by means of Latent Semantic Analysis (LSA), cf. [16].

---

[7]In fact we used the polynomials of the Taylor series for $ln(1 + x)$. Note also that k is actually restricted by the number of chunks in a snippet.

7.5.1 Potential Sense Identification

An important feature of CE is a module that attempts to group descriptive utterances by potential senses, checking their correlation in the semantic space supplied by LSA. Note that this means that CE can also be used for disambiguation of the concept in question by clustering the extracted facts according to some hidden semantic relationship. Currently, we are assuming that final disambiguation is done by the user, but we are also exploring automatic methods, e.g., by taking the complete topic graph into account.

There are many-to-many mappings between names and their concepts. On the one hand, the same name or word can refer to several meanings or entities. On the other hand, different names can indicate the same meaning or entity. To illustrate this, consider the next set S of descriptive utterances recognized by the system:

1. John Kennedy was the *35th President* of the United States.
2. John F. Kennedy was the most anti-communist US *President*.
3. John Kennedy was a Congregational minister born in *Scotland*

In these sentences, "*US President John Fitzgerald Kennedy*" is referred to as "*John Kennedy*" and "*John F. Kennedy*", while "*John Kennedy*" also indicates a Scottish congregational minister. In the scope of this work, a *sense* is one meaning of a word or one possible reference to a real-world entity.

CE disambiguates senses of a topic δ by observing the correlation of its neighbors in the reliable semantic space provided by LSA. This semantic space is constructed from the term-sentence matrix M (by considering all snippets as a single document, and each snippet as a sentence), which considers δ as a *pseudo-sentence* which is weighted according to the traditional *tf-idf*. CE builds a dictionary of terms W from normalized elements in the snippet document S, with uppercasing, removal of html-tags, and the isolation of punctuation signs. Then CE distinguishes all possible unique *n-grams* in S together with their frequencies. The size of W is then reduced by removing *n-grams*, which are substrings of another equally frequent term. This reduction allows the system to speed up the computation of M as UDV' using the *Singular Value Decomposition*. Furthermore, the absence of syntactical information of LSA is slightly reduced by taking strong local syntactic dependencies into account, following our approach described in Sect. 7.4.

7.5.2 Experiments

A baseline system was implemented in which 300 snippets were retrieved by processing the input query (the topic in question) using the same query processing module as the one used in CE. The baseline splits snippets into sentences and accounts for a strict matching of the topic in question. In addition, a random sentence from a pair that shares more than 60% of its terms, and sentences that are a substring of another sentence were discarded. The baseline and CE were then tested with 606 definition questions from the TREC 2003/2001 and CLEF 2006/2005/2004 tracks.

Overall, CE consistently outperformed the baseline. The baseline discovered answers to 74% of the questions and CE up to 94%. For 41.25% of the questions, the baseline found one to five descriptive sentences, whereas CE found 16–25 descriptive sentences for 51.32% of the questions. More specifically, results show that CE finds nuggets (descriptive phrases) for all definition questions in the TREC 2003 set, contrary to some state-of-the-art methods, which found nuggets for only 84%. Furthermore, CE finds nuggets for all 133 questions in TREC 2001 question set, in contrast with other techniques, which found a top five ranked snippet that conveys a definition for only 116 questions within the top 50 downloaded full documents.

Concerning the performance of the sense disambiguation process, CE was able to distinguish different potential senses for some topic δs, e.g., for "*atom*", the particle–sense and the format–sense. On the other hand, some senses were split into two separate senses, e.g., "*Akbar the Great*", where "*emperor*" and "*empire*" indicated different senses. This misinterpretation is due to the independent co-occurrence of "*emperor*" and "*empire*" with δ, and the fact that it is unlikely that they share common words. In order to improve this, some external sources of knowledge are necessary. This is not a trivial problem, because some δs can be extremely ambiguous like "*Jim Clark*", which refers to more than ten different real-world entities. CE recognized the racing car driver, the Netscape founder and the president of the Arizona Western Heritage Foundation. Independently of that, we found that entities and the correlation of highly closed terms in the semantic space provided by LSA can be important building blocks for a more sophisticated strategy for the disambiguation of δ.

7.6   Relation Extraction and Background Knowledge

In the previous sections we extracted concepts and collected descriptive information like definitions and further explanations. Now we want to take a closer look at the relationships holding between those concepts, i.e. we want to semantically label the edges of our topic graph.

In our approach we divide the task into two different but complementary steps:

1. In order to provide query specific background knowledge we make use of Wikipedia's infoboxes. These infoboxes contain facts and important relationships related to articles. We also tested DBpedia as a background source [3]. However, it turned out that currently it contains too much and redundant information. For example, the Wikipedia infobox for "Justin Bieber" contains 11 basic relations whereas DBpedia has 50 relations containing lots of redundancies. In our current prototype, we followed a straightforward approach for extracting infobox relations: We downloaded a snapshot of the whole English Wikipedia database

(images excluded), extracted the infoboxes for all articles if available and built a Lucene Index running on our server. We ended up with 1.124.076 infoboxes representing more than two million different searchable titles. The average access time is about 0.5 s. Currently, we only support exact matches between the user's query and an infobox title in order to avoid ambiguities. We plan to extend our user interface so that the user may choose different options. Furthermore we need to find techniques to cope with undesired or redundant information (see above). This extension is not only needed for partial matches but also when opening the system to other knowledge sources like DBpedia, newsticker, stock information and more.

2. In case of important relations missing in the infoboxes we try to extract them from the previously retrieved snippets.[8] As mentioned earlier snippets often do not contain complete sentences but only parts of them or they may contain dots which means parts of the text have been omitted. Hence the relation extraction algorithm should not rely on techniques that require more than very shallow linguistic analysis.

Bunescu and Mooney [4] perform a minimal supervised process for extracting binary relations on snippets (see also Sect. 7.3). The approach we describe here is an extension to this approach that relies on a new multiple instance learning algorithm specially developed for binary relation extraction and its extension to n-ary relations.

7.6.1 Binary Relation Extraction

This process starts with some positive and negative instances of a relation created by search queries from a standard search engine like Google or Bing. For the relation *origin(person, country)* one could formulate the following set of positive examples:

Example 7.1.

- Arnold Schwarzenegger * * * * * * * Austria (1375)
- Albert Einstein * * * * * * * Germany (622)
- Dirk Nowitzki * * * * * * * Germany (323)
- Detlef Schrempf * * * * * * * Germany (311)
- Brigitte Bardot * * * * * * * France (163)

and negative examples could be:

- Berlusconi * * * * * * * Germany (1375)
- Franz Beckenbauer * * * * * * * USA (622)
- Helmut Kohl * * * * * * * England (323)

---

[8]For "Jim Clark", e.g., wikipedia's infoboxes do not provide information for the relations: birthplace, place_of_death, or cause_of_death.

The numbers in brackets denote the number of snippets found by the search engine – in this case we used the Bing search engine. Unfortunately you cannot count on all positive examples to be really positive for the envisaged relation. For example, Arnold Schwarzenegger could have visited his parents in Austria, so in that case the relation would be *visit(person1, person2, country)*. In the negative results there is probably no snippet expressing a person's origin. Thus for this example we construct five bags of "pseudo" positive snippets and label them as positive, and three bags of negative snippets which are labeled as negative.

In the next step we apply a new multiple instance learning algorithm optimized for relation extraction to the positive and negative bags. This algorithm is specially designed to work with positive and negative bags of examples including problems which occur because of incomplete knowledge about the labels of single training examples. A bag is labeled positive if at least one example can be labeled positive and negative if all examples in the bag are negative. Our learning algorithm works as follows:

1. For each bag, merge all instances into one document and build a feature vector for this document. The feature vector may consist either of words and tokens occurring in the text or of letter n-grams. A combination of both is also possible.
2. For each feature vector compute the norm according to its size:

$$norm(bag_i) = \sqrt{\sum |x_{k_i}|} \tag{7.1}$$

   where x_{k_i} is the k-th document in the i-th bag
3. Weight the features in the vectors according to their number of occurrences in the bag:

$$w_{training}(x_{k_i}) = \#x_k \in bag_i \tag{7.2}$$

   The learning phase has now been completed and no further computations are necessary in this step, thus, making algorithm performance very fast.
4. Compute a leave-one-out calculation to identify false positive examples. This is done by removing such an instance from the feature vector of the bag and classifying it:

   • Build a feature vector of the example to be classified: The preprocessing should be the same as in the learning phase
   • Weight the features according to their relevances in the categories. This gives the mutual information gain for each token and each bag:

$$w'(x_{k_i}) = \max\left(0, \frac{\#bag_i + 1}{\sum(x_k \in bag_i)}\right) \tag{7.3}$$

$$w(x_{k_i}) = w'(x_{k_i}) * w_{training}(x_{k_i}) \tag{7.4}$$

- Compute the classification relevances for each bag:

$$\forall bag_i : relevance(bag_i) = \sum \frac{w(x_{k_i}) * \kappa_i}{norm(bag_i) * length(x)} \qquad (7.5)$$

where κ_i denotes a smoothing factor for the bags.
- The classification is done according to the category with the highest relevance

For the set of false positive examples identify the features with the highest relevance as they have the highest impact on the incorrect classification. These are removed from the feature vectors of the positive bags and added to the feature vector of one of the negative bags. Afterwards perform the steps (1) and (2) again.
5. Recompute the leave-one-out calculation using the newly weighted bags and relabel false negative examples, i.e. examples that are labeled as positive but classified as negative. As side effect, the number of false positives inside the positive bags is also reduced automatically.
6. Repeat the process from step (2) until the process converges.

The main properties of this classification method are simplicity, fastness and robustness even in unbalanced or noisy training data. The classification result has reliable confidence values which have been created by computing a final leave one out classification for all data in the training set. Furthermore it allows direct control of the mutual information gain of each feature in the feature space.

The result of the learning process on these bags is a model that can be used to classify unseen snippets and if they are classified as positive, to extract the relation predicted by the model. For the example three above, the Precision–Recall graph (cf. Fig. 7.7) shows our success. The curve shows an F-measure of about 0.76. With this method we have successfully trained several models for different relations (Table 7.1).

7.6.2 Inferring Binary and N-ary Relations

Learning of n-ary relations using the described technique is not possible for quite a few reasons. The main problem in learning n-ary relations, especially for $n > 3$, is to create enough training data by formulating useful search queries. For some relations, it might be possible to generate positive training data but it is very difficult to formulate search queries that deliver meaningful negative examples and to do so requires a good understanding of the learning technique itself. Furthermore, whenever the argument types of the relation contain times and dates, e.g. *birth(person, city, date)*, money amount, e.g. *company_acquisition(company1, company2, money amount)*, or just numbers, it is impossible to find examples for the negative training bags. But even for binary relations, we may need an additional technique because although the learning algorithm relies on a minimal amount of supervision, it still

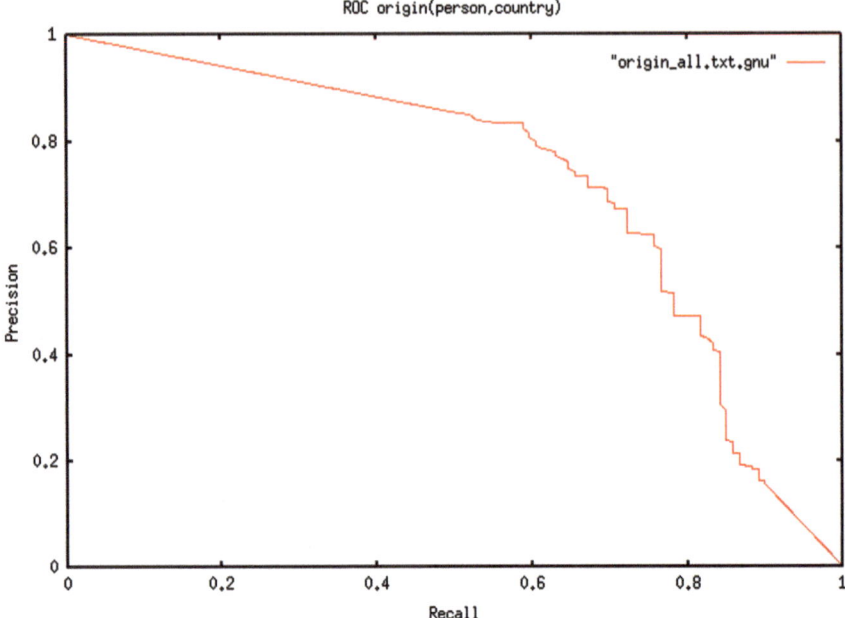

Fig. 7.7 Precision–Recall graph for origin (person,country)

Table 7.1 F-Measures for different relations

Relation	# train bags positive	# train bags negative	F–measure
birthplace(person,city)	5 (1,490 snippets)	3 (362 snippets)	0.86
origin(person,country)	5 (2,794 snippets)	3 (2,320 snippets)	0.76
place_of_death(person,city)	5 (2,466 snippets)	3 (951 snippets)	0.68
cause_of_death(person,cause)	5 (2,533 snippets)	3 (569 snippets)	0.58
married_with(person,person)	4 (388 snippets)	2 (1,025 snippets)	0.87
father_of(person,person)	4 (1,849 snippets)	3 (939 snippets)	0.72

requires some background work like finding adequate search queries or training of an optimal model, i.e. determining the desired precision/recall rate.

McDonald et al. [18] have shown that factoring n-ary relations into a set of binary relations, and reconstructing the instantiated binary relations into n-ary again by using maximal cliques in the relation graphs has been successful on a large set of biomedical data. In contrast our solution infers n-ary relations from binary ones by using manually constructed relation hierarchies (see Fig. 7.8). Starting from the binary relations using the approach described in the previous section (squared boxes) the relation hierarchy offers the possibility to infer n-ary relations (parallelograms) as well as new binary relations (rounded-corner boxes). Note that relations may only be derived if the binary relations (squared boxes) have been successfully extracted. The resulting relation candidates are then validated by generating a new search request containing the matched arguments of the relation

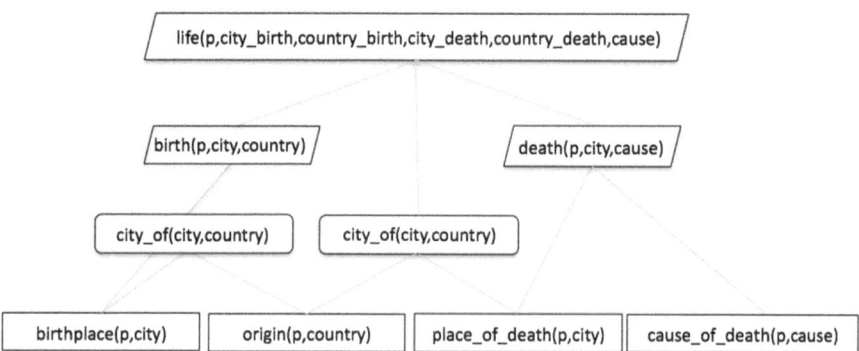

Fig. 7.8 An example for a relation hierarchy. The *squared boxes* show relations extracted from the snippets. The *rounded-corner boxes* show inferred binary relations, the *parallelograms* show inferred n-ary relations

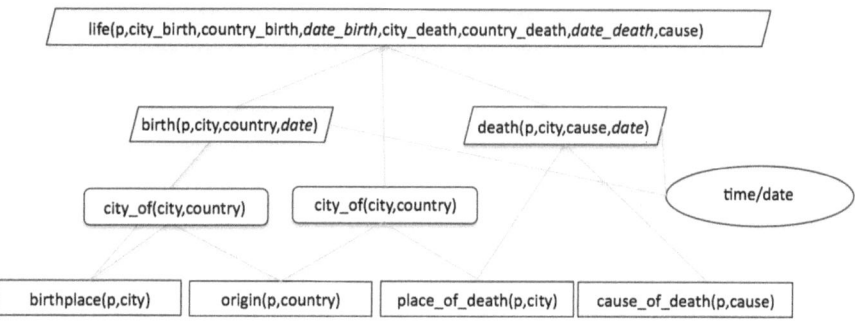

Fig. 7.9 New relations are created by attaching the topic "date"

with *AND*, and sending it to the search engine. If the number of search results is above a certain threshold we conclude that the inferred relation is validated. For argument types like times and dates, amounts of money, etc.[9] we use an extended relation hierarchy as shown in (see Fig. 7.9) because they never occur as arguments of our original or derived relations. Again we utilize search queries for validation of the resulting relation candidates.

However, general evaluation for this technique is very difficult as it depends heavily on the instances of the relations themselves and their existence on the Web. So we need to define new evaluation strategies for this special kind of relation extraction (see Sect. 7.8). Generally speaking, the success of the n-ary relation extraction as described here depends on two basic factors:

---

[9]The classification of NP chunks to argument types like times and dates is currently done by using simple regular expressions.

1. Confidence of the classified binary relations: With higher confidence figures of these relations the probability of the inferred n-ary and binary relations increases.
2. Verification on the Web: In our experiments we observed that the precision rate of inferred relations increases if they can be found in different places on the Web, i.e. if they are well supported by the Web. We also experimented for n-ary relations (n > 3) with high Web support using classification models possessing higher recall and lower precision for their underlying binary relations (this means with lower confidences). This resulted in increased recall with constant precision rates for the n-ary relations.

7.7 Evaluation

For an initial evaluation we had 20 testers: 7 came from our lab and 13 from non–computer science related fields. Fifteen persons had never used an iPad before. After a brief introduction to our system (and the iPad), the testers were asked to perform three different searches (using Google, i–GNSSMM and i–MILREX) by choosing the queries from a set of ten themes. The queries covered definition questions like *EEUU* and *NLF*, questions about persons like *Justin Bieber*, *David Beckham*, *Pete Best*, *Clark Kent*, and *Wendy Carlos*, and general themes like *Brisbane*, *Balancity*, and *Adidas*. The task was not only to get answers on questions like "Who is ..." or "What is ..." but also to acquire knowledge about background facts, news, rumors (gossip) and more interesting facts that come into mind during the search. Half of the testers were asked to first use Google and then our system in order to compare the results and the usage on the mobile device. We hoped to get feedback concerning the usability of our approach compared to the well known internet search paradigm. The second half of the participants used only our system. Here our research focus was to get information on user satisfaction of the search results. After each task, both testers had to rate several statements on a Likert scale and a general questionnaire had to be filled out after completing the entire test. Tables 7.2 and 7.3 show the overall result.

The results show that people in general prefer the result representation and accuracy in the Google style. Especially for the general themes the presentation of Web snippets is more convenient and more easy to understand. However when it comes to interesting and suprising facts users enjoyed exploring the results using

Table 7.2 Google

#question	V.good	Good	Avg.	Poor
Results first sight	55%	40%	15%	–
Query answered	71%	29%	–	–
Interesting facts	33%	33%	33%	–
Suprising facts	33%	–	–	66%
Overall feeling	33%	50%	17%	4%

Table 7.3 i-GNSSMM and i-MILREX

#question	V.good	Good	Avg.	Poor
Results first sight	43%	38%	20%	–
Query answered	65%	20%	15%	–
Interesting facts	62%	24%	10%	4%
Suprising facts	66%	15%	13%	6%
Overall feeling	54%	28%	14%	4%

the topic graph. The overall feeling was in favor of our system which might also be due to the fact that it is new and somewhat more playful.

The replies to the final questions: *How successful were you from your point of view? What did you like most/least;? What could be improved?* were informative and contained positive feedback. Users felt they had been successful using the system. They liked the paradigm of the explorative search on the iPad and preferred touching the graph instead of reformulating their queries. The presentation of background facts in i–MILREX was highly appreciated. However some users complained that the topic graph became confusing after expanding more than three nodes. As a result, in future versions of our system, we will automatically collapse nodes with higher distances from the node in focus. Although all of our test persons make use of standard search engines, most of them can imagine to using our system at least in combination with a search engine even on their own personal computers.

7.8 Conclusion and Future Work

Above, we presented an approach of interactive topic graph extraction for exploration of Web content. The initial information request is issued online by a user to the system in the form of a query topic description. The topic query is used for constructing an initial topic graph from a set of Web snippets returned by a standard search engine. At this point, the topic graph already displays a graph of strongly correlated relevant entities and terms. The user can then request further detailed information for parts of the topic graph in the form of definitions and relations from the Web and Wikipedia infoboxes.

A prototype of the system has been realized on the basis of a mobile touchable user interface for operation on an iPad. We believe that our approach of interactive topic graph extraction and exploration, together with its implementation on a mobile device, helps users explore and find new interesting information on topics about which they have only a vague idea or even no idea at all.

The next steps will be the development of a sophisticated evaluation of n-ary relation extraction and systematic field tests for complete system cycles under different user profiles. For this, we also plan to improve the concept disambiguation by implementing a stronger integration of concept and relation extraction. Furthermore, we plan to explore contextual user interaction, e.g., by taking into account all available information of the neighboring nodes and edges of the parts of the topic graph selected by the user. A particular challenge will be the exploration of multi–lingual and cross–lingual methods, which support comparison and merging of extracted information from Web snippets for different languages.

Acknowledgements The presented work was partially supported by grants from the German Federal Ministry of Economics and Technology (BMWi) to the Theseus project (FKZ: 01MQ07016).

References

1. Banko, M., Cafarella, M.J., Soderland, S., Broadhead, M.S., Etzioni, O.: Open information extraction from the Web. In: Proceedings of the International Joint Conference on Artificial Intelligence (IJCAI), Hyderabad, pp. 2670–2676. (2007)
2. Baroni, M., Evert, S.: Statistical methods for corpus exploitation. In: Lüdeling, A., Kytö, M. (eds.) Corpus Linguistics. An International Handbook. Mouton de Gruyter, Berlin (2008)
3. Bizer, C., Lehmann, J., Kobilarov, G., Auer, S., Becker, C., Cyganiak, R., Hellmann, S.: DBpedia – a crystallization point for the Web of Data. Web Semant. **7**(3), 154–165 (2009)
4. Bunescu, R.C., Mooney, R.J.: Learning to extract relations from the Web using minimal supervision. In: Proceedings of ACL'07, Prague, pp. 576–583. (2007)
5. Cui, H., Kan, M.Y., Chua T.S., Xiao, J.: A comparative study on sentence retrieval for definitional question answering. SIGIR Workshop on Information Retrieval for Question Answering (IR4QA), Sheffield (2004)
6. Downey, D., Schoenmackers, S., Etzioni, O.: Sparse information extraction: unsupervised language models to the rescue. In: Proceedings of ACL, Prague, pp. 696–703. (2007)
7. Eichler, K., Hemsen, H., Löckelt, M., Neumann, G., Reithinger, N.: Interactive dynamic information extraction. In: Proceedings of KI'2008, Kaiserslautern, pp. 54–61. (2008)
8. Etzioni, O.: Machine reading of Web text. In: Proceedings of the 4th International Conference on Knowledge Capture, Whistler, pp. 1–4. (2007)
9. Figueroa, A., Neumann, G.: Language independent answer prediction from the Web. In: Proceedings of the 5th FinTAL, Turku (2006)
10. Figueroa, A., Neumann, G., Atkinson, J.: Searching for definitional answers on the Web using surface patterns. IEEE Comput. **42**(4), 68–76 (2009)
11. Giesbrecht, E., Evert, S.: Part-of-speech tagging – a solved task? An evaluation of PoS taggers for the Web as corpus. In: Proceedings of the 5th Web as Corpus Workshop, San Sebastian (2009)
12. Giménez, J., Màrquez, L.: SVMTool: a general PoS tagger generator based on Support Vector Machines. In: Proceedings of LREC'04, Lisbon (2004)
13. Greenwood, M.A., Stevenson, M.: Improving semi-supervised acquisition of relation extraction patterns. In: Proceedings of the Workshop on Information Extraction Beyond the Document, Sydney, pp. 12–19. (2006)
14. Hildebrandt, W., Katz, B., Lin, J.: Answering definition questions using multiple knowledge sources. In: Proceedings HLT-NAACL, Boston, pp. 49–56. (2004)
15. Joho, H., Liu, Y.K., Sanderson, M.: Large scale testing of a descriptive phrase finder. In: Proceedings 1st Human Language Technology Conference, San Diego, pp. 219–221. (2001)
16. Landauer, T., McNamara, D., Dennis, S., Kintsch, W.: Handbook of Latent Semantic Analysis. Lawrence Erlbaum, Mahwah (2007)
17. Manning, C.D., Raghavan, P., Schütze, H.: Introduction to Information Retrieval. Cambridge University Press, New York (2008)
18. McDonald, R., Kulick, S., Pereira, F., Winters, S., Jin, Y., White, P.: Simple algorithms for complex relation extraction with applications to biomedical IE. In: Proceedings of the 43rd

Annual Meeting on Association for Computational Linguistics, University of Michigan, pp. 491–498. (2005)

19. Rosenfeld, B., Feldman, R.: URES: an unsupervised Web relation extraction system. In: Proceedings of the COLING/ACL on Main Conference Poster Sessions, Sydney, pp. 667–674. (2006)

20. Shinyama, Y., Sekine, S.: Preemptive information extraction using unrestricted relation discovery. In: Proceedings of the Proceedings of the Human Language Technology Conference of the NAACL, Main Conference, New York City, pp. 304–311. (2006)

21. Sekine, S.: On-demand information extraction. In: Proceedings of the COLING/ACL, Sydney, pp. 731–738. (2006)

22. Sudo, K., Sekine, S., Grishman, R.: An improved extraction pattern representation model for automatic IE pattern acquisition. In: Proceedings of ACL, Sapporo, pp. 224–231. (2003)

23. Turney, P.D.: Mining the Web for synonyms: PMI-IR versus LSA on TOEFL. In: Proceedings of the 12th European Conference on Machine Learning. Freiburg, pp. 491–502. (2001)

24. Yates, A.: Information extraction from the Web: techniques and applications. Ph.D. Thesis, University of Washington, Computer Science and Engineering (2007)

Chapter 8
Predicting Relevance of Event Extraction for the End User

Silja Huttunen, Arto Vihavainen, Mian Du, and Roman Yangarber

Abstract We present work on estimating the *relevance* of the results of an Event Extraction system to the end-user's needs. Our aim is to develop user-oriented measures of utility of the extracted events, i.e., how useful is the factual information found in the document for the end user. We introduce *discourse* and *lexical* features, and build classifiers that learn from the users' ratings of the relevance of the extraction results. Traditional criteria for evaluating the performance of Information Extraction (IE) focus on the correctness of the extracted information, e.g., in terms of recall, precision, F-measure, etc. We rather focus on subjective criteria for evaluating the quality of the extracted information: utility of results to the end-user. To measure utility, we use methods from text mining and linguistic analysis to identify features that are good predictors of the relevance of an event or a document. We report on experiments in two real-world event extraction domains: corporate activities reported in business news, and health threats in news about infectious epidemics.

8.1 Introduction

In this paper we present on-going work within the PULS project,[1] aimed at developing *user-oriented* measures of relevance of information extracted from plain-text news articles. The relevance measures have been built in collaboration with actual end users of our systems. End users view and rate the utility of extracted events using an online news surveillance server.

---

[1]The Pattern Understanding and Learning System: http://puls.cs.helsinki.fi

S. Huttunen · A. Vihavainen · M. Du · R. Yangarber (✉)
Department of Computer Science, University of Helsinki, Finland
e-mail: roman.yangarber@cs.helsinki.fi

T. Poibeau et al. (eds.), *Multi-source, Multilingual Information Extraction and Summarization 11*, Theory and Applications of Natural Language Processing, DOI 10.1007/978-3-642-28569-1__8, © Springer-Verlag Berlin Heidelberg 2013

Our aim is to show that by utilizing domain-specific and domain-independent sets of features we can build and train a system that is able to predict the utility of new information obtained by an Information Extraction system. We apply the methods on two domains in order to demonstrate that the approach is, in principle, domain independent, and easily adapted to different domains.

Our target domains are business news, with the focus on analyzing reports about corporate acquisitions and new product launches, and medical news, with the focus on outbreaks and spread of infectious epidemics. These topics are actively researched in the IE community, e.g. [6, 7, 9, 15].

The news extraction and relevance prediction works in three phases. The first phase identifies articles that may be potentially relevant to a target domain using a broad Web search, based on queries that are boolean combinations of keywords—this is done continuously on an on-going basis. The second phase employs IE to extract events from the acquired articles. The final phase then determines the *relevance* of the extracted events or articles for the end-user.

For the business domain the system extracts the names of the companies involved in the target activities. Target activities include corporate mergers and acquisitions, product launches and advertising campaigns, investments of capital and projects, deals and contracts, major hirings, firings, and layoffs. For each event, the system tries to identify the date, location, value of the transaction (if any) and, for the product-launch scenario, the product type. An example of a sentence reporting a corporate acquisition event: *"Air New Zealand said Friday it has bought 14.9% of Australia's Virgin Blue for $143 million."* A product-launch event is found, e.g., in a sentence like: *"An executive at T-Mobile said the company was introducing its new DriveSmart service at the request of customers.*

In the medical domain, the system extracts what victims were affected by what disease, where and when. For example, the sentence *"The HSE in Ireland has said that there have been a further four deaths from human swine flu in the past week"* will induce an event, with attributes country, disease, number of casualties, and the time of occurrence, [19].

In the next section, we briefly present the criteria for judging the *quality* of extracted events, and present the approach taken in our system. Section 8.3 introduces the features we use for predicting utility. Section 8.4 discusses our experimental setup and gives a short system description of the relevance generation process. Section 8.5 presents our current experiments and results with automatic assignment of relevance scores. In the final section we discuss the results and outline next steps.

8.2 Quality Measures

In IE research, performance has been traditionally measured in terms of *correctness*, counting how many of the fields in each record were correctly extracted by comparing the system's answers to a set of answers pre-defined by human annotators.

Table 8.1 Guidelines for relevance scores

Criteria	Score
New information; highly relevant	5
Important updates; on-going developments	4
Review of current events; hypothetical, predictions	3
Historical/non-current, background information	2
Non-specific, *non-factive* events; secondary topics	1
Unrelated to target domain; useless	0

In the MUC and ACE initiatives, e.g., this was computed mainly in terms of recall and precision, and F-measure, [1, 11].

We would like to distinguish *objective* vs. *subjective* measures of quality. Objective measures take the perspective of the system in evaluating the obtained IE results in terms of correctness and confidence. Confidence has been studied to estimate the probability of the correctness of the system's answer, in e.g. [5]. Our IE system, PULS, computes *confidence* using *discourse-level cues*, [16], such as: confidence decreases as the distance between the sentence containing the event and event attributes increases; confidence increases if a document mentions only one country.

Subjective measures reflect the end users' perspective, that is the relevance (or utility) of the extracted information, and the reliability of the information found, [17]. *Reliability* measures whether the reported event is "true", or trustworthy. We mention this criterion for completeness, since it is the ultimate goal of any news surveillance process. However, this judging reliability requires pragmatic knowledge, including information that is obtainable by the user only through downstream verification, and is thus beyond the scope of this paper.

Utility measures how *useful* the result is to the user *irrespective* of its correctness. An event may be correctly extracted, and yet be of low utility to the user; for example, historical or hypothetical events may not be useful for an analyst concerned with the current state of affairs. Conversely, an event may have many incorrectly extracted attributes, and yet be of great *value* and interest to the user.

We focus here specifically on relevance vs. correctness. The relevance rating scale currently used in our work are shown in Table 8.1. Our goal is, specifically to devise methods for automatic assignment of relevance scores to extracted events, and to the documents in which they are found.

The users assign the scores as presented in the Likert-like scale, Table 8.1. Further, in the work and experiments reported in this paper, these scores are reduced for simplicity, into either a three-way classification—*high* (4–5), *low* (1–3) and *irrelevant* (0) events,—or a binary classification—where events with a score of 4–5 are considered high relevance, and those with a score of 0–3 are considered low-relevance. The binary classification is useful because one practical purpose of introducing the relevance score is for the system to determine whether a given extracted event ought to be present to the user on the main page of the site—which is a binary decision.

8.3   Linguistic Features

In this section we describe the features that we use in our system for predicting the relevance of an event. These features were devised through a detailed analysis of the domains and user-evaluated events, and were chosen based on their potential for relevance prediction.

Many features are characterized in terms of the event *trigger* and its *attributes*. The PULS IE system operates by pattern matching, [8]. The system employs a pattern base—a large set of domain-specific linguistic patterns, which map from surface-syntactic representation of the facts in the sentence into the semantic representation in the database. A trigger is a sentence where an event pattern fires, signaling a mention of an event at that point in the document. For example, in the sentence "... *Department says there have been eight confirmed cases of measles, after an outbreak at Royal Perth Hospital.*" a pattern is triggered by the phrase "cases of *disease*". The attributes of the event correspond to the fills in the database record, in this example, the name of the disease, the location, date, the number of cases, etc. Several events may appear in a news article.

At this point, we introduce a distinction between *discourse features* and *lexical features*. Discourse features are based on properties of the article text and of the events extracted from it. Lexical features are simpler, low-level features based on bags of words, discussed in Sect. 8.3.2. In essence, lexical features capture local information, while discourse features capture longer-range relationships within the document.

8.3.1   Discourse Features

Discourse features operate on information about the article such as the number of events found in the article, the positioning of the events in the document, the compactness of the placement of the event's attributes [2, 12], and the recency of the events' occurrence.

8.3.1.1   Layout and Positioning

We use a set of features describing the (relative) position of the trigger sentence within the document. These features allow us to quantify the assumption that important details of news topics are placed in the beginning of an article, whereas less important details are stated later.[2] Layout features also include the length of the document and the position of the trigger sentence in the document.

---

[2]The so-called "inverted pyramid" principle, [3]

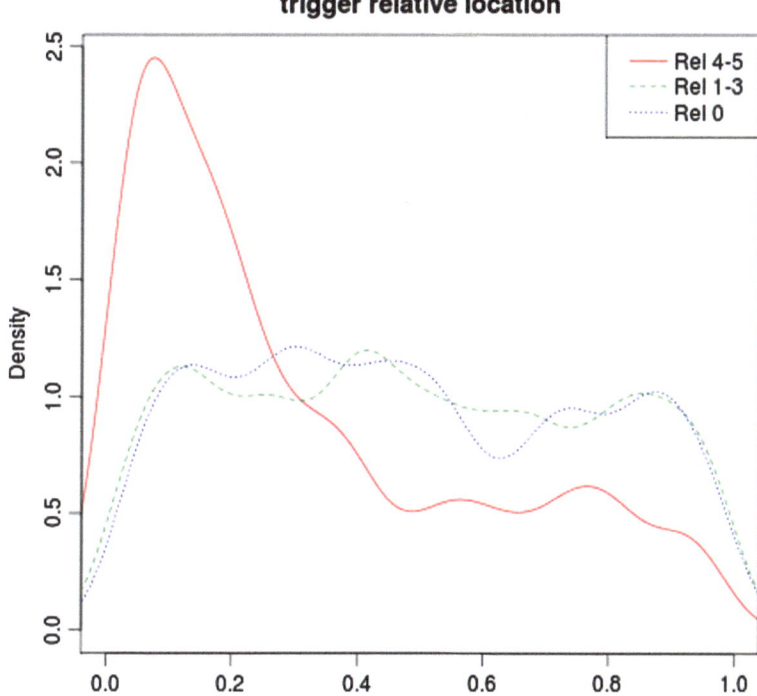

Fig. 8.1 Medical Domain: Conditional distribution of the relative position of the trigger sentence in the document, given the relevance class/rating: high vs. low vs. zero. The relative position is the X-axis, the y-axis shows the probability density

Figure 8.1 shows the conditional distribution of the relative location of the event in the text, given that the event has a high relevance score (4–5), low relevance (1–3), or is completely irrelevant (score 0).

Figure 8.2 shows that high-relevance events favor the placement of the trigger sentence earlier on in the document. We use the term document *header* to mean the article's headline and the first two sentences of the news article.

8.3.1.2 Event Compactness

In a compact event, all the event attributes are situated near the trigger in the text. The *compactness features* track the distance of mentions of the attributes of an event from the event's trigger. We model the effect of compactness of an event on its relevance by, e.g., measuring the distance between the trigger and the disease name (for the medical domain) or a company name (for the business domain). The distance may be measured as the number of bytes, words, or sentences. The "active" participating attributes of an event are here called *actors*.

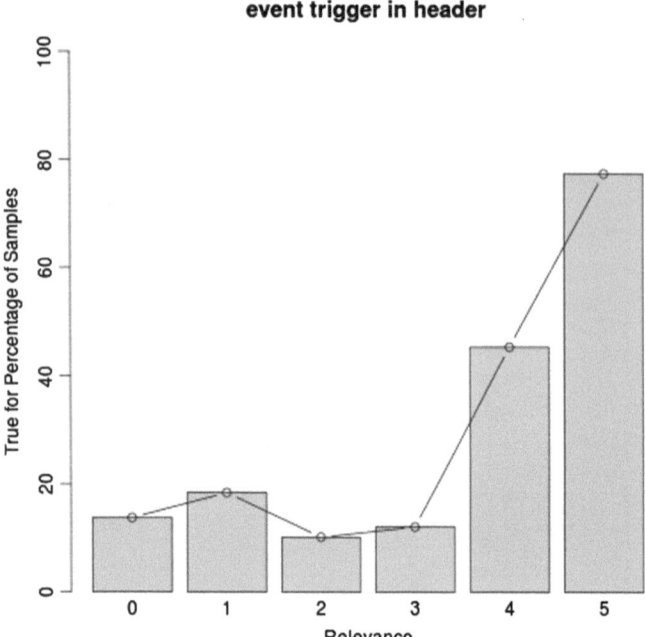

Fig. 8.2 Medical Domain: Probability that the trigger of the event is in the header, given the relevance score

Figure 8.3 shows the distribution of the distance in sentences (horizontal axis) between the disease name and the trigger given high, low and zero relevance. For events that contain no actor at all, the feature receives a special value *NA*. If the disease name is totally missing in document, the NA value is mapped to a high value on the x-axis, (simulating the effect that the disease name is "far away" from the trigger.). The name of a disease is more likely to occur in a trigger sentence of a high-relevance event, than in the trigger sentence of low-relevance event.

Content-repetition features measure whether an important fill, such as an actor, is repeated inside the document. These features are based on the assumption that repeated mentions of a key actor should positively affect the relevance. Conversely, features that count the number *distinct* actors mentioned in the text may serve as good indicators of *lower* relevance. For example, an article mentioning many different diseases or companies is less likely to be of high relevance, containing very specific and topical news, and may be more likely to be a broader review or overview article.

8.3.1.3 Time and Recency

Time features relate to the recency of an event, comparing the time attributes of an event with the publication date of the news article, i.e., the difference between

trigger disease distance

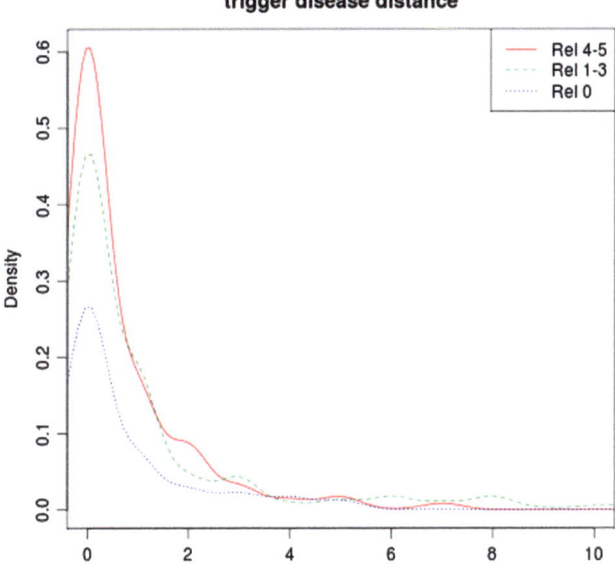

Fig. 8.3 Medical domain: distribution of distance from the trigger sentence to the mention of the disease name. Instances with no disease mention, or with distance > 10 are mapped to the point distance $= 20$, not shown in the graph

publication date and the reported event date. The PULS system may extract different kinds of events, including hypothetical events, and events which have a projected or expected date in the future. Recency is a good indicator for relevance, as can be seen in Fig. 8.4. Highly relevant articles, usually describe more recent events.

8.3.1.4   Indicators of Irrelevant Domains

Blacklist features signal that the extracted event is potentially part of a wrong topic. Blacklist features are small sets of manually selected terms that commonly appear in irrelevant documents. In epidemic surveillance, for example, terms such as "vaccination campaign" and "obituary" are strong indicators low relevance. A common source of false positives in the medical domain is obituaries, when, e.g., a death after an illness is reported. The extraction patterns may not be able to distinguish those from deaths due to infectious diseases, based on *local* cues alone.

In the business domain, an indicator of low relevance is, e.g., "President", possibly followed by a proper name, and a country. This mostly refers to heads of state, rather than heads of a company.

PULS extracts "negative events", (called here *harm* events), as well, to capture events that frequently interfere with events of interest. For example, in the business domain, satellite/rocket launches may trigger patterns for finding product launches, since they are syntactically similar; natural disasters (flooding, earthquake, etc.) with

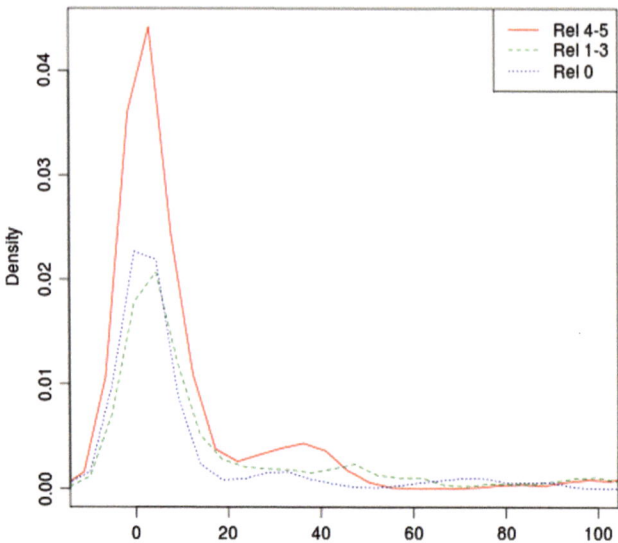

Fig. 8.4 Medical Domain: distribution of the difference days from publication to event date. Negative values indicate events in the future

casualties often interfere with patterns in for medical domain. The number of found harm events in a document is a discourse feature.

A missing attribute may also be an indication of an irrelevant event. Events rejected or marked irrelevant by the user are more likely to be missing the name of a disease. The system also extracts victim names where possible, since obituaries, stories about public figures, and other items irrelevant from the epidemiological perspective, tend to name the victims. It is important to keep in mind that all these features only capture tendencies and probabilities, and are not deterministic. For example, some news articles about genuine epidemic outbreaks may name the victims as well—in an attempt to personalize them for the reader and make the experience more immediate.

The number of unique actors preceding the trigger sentences is potentially correlated with irrelevance.[3] For example, if no disease names exist before the trigger sentence, then the document is likely to be irrelevant. On the other hand, important news events often mention only a single disease or company.

A more complete list of the discourse features used in the experiments is given in Table 8.2.

---

[3]PULS system normalizes and unifies variants of disease names and organization names, e.g., Swine Flu with H1N1; company full-names and acronyms.

Table 8.2 Examples of discourse features

Layout features
Event trigger in header or headline?
Any event trigger found in header/headline?
Trigger's relative location in document
Actor in trigger/header/headline?
Country in trigger/header/headline?
Document length
Compactness
Distance from trigger to actor
Actor found before end of trigger sentence?
Number of unique countries/actors in trigger sentence
Number of unique countries/actors until end of trigger sentence
Number of unique countries/actors found in document events
Event contains a valid country/actor?
Is content repeated in the header/in the document?
Number of events found in document
Time
Event has time of occurrence?
Distance between trigger sentence and mention of event date
Time difference between publication date and event start/end date
Low relevance indicators
Is blacklisted data found in Headline/Header/Document?
Number of *harm* events in the document (medical domain)
Victim count
Is victim named?
Is illness unspecified?

8.3.2   Lexical Features

Lexical features for an event consists of bags of words in the trigger sentence, and in the sentences immediately preceding and following the trigger sentence. The surrounding sentences provide additional context for disambiguation. For example, the trigger sentence may include deaths and injuries, but in principle the article could be about any kind of casualties.

8.3.3   Domain Specificity

Some features are applicable directly to different domains. An example of such features are the recency features, which compare the event date to the publication date of the document. Other features are domain-specific, and make use of the domain-specific attributes. For example, we may check the position of an actor attribute, and see whether it appears in the headline. In the medical domain, such an attribute would be the disease name, in the business domain, we use the company name in an analogous fashion.

8.4   Experimental Setup and System Description

Next, we briefly describe how the relevance classifiers are built. We have an online news surveillance system that allows users to review, rate and correct events extracted from news articles. The work-flow for finding relevant events is as follows:

The system's information retrieval (IR) component continuously polls news sites, [18]. News filtering is done using Boolean keyword-based queries. The result is a continuous stream of potentially relevant documents, that is forwarded to the information extraction system.

Our IE system then extracts events of potential relevance from this stream of articles. The extracted information, i.e., the structured events with their attributes, is stored into a database. The IE component uses a large set of linguistic patterns, which in turn utilize general and domain-specific concepts, such as diseases, locations and organizations.

Once the attributes of an event have been extracted, the relevance classifier is invoked. Each event is converted to a feature vector, to which a classifier assigns a relevance score.

After the event receives a relevance score, it appears on the on-line server. Relevance predictions are highlighted with different colors, which enhances the user experience and allows easy notification of high relevance events.

The system's user interface (UI) provides a simple editing facility for the extracted events. In case of errors in the automatic extraction, the UI allows the user to correct erroneous fills, e.g., if a company name, country, or a disease name was extracted incorrectly. In addition to editing the event fills, the users can also assign or edit the relevance labels to the extracted events. The set of events that have been corrected and/or relevance-labeled manually by human users are used for training and testing the relevance classifiers.

In the business domain, we use a set of hand-labeled data, in which currently roughly 45% of the events are high-relevance, and 55% are low-relevance. In the medical domain, about 80% of examples are labeled with lower relevance. We experimented with building balanced and unbalanced classifiers for the medical domain; we took a sample from the complete labeled set, so that the class distribution in the sample is about even—i.e., the randomly sampled training subset is balanced so it contains about the same amount of low- and high-relevance events.

Since parts of the labeled data are actually corrected by the user, we obtain *two* parallel sets of events with relevance labels: the "raw" events, as extracted by the system, and corresponding "cleaned" events, i.e., the same events with corrections. The raw set is more noisy, since it contains the errors that were introduced by the system.

The relevance classifiers are built using the cleaned labeled data. For evaluation, we test the classifier performance against both the cleaned and the raw events. We focus on classification performance on the raw events, because ultimately the goal is to build a classifier that can be applied to the extracted event stream, which are not validated or corrected by the end-user. In any case, the IE system must assign the relevance score to each event, before a user examines it, and possibly validates it. Therefore, the "raw" scores in the evaluation give us an indication of what performance we can expect in the real-world setting.[4]

8.5   Evaluation Methodology and Results

The predictive power of our features is evaluated by using three different classifiers: Naive Bayes [13], SVM [14] and BayesNet [4]. We used the implementations from the WEKA toolkit [10], which provides a collection of machine learning algorithms.

Evaluations are done using a ten-fold cross-validation. We evaluated the results using precision, recall, F-measure and accuracy for high/low-relevance classification. It is important to note that when we split the corpus into ten parts. we were careful not to split examples that belong to the same document across the training and test sets. That is, we make sure that for any given document, either *all* events found in that document fall within the training set or they all fall within the test set—to assure that a document never contributes events to both the training and the testing set. This is because there may be a great deal of redundancy within a document, in which case the examples would give away much information about each other if they were split across the training/test boundary. Failing to do so, would produce overly optimistic results.

8.5.1   Business Domain

In the business domain, we use about 213 user-labeled events, in 127 documents. Table 8.3 shows classification performance achieved on discourse, lexical and combined features. Discourse feature construction is as described in Sect. 8.3. We currently utilize roughly 40 discourse-level features.

In the table, we report the system's performance on *all* events in our labeled corpus, as well as only on events that appear *first* within a document (which may contain more than one event). The first-event evaluation is interesting since we can view it as an additional document-level *text-filtering* task, where the relevance of the first event is used to define the relevance of the entire document.

We train two types of binary classifiers: the high-vs.-low classifiers separate between events labeled 4–5 and 0–3. The zero-vs.-rest classifiers separate the zero-relevance (i.e., completely useless) events from the rest. In each case, the F-measure is calculated for predicting the higher-relevance class.

For each classifier, we show the performance using discourse features only, lexical features only, and the combined set of features. The classifiers are trained

---

[4]Note that, on the other hand, it makes less sense to train the classifier on raw data, since it is inherently more noisy, degrades the classifier performance.

Table 8.3 Relevance classification results on business domain: accuracy and F-measure (in parentheses) for discourse features, lexical features, and combined features

High-vs.-low	Business Domain					
	All events			First events only		
	Lexical	Discourse	Combined	Lexical	Discourse	Combined
SVM	72.2 (0.696)	84.6 (0.83)	**85.3 (0.833)**	70.4 (0.738)	81.8 (0.826)	81.4 (0.818)
Naive Bayes	74.3 (0.73)	75.7 (0.753)	82.5 (0.814)	70.3 (0.731)	81.7 (0.823)	82.2 (0.825)
Bayes net	75.3 (0.73)	84.2 (0.823)	84.5 (0.823)	71.6 (0.718)	81.5 (0.822)	**82.8 (0.834)**
Zero-vs.-rest						
SVM	81.0 (0.894)	84.8 (0.916)	82.6 (0.904)	84.0 (0.912)	84.4 (0.914)	84.7 (0.916)
Naive Bayes	84.8 (0.915)	83.0 (0.906)	**85.5 (0.92)**	**89.2 (0.94)**	83.6 (0.91)	86.2 (0.924)
Bayes net	83.0 (0.908)	82.4 (0.903)	81.7 (0.899)	84.2 (0.915)	84.2 (0.915)	84.1 (0.914)

Table 8.4 Initial relevance classification results on Medical domain. Accuracy and F-measure (in parentheses) for discourse features, lexical features, and combined features

High-vs.-low	Medical domain					
	All events			First events only		
	Lexical	Discourse	Combined	Lexical	Discourse	Combined
SVM	82.2 (0.537)	**85.1 (0.618)**	84.2 (0.613)	87.2 (0.625)	88.5 (0.664)	**89.6** (0.71)
Naive Bayes	79.7 (0.64)	80.7 (0.598)	84.6 **(0.702)**	85.8 (0.679)	85.0 (0.639)	89.2 **(0.728)**
Bayes net	80.6 (0.558)	79.1 (0.615)	79.5 (0.64)	82.6 (0.529)	82.0 (0.612)	82.5 (0.619)
Zero-vs.-rest						
SVM	83.9 (0.907)	84.8 (0.913)	**85.9 (0.917)**	80.6 (0.888)	81.6 (0.895)	83.0 (0.897)
Naive Bayes	85.3 (0.915)	84.1 (0.908)	85.7 **(0.918)**	82.7 (0.898)	82.5 (0.895)	**83.8 (0.902)**
Bayes net	82.4 (0.903)	81.7 (0.891)	82.1 (0.893)	78.3 (0.876)	78.8 (0.868)	78.2 (0.864)

with feature selection using information gain. In the table, the bold score indicates the best score achieved for the given column.

8.5.2 Medical Domain

Table 8.4 shows the classification results using the same strategy as in business domain. In most cases, discourse features perform better than lexical features, and combining the discourse and lexical features improves the predictive performance over both discourse and lexical features alone. These classifications were obtained on approximately 900 events, in 530 documents.

8.6 Discussion and Conclusions

As the quantity of information available from different news services increases rapidly, the capability to extract and highlight relevant news items becomes important. For intelligence officers such as business analysts and epidemiologists, it is important that they can limit the amount of time used to monitor extracted facts.

The relevance classifiers form a component of the on-line news monitoring, to predict the relevance of extracted events to the users. In the experiments in the business domain, the discourse features alone perform better than lexical features. In most cases for the business domain, combining discourse and lexical features helps the classifier. The nature of product launch and corporate acquisition news is typically such that most of the information is available in the first few sentences.

In medical domain, combining discourse and lexical features also generally helps classification performance. Information such as disease type, adjectives related to the event and other subtle hints (e.g. female victims are often described through their family relations) are missing from the discourse features, but have an effect on the classifier performance.

In certain knowledge-intensive domains—such as the ones studied here— missing a relevant news item carries a higher cost to the end-user. In our future work, we will also test classification with different precision-recall-ratio, by adjusting the classification threshold, to model the utility of the results to the users with a preference for high- or low-relevance news items.

To summarize the points addressed in this paper:

- We present prediction of relevance in the task of event extraction in the domains of public health and business intelligence, that we believe to be generalizable to different domains.
- We emphasize the importance of the user's perspective when estimating quality, not just the system's performance. *Relevance* to the user is at least as important as (if not more important than) correctness.
- For the present, we assume that users have the same notion of relevance of an event in a given domain. We do not model differences between individual users (as with collaborative filtering), and treat them as a single group with a shared perspective.
- We have presented experiments and an initial evaluation of assignment of relevance scores.
- Our experiments indicate that relevance is a *tractable* measure of quality, at least in the studied domains.

Our on-going work includes refining the classification approaches, especially exploring feature dependencies using Bayesian networks, extending the system to cover multiple languages, and exploring collaborative filtering to address users' and user-groups differing interests. We are currently working on applying our approach to other domains as well.

References

1. ACE: Automatic content extraction. http://www.nist.gov/speech/tests/ace/ (2004)
2. Bagga, A., Biermann, A.W.: Analyzing the complexity of a domain with respect to an information extraction task. In: Proceeding of the 10th International Conference on Research on Computational Linguistics (ROCLING X), Taipei (1997)

3. Bell, A.: The Language of News Media. Language in Society/Blackwell, Oxford (1991)
4. Bouckaert, R.: Bayesian network classifiers in Weka. Technical Report (2004)
5. Culotta, A., McCallum, A.: Confidence estimation for information extraction. In: Proceedings of Human Language Technology Conference and North American Chapter of the Association for Computational Linguistics, Boston (2004)
6. Cvitas, A.: Information extraction in business intelligence systems. In: MIPRO, 2010 Proceedings of the 33rd International Convention, Opatija, May 2010, pp. 1278–1282
7. Freifeld, C., Mandl, K., Reis, B., Brownstein, J.: HealthMap: global infectious disease monitoring through automated classification and visualization of internet media reports. J. Am. Med. Inf. Assoc. **15**(1), 150–157 (2008)
8. Grishman, R., Huttunen, S., Yangarber, R.: Event extraction for infectious disease outbreaks. In: Proceedings of the 2nd Human Language Technology Conference (HLT 2002), San Diego, March 2002
9. Grishman, R., Huttunen, S., Yangarber, R.: Information extraction for enhanced access to disease outbreak reports. J. Biomed. Inf. **35**(4), 236–246 (2003)
10. Hall, M., Frank, E., Holmes, G., Pfahringer, B., Reutemann, P., Witten, I.H.: The WEKA data mining software: an update. SIGKDD Explor. Newsl. **11**(1), 10–18 (2009). http://dx.doi.org/10.1145/1656274.1656278
11. Hirschman, L.: Language understanding evaluations: lessons learned from MUC and ATIS. In: Proceedings of the First International Conference on Language Resources and Evaluation (LREC), Granada, May 1998, pp. 117–122
12. Huttunen, S., Yangarber, R., Grishman, R.: Complexity of event structure in information extraction. In: Proceedings of the 19th International Conference on Computational Linguistics (COLING 2002), Taipei, August 2002
13. John, G.H., Langley, P.: Estimating continuous distributions in bayesian classifiers. In: Eleventh Conference on Uncertainty in Artificial Intelligence, Montreal, pp. 338–345. Morgan Kaufmann, San Mateo (1995)
14. Platt, J.C.: Fast training of support vector machines using sequential minimal optimization. In: Advances in kernel Methods: Support Vector Learning, pp. 185–208. MIT, Cambridge (1999)
15. Saggion, H., Funk, A., Maynard, D., Bontcheva, K.: Ontology-based information extraction for business intelligence. In: Proceedings of the 6th International Semantic Web Conference and 2nd Asian Semantic Web Conference. ISWC'07/ASWC'07, Busan, pp. 843–856. Springer, Berlin/Heidelberg (2007). http://portal.acm.org/citation.cfm?id=1785162.1785225
16. Steinberger, R., Fuart, F., van der Goot, E., Best, C., von Etter, P., Yangarber, R.: Text mining from the web for medical intelligence. In: Perrotta, D., Piskorski, J., Soulié-Fogelman, F., Steinberger, R. (eds.) Mining Massive Data Sets for Security. OIS, Amsterdam (2008)
17. von Etter, P., Huttunen, S., Vihavainen, A., Vuorinen, M., Yangarber, R.: Assessment of utility in Web mining for the domain of public health. In: Proceedings of the NAACL HLT 2010 Second Louhi Workshop on Text and Data Mining of Health Documents. Association for Computational Linguistics, Los Angeles, June 2010, pp. 29–37. http://www.aclweb.org/anthology/W10-1105
18. Yangarber, R., Best, C., von Etter, P., Fuart, F., Horby, D., Steinberger, R.: Combining information about epidemic threats from multiple sources. In: Proceedings of the MMIES Workshop, International Conference on Recent Advances in Natural Language Processing (RANLP 2007), Borovets, September 2007
19. Yangarber, R., Steinberger, R.: Automatic epidemiological surveillance from on-line news in MedISys and PULS. In: Proceedings of IMED-2009: International Meeting on Emerging Diseases and Surveillance, Vienna (2009)

Chapter 9
Open-Domain Multi-Document Summarization via Information Extraction: Challenges and Prospects

Heng Ji, Benoit Favre, Wen-Pin Lin, Dan Gillick, Dilek Hakkani-Tur, and Ralph Grishman

Abstract Information Extraction (IE) and Summarization share the same goal of extracting and presenting the relevant information of a document. While IE was a primary element of early abstractive summarization systems, it's been left out in more recent extractive systems. However, extracting facts, recognizing entities and events should provide useful information to those systems and help resolve semantic ambiguities that they cannot tackle. This paper explores novel approaches to taking advantage of cross-document IE for multi-document summarization. We propose multiple approaches to IE-based summarization and analyze their strengths and weaknesses. One of them, re-ranking the output of a high performing

H. Ji (✉)
Computer Science Department, Queens College and Graduate Center, City University
of New York, New York, NY, USA
e-mail: hengji@cs.qc.cuny.edu

B. Favre
LIF, Aix-Marseille Université, France
e-mail: benoit.favre@lif.univ-mrs.fr

W.-P. Lin
Computer Science Department, Queens College and Graduate Center, City University
of New York, NY, USA

D. Gillick
Computer Science Department, University of California, Berkeley, CA, USA
e-mail: dgillick@berkeley.edu

D. Hakkani-Tur
Speech Labs, Microsoft, Mountain View, CA, USA
e-mail: dilek@ieee.org

R. Grishman
Computer Science Department, New York University, New York, NY, USA
e-mail: grishman@cs.nyu.edu

T. Poibeau et al. (eds.), *Multi-source, Multilingual Information Extraction
and Summarization 11*, Theory and Applications of Natural Language Processing,
DOI 10.1007/978-3-642-28569-1__9, © Springer-Verlag Berlin Heidelberg 2013

summarization system with IE-informed metrics, leads to improvements in both manually-evaluated content quality and readability.

9.1 Introduction

Since about one decade ago Information Extraction (IE) and Automated Text Summarization have been recognized as two tasks sharing the same goal—extract accurate information from unstructured texts according to a user's specific desire, and present the information to the user in a compact form [14, 23]. Summarization aims to formulate this information in natural language sentences, whereas IE aims to convert the information into structured representations (e.g., databases). These two tasks have been studied separately and quite intensively over the past decade. Various corpora have been annotated for each task, a wide range of models and machine learning methods have been applied, and separate official evaluations have been organized. There has clearly been a great deal of progress on the performance of both tasks.

Because a significant percentage of queries in the summarization task involve facts (entities, relations and events), it is beneficial to exploit facts extracted by IE techniques to improve automatic summarization. Some earlier work (e.g., [28, 36]) used IE as defined in the Message Understanding Conferences (MUC) [15] to generate or improve summaries. The IE task has progressed from MUC-style single template extraction to more comprehensive extraction tasks that target more fine-grained types of facts, such as the 18 types of relations and 33 types of events defined in NIST Automatic Content Extraction (ACE2005)[1] and the 42 types of slots defined in the Knowledge Base Population (KBP) track at the Text Analysis Conference (TAC2010) [20]. IE methods have also advanced from single-document IE to cross-document dynamic event chain extraction (e.g., [19]) and attribute extraction in KBP. In addition, recent progress on open-domain IE [1, 2] and on-demand IE [33] can address the portability issue of IE systems and makes IE results more widely applicable. Furthermore, many current IE systems exploit supervised learning techniques, which enable them to produce reliable confidence values (e.g., [18]). Therefore allowing the summarization task to choose using IE results according to confidence values would improve the flexibility of this task. For these reasons we feel the time is now ripe to explore some novel methods to marry these two tasks again and improve the performance of the summarization task.

In this study, we test the following scenarios for combining these two tasks: IE-only based template filling and sentence compression for abstractive summary generation, IE for sentence re-ranking and redundancy removal, and IE-unit based coverage maximization. We start from a more ambitious paradigm which can generate abstractive summaries entirely based on IE results. Given a collection

---

[1]http://www.nist.gov/speech/tests/ace/

of documents for a specific query, we extract facts in both the queries and the documents. We implement two different approaches of utilizing these facts: template-filling and fact stitching based sentence compression. Both approaches obtain poor content and readability/fluency scores because IE still lacks coverage, accuracy and inference. Then we take a more conservative framework. We use a high-performing multi-document extractive summarizer as our baseline, and tightly integrate IE results into its sentence ranking and redundancy removal. Experiments on the NIST Text Analysis Conference (TAC) multi-document summarization task [10] show this integration method can achieve significant improvement on both standard summarization metrics and human judgment. In addition, we also provide extensive analysis on the strengths and weaknesses of these approaches.

9.2 Related Work

Our work re-visits the idea of exploiting IE results to improve multi-document summarization proposed by Radev et al. [28] and White et al. [36]. In [28], IE results such as entities and MUC events were combined with natural language generation techniques in summarization. White et al. [36] improved Radev et al.'s method by summarizing larger input documents based on relevant content selection and sentence extraction. They also formally evaluated the performance of this idea. More recently, Filatova and Hatzivassiloglou [12] considered the contexts involving any pair of names as general 'events' and used them to improve extractive summarization. Vanderwende et al. [34] explored an event-centric approach and generated summaries based on extracting and merging portions of logical forms. Biadsy et al. [4] exploited entity and time facts extracted from IE to improve sentence extraction for biographical summaries. Hachey [17] used generic relations to improve extractive summarization. Compared to these previous methods, we extend the usage of IE from a single template to wider types of relations and events. To the best of our knowledge our approach is the first work to apply KBP slot filling and event coreference resolution techniques to remove summary redundancy.

Recently there has been increasing interest in generating abstractive multi-document summaries based on template filling (e.g., [31]) and sentence compression (e.g., [13, 22]). In this paper, we explore both of these methods entirely based on IE results. Rusu et al. [30] performed entity coreference resolution and generated a semantic graph with subject-verb-object triplets. Then they predicted which triplets should be included in the summary using Support Vector Machines based on diverse features including words, part-of-speech tags, sentence location, named entities, cosine similarity to centroid, pagerank scores and other graph-derived features. Our work is also related to the summarization research that incorporates semantic role labeling (SRL) results (e.g., [24, 25]). Semantic roles cover more event categories than IE, while IE can provide additional annotations such as entity resolution and event resolution which are beneficial to summarization. Furthermore, our approach of selecting informative concepts is similar to defining Summarization Content

Units (SCUs) in the Pyramid Approach [27] because both methods aim to maximize the coverage of logical 'concepts' in summaries.

9.3   Cross-Document IE Annotation

We apply two Englishcross-document IE systems to extract facts from the query and source documents. These IE systems were developed for the NIST Automatic Content Extraction Program (ACE 2005) and the NIST TAC Knowledge Base Population (KBP 2010) Program [20]. ACE2005 defined seven types of entities (persons, geo-political entities, locations, organizations, facilities, vehicles and weapons), 18 types of relations (e.g., *"a town some 50 miles south of Salzburg"* indicates a *"located"* relation.); and 33 distinct types of relatively 'dynamic' events (e.g., *"Barry Diller on Wednesday quit as chief of Vivendi Universal Entertainment."* indicates a *"personnel-start"* event). The KBP Slot Filling task involves learning a pre-defined set of attributes for person and organization entities. KBP 2010 defined 26 slot types for persons and 16 slot types for organizations. For example, *"Ruth D. Masters is the wife of Hyman G. Rickover"* indicates that the *"per:spouse"* of *"Hyman G. Rickover"* is *"Ruth D. Masters"*. Both systems produce reliable confidence values.

9.3.1   ACE IE System

The ACE IE pipeline [16, 18, 19] includes name tagging, nominal mention tagging, entity coreference resolution, time expression extraction and normalization, relation extraction and event extraction. Names are identified and classified using a Hidden Markov Model. Nominals are identified using a Maximum Entropy (MaxEnt)-based chunker and then semantically classified using statistics from the ACE training corpora. Entity coreference resolution, relation extraction and event extraction are also based on MaxEnt models, incorporating diverse lexical, syntactic, semantic and ontological knowledge. At the end an event coreference resolution component is applied to link coreferential events, based on a pair-wise MaxEnt model and a graph-cut clustering model. Then an event tracking component is applied to link relevant events on a time line.

9.3.2   KBP Slot Filling System

In addition, we apply a state-of-the-art slot filling system [8] to identify KBP slots for every person or organization entity which appears in the query and source documents. This system includes a bottom-up pattern matching pipeline and a top-down question answering (QA) pipeline. In pattern matching, we extract and rank patterns based on a distant supervision approach [26] using entity-attribute pairs from Wikipedia Infoboxes and Freebase [5]. Then we apply these patterns to extract attributes for unseen entities. We set a low threshold to include more candidate attribute answers, and then apply several filtering steps to remove wrong answers. The filtering steps include removing answers which have inappropriate entity types or involve inappropriate dependency paths to the entities. We also apply an open domain QA system, OpenEphyra [32] to retrieve more candidate answers. To estimate the relevance of a query and answer pair, we use the Corrected Conditional Probability (CCP) for answer validation. Finally we exploit an effective MaxEnt based supervised re-ranking method to combine the results from these two pipelines. The re-ranking features include confidence values, dependency parsing paths, majority voting values and slot types.

In the slot filling task, each slot is often dependent on other slots. For example, if the age of X is *"2 years old"*, we can infer that there are unlikely any *"employer"* attributes for X. Similarly, we design propagation rules to enhance recall, for example, if both X and Y are children of Z, then we can infer X and Y are siblings. Therefore we develop a reasoning component to approach a real world acceptable answer in which all slot dependencies are satisfied. We use Markov Logic Networks (MLN) [29], a statistical relational learning language, to model these inference rules more declaratively. Markov Logic extends first order logic in that it adds a weight to each first order logic formula, allowing for violation of those formulas with some penalty.

The general architecture of these two IE systems is depicted in Fig. 9.1. Based on the assumption that the documents for a given query in a summarization task are topically related, we apply the extraction methods to each 'super-document' that includes the query and the source documents. As a result we can obtain a rich knowledge base including entities, relations, events, event chains and coreference links.

9.4   Motivation for Using IE for Summarization

Using the combination of fact types in ACE and KBP, we can cover rich information in source documents. For example, among the 92 queries in the NIST TAC multi-document summarization task [10], 28 queries include explicit ACE events and their corresponding source documents include 2,739 event instances. Some queries include specific events. For example, the query *"Provide details of the **attacks** on Egypt's Sinai Peninsula resorts targeting Israeli tourists."* specifies *"attack"* events, one of the ACE event types. Some other queries inquire about general event series, such as *"Describe the views and **activities** of John C. Yoo."* Previous research extensively focused on using entity extraction to improve summarization, so in this section we only present some concrete examples of using relations and events to improve summarization quality.

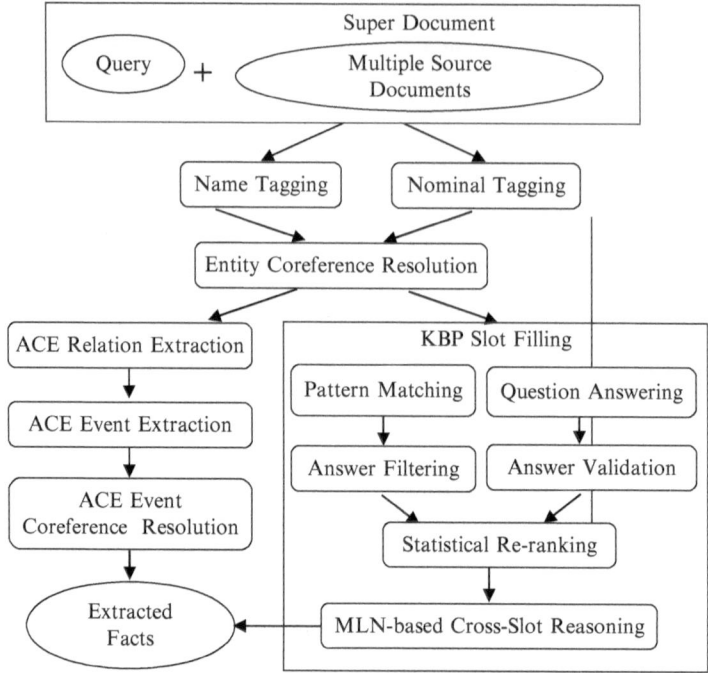

Fig. 9.1 Overview of IE systems

9.4.1 Relations/Events can Promote Relevant Sentences

Traditional sentence ranking methods in summarization used keyword matching, so the knowledge acquisition bottleneck [37] still remains due to sparse data. In order to learn a more robust sentence ranker, the method of matching query and sentences should go beyond the lexical and syntactic level in order to capture semantic structures. Several extractive summarizers (e.g., [3,7,9,35]) used semantic relations in WordNet [11]. This approach has two main limitations: (1) It cannot address broader semantic relatedness; (2) It cannot address the semantic relations between two words with different part-of-speech tags. Semantic relation and event classification can provide a more flexible matching framework. Our basic intuition is that a sentence should receive a high rank if it involves many relations and events specified in the query, regardless of the different word forms indicating such relations and events. For example, for the following query sentences 1, 2 and 3 should receive high ranks according to the gold-standard summary:

[**Query**]
Describe the July 7, 2005 **bombings** in **London, England** and the events, casualties and investigation resulting from the attack.
[**High-Rank Sentence 1**]

> The **attacks**, the deadliest ever carried out on **London** in peacetime, coincided with a summit of the Group of Eight in Gleneagles, Scotland.
> **[High-Rank Sentence 2]**
> A group called Secret al-Qaida Jihad Organization in Europe claimed responsibility, saying the **attacks** were undertaken to avenge **British** involvement in the wars in Afghanistan and Iraq.
> **[High-Rank Sentence 3]**
> The **bomb exploded** in the lead car moments after the train pulled out of the **King's Cross station**, blowing apart the car and making it impossible to reach the dead and injured from the rear.

In sentences 1 and 2, a summarizer without using IE may not able to detect *"attacks"* as the same event as *"bombings"* because they have different lexical forms. However, the IE system extracts *"conflict-attack"* events and labels *"London/British"* as *"place"* arguments in both sentences. This provides us much stronger confidence in increasing the ranks of sentences 1 and 2. Furthermore, even if the event triggers in sentence 3 *"bomb"* can be matched with *"bombings"* in the query, a summarizer may still assign a low weight to sentence 3 if it cannot detect the *"Located"* relation between *"King's Cross station"* and *"London"*. But IE can successfully identify this *"Located"* relation from another sentence in the same document set: *"London – The subway tunnel between **King's Cross** and Russell Square is one of several 'deep tubes' bored through **London**'s bedrock and clay more than a century ago"*.

9.4.2 Relations/Events can Demote Irrelevant Sentences

Relations and events can also filter some irrelevant sentences by deep semantic structure analysis. For example,

> **[Query]**
> Describe the **murders** of **Judge Joan Lefkow**'s husband and mother, and the subsequent investigation. Include details about any evidence, witnesses, suspects and motives.
> **[Low-Rank Sentence 4]**
> They remembered that he would sometimes show up at the federal courthouse to take his wife, U.S. District **Judge Joan Humphrey Lefkow**, to lunch and brought her flowers.

A summarizer without using IE may mistakenly assign a high rank to sentence 4 because it involves a name *"Joan Humphrey Lefkow"*. However, event extraction can be used to decrease the rank of this sentence because it does not include any *"Conflict-attack (murder)"* events as specified in the query.

9.4.3 Event Coreference can Remove Redundancy

What we have presented above is advancing summaries in terms of their content quality. Another central track of summarization research is the issue of readability, especially how to remove redundancy from multiple documents. In this paper we propose a novel approach based on event coreference resolution to reach this goal. Compared to similarity computation methods based on lexical features, our method can detect similar pairs of sentences even if they use completely different expressions. For example, we can fuse the following sentences because they include coreferential *"Conflict-attack"* events, with *"blasts/bombings"* as indicative words and *"London"* as their place arguments:

> **[Sentence 5]**
> It was the deadliest of the four bomb **blasts** in **London** last week.
> **[Sentence 6]**
> The bus explosion was one of four co-ordinated **bombings**, the others on **London** Underground subway trains.

It will be challenging for a summarizer without using IE to detect this redundancy because most words don't overlap in these two sentences.

9.4.4   Integrating IE and Summarization

Methods for incorporating IE into summarization range from using IE alone to using IE to modify the behavior of existing summarization systems. Here we list five general approaches using facts extracted by IE, entities, relations and events, which we call IE units. These methods are schematized in Fig. 9.2.

Template-based generation consists in detecting IE units, such as events and the entities involved in them, and feeding a generation module which uses templates for building summary sentences. Such templates could be: *"Attack: [Attacker] attacked [Place] on [Time-Within]"*, which would result in, for instance, *"[Terrorists] attacked [the police station in South Bagdad] on [Friday]"*. Such an approach is known for being susceptible to the coverage of the template rules.

IE-based compression is similar to template-based generation but it does not use pre-existing templates but rather takes advantage of the support of the IE units in the original documents. The process is less prone to the lack of good template coverage and generates sentences closer to the source but it requires very accurate detection of IE unit spans.

If an existing summarization system is available, one unobtrusive way to take advantage of IE is to pre-filter the input documents. For instance, if the summarization system is extractive, sentences that do not contain any IE elements can be dropped from its input. Various ways of performing such filtering can be devised depending on the type of summarization system.

The next step applies specifically to summarization systems that compute sentence-level similarities (like Maximal Marginal Relevance) by infusing these similarities with IE units. For instance, if two sentences involve coreferential events, then they may be marked as redundant and not used together in the summary. Various graph-based methods, unsupervised and supervised sentence relevance prediction methods can be used to reach this goal.

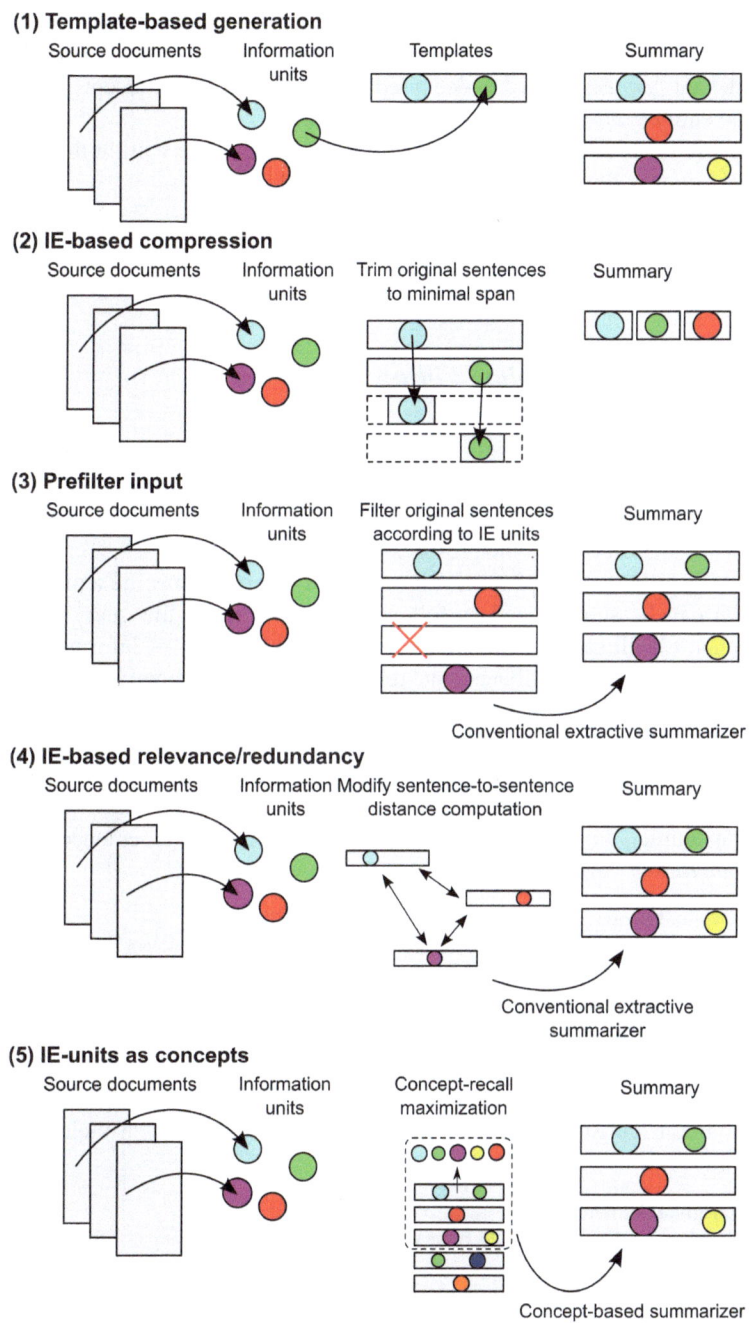

Fig. 9.2 Methods for integrating IE with summarization

Finally, for coverage-based methods that account for *"concepts"* or *"information units"* present in the summary instead of sentence-level scores, IE can be used to infer relevant *"concepts"* and to deem redundant, for instance concepts that refer to the same entity (*"the president"* and *"Mr Obama"*).

All these approaches can be used in conjunction, and we will, in the following sections, review a few possibilities.

9.5 Proposed Approaches

9.5.1 IE-Based Template Filling

Traditional IE by definition is a task of identifying 'facts' from unstructured documents, and filling these facts into structured representations (e.g., templates). Therefore, the most natural way of using IE results for summarization is to fill templates as described in some previous work [28, 31, 36].

For ACE event types and some KBP slots, we construct specific templates. This approach can be summarized as follows: (1) apply IE to the query and source documents; (2) fill sentence templates with available pieces of information (event arguments, entities and attributes), and replace pronouns and nominal mentions with their coreferential proper names; (3) arrange sentences using temporal relations (if there are no explicit temporal arguments, then using the text reporting order) up to the summary length constraint. Some examples of the templates generated based on ACE events are shown in Table 9.1.

For example, we can generate the following summary sentence using the *"Personnel/Elect"* event template *"[Person] was elected in [Place]."*:

[Original Sentence 7]
After a bitter and protracted recount fight in the **Washington** governor's race, **elections** officials announced Wednesday that the Democratic candidate, **Christine O. Gregoire**, was leading her Republican opponent by 10 votes, a minuscule margin but a stunning reversal of the Nov. 2 election results.
[Summary Sentence]
Christine O. Gregoire was elected in **Washington**.

In addition, the summary sentences can be ordered based on their time arguments:

[Original Sentence 8]
Charles announced he would **marry Camilla** in a civil ceremony at **Windsor Castle** on **April 8**.
Charles was previously **married** to Princess **Diana**, who **died** in a car crash in **Paris** in **1997**.
[Summary Sentence]
Charles and **Diana married**. **Diana** died in **Paris** on **1997**. **Charles** and **Camilla** married in **Windsor Castle** on **April 8**.

Table 9.1 IE-based template examples

Event type	Event subType	Templates
Movement	Transport	[Agent] left [Origin]
Personnel	Elect	[Entity] elected [Person] as [Position] of [Place] on [Time-Within]
		[Person] was elected in [Place]
		[Person] was elected in [Place] on [Time-Within]
Personnel	Start-position	[Person] was hired
	End-position	[Person] was fired
Life	Die	[Agent] killed [Victim]
		[Victim] died.
		[Victim] died on [Time-Within]
	Marry	The marriage took place in [Place].
		[Person] and [Person] married
		[Person] and [Person] married in [Place] on [Time-Within]
Conflict	Demonstrate	There was a demonstration in [Place] on [Time-Within]
	Attack	[Attacker] attacked [Place] on [Time-Within]
		[Attacker] attacked [Place]
		[Attacker] attacked [Target]
		[Target] was attacked on [Time-Within]
		The attack occurred on [Time-Within]
		[Attacker] attacked
Transaction	Transfer-ownership	[Buyer] made the purchase on [Time-Within]
Justice	Arrest-jail	[Person] was arrested
	Trial-hearing	[Defendant] was tried

Ordering sentences based on event time arguments can produce summaries with better readability because the text order by itself is a poor predictor of chronological order (only 3% temporal correlation with the true order) [19].

9.5.2   IE-Based Sentence Compression

IE-based template filling can only be fruitful if the template database has a large-enough coverage. In order to build a template-independent approach, we define IE-based sentence compression: instead of filling templates, use source words found in event and relation mentions to build summary sentences. For example, the following summary can be generated by compressing sentence 9:

[**Original Sentence 9**]
Four bombers were among those **killed** in **Thursday**'s **attacks on a double-decker bus**, Sky News television reported **Tuesday**, quoting police sources.
[**Summary Sentence**]
Four bombers were among those **killed** in **Thursday**'s **attacks on a double-decker bus**, Sky News television reported **Tuesday**.

Using 'mentions' (the maximum span that covers the trigger and all arguments of a relation/event instance) to build summary sentences results in more faithful wording with respect to the source. However, the syntactic structure of source sentences is not always favorable to these kinds of extractions, generating verb-less sentences or decontextualized entity references. For example, the following summary was created from multiple sentences:

> **[Original Sentences 10]**
> **Forty-four victims** of the London subway and bus **bombings** remained in hospitals Friday.
> **[Summary Sentence]**
> **Forty-four victims** of the London subway and bus **bombings**.
> **[Original Sentences 11]**
> British police said Friday they were "aware" of an **arrest** in **Egypt** in connection with the investigation into **last week**'s London **bombings**.
> **[Summary Sentences]**
> **Arrest** in **Egypt** in connection with the investigation into **last week**'s London **bombings**.

9.5.3   IE-Based Relevance Estimation

IE provides an effective way of modeling the central information described in the source documents. Even if the IE model describes such information perfectly, it does not tell us what subset of IE units should appear in a summary. As discussed earlier, IE can be integrated to existing summarization systems at the sentence similarity level to characterize relevance or redundancy. For the purpose of this work, we focus on a linear model to re-rank the relevance scores of a baseline summarizer with sentence-level IE scores.

For a given query Q and a collection of source documents D that includes N sentences $(s_1, \ldots, s_N)$, we generate a summary based on an integrated approach as follows.

Each IE component includes a statistical classifier which generates reliable confidence values. For example, for each event mention in D, the baseline Maximum Entropy based classifiers produce three types of confidence values:

- *Conf(trigger,etype)*: The probability of a string *trigger* indicating an event mention with type *etype*.
- *Conf(arg, etype)*: The probability that a mention *arg* is an argument of some particular event type *etype*.
- *Conf(arg, etype, role)*: If *arg* is an argument with event type *etype*, the probability of *arg* having some particular role.

For any sentence s_i in D, we extract the confidence values presented in Table 9.2. We then linearly combine them to form the final IE confidence for si as follows.

Table 9.2 IE confidence values

Confidence	Description
$c_1(s_i,e_j)$	Confidence of si including an entity ej which is coreferential with an entity in Q
$c_2(s_i,r_k)$	Confidence of si including a relation mention rk which shares the same type and arguments with a relation mention in Q
$c_3(s_i,ev_l)$	Confidence of si including an event mention evl which shares the same type and arguments with an event mention in Q
$c_4(s_i,kbp_m)$	Confidence of si including a KBP relation kbpm which shares the same slot type, entity and slot value with a KBP relation in Q

$$c_{ie}(s) = \alpha_1 \times \sum_j c_1(s_i, e_j) + \alpha_2 \times \sum_k c_2(s_i, r_k) + \alpha_3 \times \sum_l c_3(s_i, ev_l)$$
$$+ \alpha_4 \times \sum_m c_4(s_i, kbp_m)$$

The α parameters are optimized using a development set. Assuming the ranking confidence from the baseline summarizer for s_i is $c_{baseline}(s_i)$, we can get the combined confidence of using si as a summary sentence:

$$c_{summary}(s_i) = (1 - \lambda) \times (c_{baseline}(s_i)/\sum_{i=1}^{N} c_{baseline}(s_i)) + \lambda \times (c_{ie}(s_i)/\sum_{p=1}^{N} c_{ie}(s_p))$$

We believe that incorporating these confidence values into a unified re-ranking model can provide a comprehensive representation of the information in the source collection of documents. Based on the combined confidence values, we select the top sentences to form a summary within some certain length constraint specified by the summarization task.

9.5.4   IE-Based Redundancy Removal

While IE extracted entities, relation mentions and event mentions which might help introduce more relevant sentences, it does not prevent identical pieces of information from being represented multiple times in the summary under different wordings. We address redundancy removal by taking advantage of coreference links to drop sentences that do not bring new content.

This approach is implemented by filtering the ranked-list of sentences generated from the baseline summarizer. In particular, we conduct the following greedy search through any sentence pair of $< s_i, s_j >$:

- If all of the entity and event mentions in s_i are coreferential with a subset of the entity and event mentions in s_j, then remove s_i;

- If all of the entity and event mentions in s_i and s_j are coreferential, and s_i is shorter than s_j, then remove s_i.

For example, the following sentences include coreferential *"Personnel/End-Position"* events, so we remove the shorter sentence 13.

[Sentence 12]
Armstrong, who **retired** after his seventh yellow jersey victory **last month**, has always denied ever taking banned substances, and has been on a major defensive since a report by French newspaper L'Equipe last week showed details of doping test results from the Tour de France in 1999.
[Sentence 13]
Armstrong retired from cycling after his record seventh straight Tour victory **last month**.

9.5.5   IE-Unit Coverage Maximization

Recent work in summarization has lead to the emergence of coverage-based models ([13] and references therein). Instead of modeling sentence-level relevance and redundancy, these models assess the value of information units, or *"concepts"*, that appear in input sentences. A summary is created by concatenating sentences according to the concepts they contain, effectively tackling the problem of redundancy in sets of more than two sentences. Finding a selection of sentences in this model corresponds to solving a set-cover problem with a knapsack constraint (the length limit of the summary).

While concepts are mostly embodied by word n-grams, we suggest to cast the problem as finding the set of sentences that cover the most important IE units. We use frequency for characterizing the importance of IE units and perform inference with the following Integer Linear Program (ILP):

$$\text{Maximize} \quad \sum_i frequency(i) \times (u_i)$$

$$\text{Subject to} \quad \sum_j length(j) \times s_j \leq \text{length\_bound}$$

$$s_j = 1 \Rightarrow u_i = 1 \ \forall units \in sentence$$

$$u_i = 1 \Rightarrow \text{at least one } s_j = 1 \forall \text{ sentences that contain } u_i$$

In this ILP, u_i is a binary variable indicating the presence of unit i in the summary; s_j is a binary variable indicating the presence of sentence j in the summary. Details on the formulation can be found in [13]. This model is particularly suited for incorporating IE results because IE extracted facts make particularly good concepts for the model. For instance, selecting the sentences that would cover all events central to the topic would make a very relevant summary. In this approach, we perform cross-document IE to detect entities, relations and events, then associate each of the detected elements to corresponding units and find the set of sentences that maximizes weighted IE-unit coverage.

9.6 Experimental Results

In this section, we describe an experimental framework for evaluating the quality of the proposed approaches.

9.6.1 TAC Summarization Task

The summarization task we are addressing is that of the NIST Text Analysis Conference (TAC) multi-document summarization evaluation [10]. This task involves generating fixed-length summaries from ten newswire documents, each related to a given query including a specific topic. While TAC also includes an update summarization task—additional summaries assuming some prior knowledge—we focus only on the standard task in this paper. For example, given a query *"Judge Joan Lefkow's Family Murdered/Describe the murders of Judge Joan Lefkow's husband and mother, and the subsequent investigation. Include details about any evidence, witnesses, suspects and motives."* and ten related documents, a summarization system is required to generate a summary about specific entities (*"Judge Joan Lefkow"*), relations (*"family"*) and events (*"murder"* and *"investigation"*).

In the TAC campaigns, the quality of system-generated summaries is evaluated through manual and automatic evaluations. In manual evaluation, judges give ratings on a Likert scale (1–5 or 1–10) on content responsiveness (informativeness given the input documents and the user query) and linguistic quality (grammaticality, non-redundancy, clarity of references, global organization. . .). Automatic evaluation compares each automatic summary to a set of expert-written summaries using distances such as word n-gram overlap (ROUGE) [21]. Multiple reference summaries are used to address the fact that there is no single good answer to the summarization problem.

9.6.2 Baseline Summarization System

As a baseline, we apply a top-performing TAC summarization system [13] using the principles of coverage-based summarization which was already described in Sect. 5.5. In this model, a summary is the set of sentences that cover the most relevant concepts in the source document set, where concepts are simply word bigrams weighted by their document frequency. The concepts that include low-frequency words or stop-words are filtered. For sentence-level experiments, the value of a sentence is the sum of the concept values it contains. In addition, a sentence compression component is used to post-process the candidate sentences. The compression step consists of dependency tree trimming using high-confidence semantic role labeling decisions. Non-mandatory temporal and manner arguments are removed and indirect discourse is reformulated in direct form.

9.6.3 Evaluation of IE-Based Template Filling and Sentence Compression Approaches and Sentence-Level Integration

We first evaluated the approaches that do not rely on an existing summarization system, IE-based Template Filling and IE-based Sentence Compression, and a system using sentence-level integration through IE-based Relevance Estimation followed by IE-based Redundancy Removal. In order to perform this evaluation, we randomly selected 31 topics from the TAC 2008 and TAC 2009 summarization tasks as our blind test set. The summaries are evaluated automatically with the ROUGE-2 and ROUGE-SU4 metrics [21]. In order to focus more on evaluating the ordering of sentences and coherence across sentences, we extend the length restriction in the TAC setting from 100 words to 20 sentences. Therefore the results are not directly comparable with those of the official evaluation.

We also asked 16 human subjects to manually evaluate summaries based on the TAC Responsiveness metrics [10] consisting of Content and Readability/Fluency measures. In order to compare different methods extensively, we asked the annotators to give a score in the [14, 19] range (1-Very Poor, 2-Poor, 3-Barely Acceptable, 4-Good, 5-Very Good).

In this evaluation, our baseline is the word-bigram based system described in Sect. 6.2 in which sentences are first valued according to the concepts they contained, and then selected in order of decreasing value. The sentences output by this baseline are then rescored by IE-based Relevance Estimation and pruned using IE-based Redundancy Removal. Parameters of the relevance re-ranking module are estimated on a development set (the documents not used for scoring) in order to maximize ROUGE-2 recall. For processing the test set, we use $\alpha_1 = 1$, $\alpha_2 = 2$, $\alpha_3 = 3$, $\alpha_4 = 1$ (IE components) and $\lambda = 0.7$ (IE weight respective to the baseline). If a query does not include any facts extracted by IE, we use the summaries generated from the baseline summarizer.

9.6.3.1 ROUGE Scores

The ROUGE-2 results of each system are summarized in Table 9.3. The IE-based Template Filling and Sentence Compression methods perform poorly in term of ROUGE score; the integrated approach using relevance estimation and redundancy

Table 9.3 TAC ROUGE-2 scores

Method	Recall	Precision
Baseline with dependency tree trimming-based sentence compression	0.1674	0.1058
IE-based template filling	0.1239	0.0825
IE-based sentence compression	0.1297	0.0901
IE-based relevance estimation and redundancy removal	0.1798	0.1084

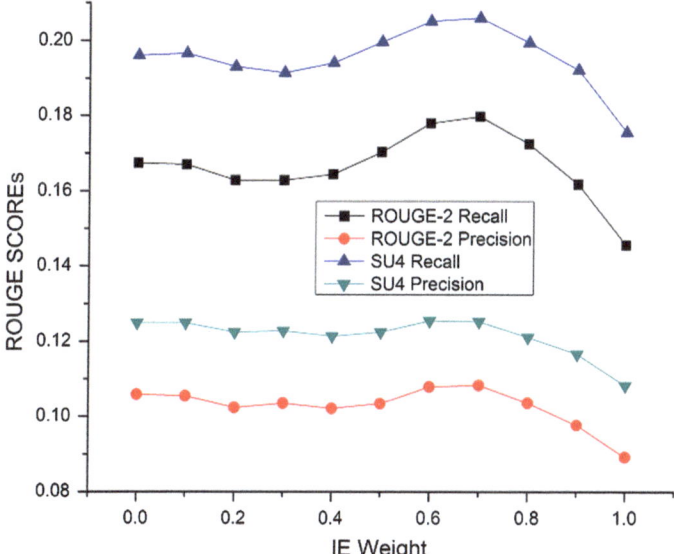

Fig. 9.3 IE-based relevance estimation and redundancy removal results

removal yields improvement over the baseline, suggesting a benefit of taking advantage of IE output.

Figure 9.3 presents the detailed ROUGE-2 and ROUGE-SU4 results of the integrated approach according to the λ parameter. It achieves significant improvement on Recall: when we use $\lambda = 0.7$, which is also the best weight optimized from the development set, our methods achieve 7.38% relative ROUGE-2 gain. In order to check how robust our approach is, we conducted the Wilcoxon Matched-Pairs Signed-Ranks Test on ROUGE scores for these 31 topics. The results show that we can reject the hypothesis that the improvements were random at a 95.7% confidence level. From these curves we can also conclude that using IE results only ($\lambda = 1$) for sentence ranking produced worse ROUGE scores than the baselines.

9.6.3.2 TAC Responsiveness Scores

Table 9.4 presents the average scores across all topics based on manual evaluation using TAC Responsiveness metrics.

Table 9.4 TAC responsiveness comparison

Method	Content	Readability	Responsiveness
Baseline with sentence compression ($\lambda = 0$)	3.11	3.56	3.39
IE-based template filling	2.24	3.08	2.64
IE-based sentence compression	2.73	2.85	2.76
IE-based relevance estimation ($\lambda = 0.7$) and redundancy removal	3.89	3.67	3.61

The IE-only methods obtain lower responsiveness, content and readability scores, which is probably a combination of lack of coverage and bad linguistic quality. But Table 9.3 also shows that the IE-integrated method receives better content scores based on human assessment and even improves over the baseline. This is probably due to document sets involving facts that are ambiguous when using words only for modeling. For example, for the query *"Provide details of the kidnapping of journalist Jill Carroll in Baghdad and the efforts to secure her release"*, the baseline summarizer received a score of *"2"* because of a mismatch between *"kidnapping"* in the query and the *"arrest"* events involving other person and place arguments in the source documents. In contrast, the IE-informed method received a score of *"4"*, because of the effective integration of the *"kidnap"* event detection results when re-ranking sentences. Furthermore, according to the user feedback, our method produced fewer redundant sentences for most topics.

Error analysis shows that for three topics IE had negative impact because of incorrect event categorization for the queries, and missing/spurious extraction errors. For example, for the query *"BTK/Track the efforts to identify the serial killer BTK and bring him to justice."*, IE mistakenly recognized *"Justice"* as the main event type while it missed the more important event type *"Investigation"* which was not defined in the 33 event types. In these and other cases, we could apply salience detection to assign weights to different fact types in the query. Nevertheless, as the above results indicate, the rewards of using the IE information outweigh the risks.

9.6.4 Evaluating the IE-Unit Coverage Maximization Approach

For this set of experiments, we compared the coverage maximization system based on word bigrams as concepts with the same system with IE-units as concepts in addition to the word-based concepts. The behavior of this system can be tuned through a λ parameter which acts as a multiplicative factor of the frequency of IE units when computing their value. Therefore, $\lambda = 0$ implies ignoring IE units and $\lambda = 1$ gives an equal weight to word-based and IE-based concepts.

This time we processed 50 document sets from the non-update part of the TAC'09 evaluation in order to compare with state-of-the-art results. Table 9.5 shows the

Table 9.5 ROUGE-2 Scores
for the Coverage-based
Systems on TAC'09

System	ROUGE-2 Recall
Baseline ($\lambda = 0$)	0.1237
IE only	0.0859
Baseline + IE ($\lambda = 1$)	0.1199

ROUGE-2 results for the IE-only system (where IE-units are used as concepts), the baseline system (Coverage maximization, words only) and the mixed system which takes advantage of both units.

Results show that neither IE-only nor the mixed system outperform the baseline in term of ROUGE. We tried with different values for the mixing parameter but none of them resulted in an improvement. However, we observed that sometimes the IE-infused system can outperform the baseline (for instance, in 10 out of 50 topics for the $\lambda = 1$ system). A careful analysis of the results demonstrated that IE does not cover enough relevant events to gain the advantage over word n-grams, and that most events are only detected once which makes frequency a bad estimator of their relevance.

These results are very interesting because one of the most common criticisms of word-based content-coverage models is that they do not model meaningful real-world entities whereas IE would provide those natural representatives of the actual information. Clearly, more work has to be pursued in this direction in order to empower those already high-performing models.

9.7 Discussion

We have seen in the experiment section that IE is not a good contender for building summarization systems when used alone. However, when it is blended with existing summarization technology, IE can bring some interesting improvements. We will discuss in this section the limitations of the current approaches and devise new avenues for future work. These limitations are annotation coverage, assessment of importance, readability and inference.

9.7.1 Content Coverage

The first problem with targeting IE for open-domain summarization is that even though IE methods apply broadly to many kinds of entities, relations and events, actual systems are developed in the framework of evaluation campaigns and rely on the annotated data produced in these campaigns.

Unfortunately, none of the IE shared tasks (e.g., ACE, KBP) has a large-enough coverage of frequent relations and event types. To demonstrate that, we compare the event types represented in a large corpus with those of the ACE evaluation campaign. We cluster event verbs based on cross-lingual parallel corpora [6],

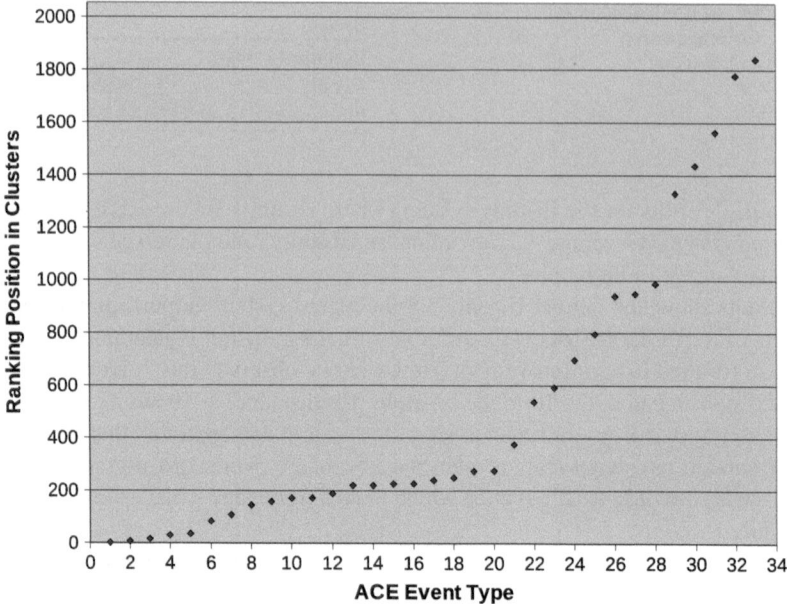

Fig. 9.4 Distribution of ACE event types among verb clusters

creating classes from verbs that often align to the same words in foreign languages, to obtain 2,504 English verb clusters supposed to represent important event types. Then we rank the clusters according to the frequency of their members appearing in the LDC English Gigaword corpus. In parallel, we look at the ranking position of each ACE event type among these clusters, which gives an idea of their coverage of the whole set.

Figure 9.4 presents the ranking results (each point indicates one event type). We can see that although most ACE event types rank very high, a lot of other important event types are missing in the ACE paradigm. The result of this observation is that IE systems trained on ACE-annotated data have a relatively poor coverage of the long tail of event verbs found in open-domain corpora like Gigaword.

It is also interesting to look at IE coverage on the TAC datasets. These datasets contain a manual annotation of the hand-written reference summaries with basic information units (called Summarization Content Units, SCUs). Whereas detecting the information represented by these SCUs is essential for generating relevant summaries, our IE system covers only 60.67% of their words, that is 25,983 out of 42,822 for the TAC'08 data. It seems clear that by being blind to a large part of the relevant information, IE cannot help summarization to its full extent.

According to our IE-unit Coverage Maximization system, the TAC'08 document-set with the fewest information units is D0827E-A. It has 241 information units, and only three of them are in its SCU annotation. The document-set that has the most information units is D0826E-A, with 3,229 units, but also only three of them overlap with the reference SCUs. This partially explains why IE does not seem to improve over the word-based coverage maximization system.

We can also make a comment on the template-based methods. They require very high accuracy in event argument role labeling, otherwise the generated sentences might contain wrong information. For example, in the sentence 8 in Sect. 5.1, our template was not able to capture the tense information (*"would"*) for the trigger word *"marry"* and so produced the wrong tense in the summary *"Charles and Camilla married..."*. For these methods, trading-off coverage for accuracy is not going to be an option.

In order to improve the coverage of current IE systems, we will have to devise new techniques for automatically expanding data from evaluation campaigns with new entity types and new events. Another approach would be to rely on semi-supervised learning to generalize knowledge learned from existing annotation to unseen types. Domain-independent techniques such as On-demand IE [33] can cover wider fact types, but have lower precision. An additional possible solution is to exploit more coarse-grained templates based on generic relation discovery [17] or semantic role labeling [25], but fine-grained annotations, such as event types and argument roles, are beneficial to select sentences that contain relevant facts specified in queries. Clearly, in order to get good-enough coverage for summarization applications, IE researchers will have to close the gap between fine-grained IE-elements for which only little training data is available and broader definitions of semantic roles and semantic frames.

9.7.2 Assessment of Importance

Summarization consists in gathering the most important pieces of information of a text in a limited space. This aspect is typically called *"relevance"* in the summarization literature, originating from the information retrieval literature. Relevance is measured as a combination of frequency and specificity. Frequent phenomena are good topical representatives (content words) unless they are part of the structure of the input (stop-words). The term frequency-inverse document frequency (TF-IDF) framework has been very successful for information retrieval and summarization, but does it directly apply to facts extracted by IE? In particular, in the word-based coverage maximization model, the document frequency of word n-grams is used to estimate their importance for the summary. We tried to apply the same framework for IE-units but it did not yield positive results. How are we supposed to deal with types of units and relational elements like the fact that a person is involved in a particular event? For instance, in the TAC'08 document sets, events occur at most once per document, making frequency a very crude proxy of importance. In addition it is difficult to infer the importance of an event or a relation according to the importance given to the involved participants. Devising good models of importance is critical in order to construct summaries focused on the most important facts if the length constraint does not allow for all facts to be presented. Using frequency from large corpora directly might not be a good solution since the documents being summarized are likely to deal with non-recurring events or events outside of the domain of available corpora. A generalization process could help for instance to infer the importance of an event from similar events, events from the same category or co-occurring events (e.g. *"what is the expected timeline following an earthquake event?"*).

9.7.3   Readability

The IE-only summarization methods like Template Filling and Sentence Compression resulted in poor readability. Both methods suffer from misdetections that crudely insert spurious elements in the templates or extract wrong elements in the compression method. When coupled with extractive summarization, the effect is not as visible due to the fact that full original sentences are used in the final summary. In TAC-involved summarization systems, IE is also often used for replacing pronominal and nominal references with their antecedent. There again, IE errors lead to reduced readability. Improving IE accuracy seems the only remedy to those problems, but confidence scores could be a good source of information for relaxing IE-induced linguistic constraints when IE output is not estimated to be of high-enough quality. We showed an example of such processing in our IE-based relevance estimation method. Another way to improve readability is to take advantage of the progress in text generation, for instance by using language models for rescoring multiple summary hypotheses.

9.7.4   Inference and Generalization

Most of the current IE techniques do not tackle sophisticated inferences; that is, for instance, the fact that if somebody was somewhere on a given date, then this person cannot be at another place at that time. In fact, IE only detects stated information, and misses information that would be implied by world knowledge. None of the systems presented in this paper perform such processing whereas it would be particularly appropriate to assess the importance of IE elements. For instance, if it can be implied that a person participated in an important meeting then the importance of that person can be increased for the final summary. In addition, inferences could provide means of detecting inconsistencies in the input documents. Textual entailment sometimes uses IE as a tool, but both should really be considered as a joint problem.

9.8   Conclusions and Future Work

We investigated the once-popular IE-driven summarization approaches in a wider IE paradigm. We proposed multiple approaches to IE-based summarization and analyzed their strengths and weaknesses. We first concluded that simply relying upon IE for abstractive summarization is not in itself sufficient to produce informative and fluent summaries. Then we demonstrated that a simple re-ranking approach with IE-informed metrics can achieve improvement over a high-performing extractive summarizer. We expect that as IE is further developed to achieve higher performance in broader domains, the summarization task can benefit more from such extended semantic frames. We hope the experiments shown in this paper can draw some interest in both the IE and summarization communities. In the future we will attempt a hybrid approach to combine abstractive and extractive summarization techniques. In addition, we plan to incorporate high-confidence results from open-domain IE to increase the coverage of information units for summarization.

Acknowledgements The first author and the third author were supported by the U.S. Army Research Laboratory under Cooperative Agreement Number W911NF-09-2-0053, the U.S. NSF CAREER Award under Grant IIS-0953149 and PSC-CUNY Research Program. The views and conclusions contained in this document are those of the authors and should not be interpreted as representing the official policies, either expressed or implied, of the Army Research Laboratory or the U.S. Government. The U.S. Government is authorized to reproduce and distribute reprints for Government purposes notwithstanding any copyright notation hereon.

References

1. Banko, M., Cafarella, M.J., Soderland, S., Etzioni, O.: Open information extraction from the web. In: Proceeding of the International Joint Conferences on Artificial Intelligence (IJCAI 2007), Hyderabad (2007)
2. Banko, M., Etzioni, O.: The tradeoffs between open and traditional relation extraction. In: Proceeding of the 46th Annual Meeting of the Association for Computational Linguistics: Human Language Technologies (ACL-HLT 2008), Columbus (2008)
3. Bellare, K., Sarma, A.D., Loiwal, N., Mehta, V., Ramakrishnan, G., Bhattacharyya, P.: Generic text summarization using wordNet. In: Proceeding of the 4th International Conference on Language Resource and Evaluation (LREC2004), Lisbon (2004)
4. Biadsy, F., Hirschberg, J., Filatova, E.: An unsupervised approach to biography production using wikipedia. In: Proceeding of the 46th Annual Meeting of the Association for Computational Linguistics: Human Language Technologies (ACL-HLT 2008), Columbus, pp. 807–815. (2008)
5. Bollacker, K., Cook, R., Tufts, P.: Freebase: a shared database of structured general human knowledge. In: Proceeding of the National Conference on Artificial Intelligence, Vancouver, vol. 2 (2007)
6. Callison-Burch, C.: Syntactic constraints on paraphrases extracted from parallel corpora. In: Proceeding of the Conference on Empirical Methods in Natural Language Processing (EMNLP 2008), Honolulu (2008)
7. Chaves, R.P.: WordNet and automated text summarization. In: Proceeding of the 6th Natural Language Processing Pacific Rim Symposium, Tokyo (2001)
8. Chen, Z., Tamang, S., Lee, A., Li, X., Lin, W., Artiles, J., Snover, M., Passantino, M., Ji, H.: CUNY-BLENDER TAC-KBP2010 entity linking and slot filling system description. In: Proceeding of the Text Analysis Conference (TAC2010), City University of New York (2010)
9. Dang, C., Luo, X., Zhang, H.: Wordnet-based summarization of unstructured document. J. WSEAS Trans. Comput. 7(9), 1467–1472 (2008)

10. Dang, H. T., Owczarzak, K.: Overview of the TAC 2009 summarization track. In: Proceeding of the Text Analysis Conference (TAC 2009), NIST (2009)
11. Fellbaum, C. (ed.). WordNet: An Electronic Lexical Database. MIT, Cambridge (1998)
12. Filatova, E., Hatzivassiloglou, V.: A formal model for information selection in multi-sentence text extraction. In: Proceeding of the 20th International Conference on Computational Linguistics (COLING 2004), Geneva (2004)
13. Gillick, D., Favre, B., Hakkani-Tur, D., Bohnet, B., Liu, Y., Xie, S.: The ICSI/UTD summarization system at TAC 2009. In: Proceeding of the Text Analysis Conference (TAC 2009), NIST (2009)
14. Grishman, R., Hobbs, J., Hovy, E., Sanfilippo, A., Wilks, Y: Cross-lingual information extraction and automated text summarization. Linguist. Comput. **XIV–XV** (1997)
15. Grishman, R., Sundheim, B.: Message understanding conference - 6: a brief history. In: Proceeding of the 16th International Conference on Computational Linguistics (COLING 1996), Copenhagen, pp. 466–471. (1996)
16. Grishman, R., Westbrook, D., Meyers, A.: NYUs Chinese ACE 2005 EDR system description. In: Proceeding of the NIST Automatic Content Extraction Workshop (ACE2005) (2005)
17. Hachey, B.: Multi-document summarisation using generic relation extraction. In: Proceeding of the Conference on Empirical Methods in Natural Language Processing (EMNLP 2009), Singapore, pp. 420–429. (2009)
18. Ji, H., Grishman, R.: Refining event extraction through cross-document inference. In: Proceeding of the 46th Annual Meeting of the Association for Computational Linguistics: Human Language Technologies (ACL-HLT 2008), Columbus (2008)
19. Ji, H., Grishman, R., Chen, Z., Gupta, P.: Cross-document event extraction, ranking and tracking. In: Proceeding of the Recent Advances in Natural Language Processing (RANLP 2009), Borovets, pp. 166–172. (2009)
20. Ji, H., Grishman, R., Dang, H. T., Griffitt, K., Ellis, J.: An overview of the TAC2010 knowledge base population track. In: Proceeding of the Text Analysis Conference (TAC2010), Gaithersburg (2010)
21. Lin, C., Hovy, E.: Automatic evaluation of summaries using N-gram co-occurrence statistics. In: Proceeding of the Human Language Technology Conference of the North American Chapter of the Association for Computational Linguistics (HLT-NAACL 2003), Edmonton, pp. 150–156. (2003)
22. Liu, F., Liu, Y.: From extractive to abstractive meeting summaries: can it be done by sentence compression? In: Proceeding of the Joint Conference of the 47th Annual Meeting of the Association for Computational Linguistics and the 4th International Joint Conference on Natural Language Processing of the Asian Federation of Natural Language Processing (ACL-IJCNLP 2009), Singapore (2009)
23. McKeown, K., Passonneau, R., Elson, D., Nenkova, A., Hirschberg, J.: Do summaries help? A task-based evaluation of multi-document summarization. In: Proceeding of the 28th Annual International ACM SIGIR Conference on Research and Development in Information Retrieval (SIGIR 2005), Salvador (2005)
24. Melli, G., Shi, Z., Wang, Y., Liu, Y., Sarkar, A., Popowich, F.: Description of SQUASH, the SFU question answering summary handler for the DUC-2006 summarization task. In: Proceeding of the Document Understanding Conference (DUC 2006), Brooklyn (2006)
25. Melli, G., Wang, Y., Liu, Y., Kashani, M.M., Shi, Z., Gu, B., Sarkar, A., Popowich, F.: Description of SQUASH, the SFU question answering summary handler for the DUC-2005 summarization task. In: Proceeding of the Document Understanding Conference (DUC2005), Vancouver (2005)
26. Mintz, M., Bills, S., Snow, R., Jurafsky, D.: Distant supervision for relation extraction without labeled data. In: Proceeding of the Joint Conference of the 47th Annual Meeting of the Association for Computational Linguistics and the 4th International Joint Conference on Natural Language Processing of the Asian Federation of Natural Language Processing (ACL-IJCNLP 2009), Singapore (2009)

27. Nenkova, A., Passonneau, R.: Evaluating content selection in summarization: the pyramid method. In: Proceeding of the Human Language Technology Conference-North American Chapter of the Association for Computational Linguistics Annual Meeting (HLT-NAACL 2004), Boston (2004)
28. Radev, D.R., McKeown, K.R.: Generating natural language summaries from multiple on-line sources. Comput. Linguist. **24**(3), 469–500 (1998)
29. Richardson, M., Domingos, P.: Markov logic networks. Mach. Learn. **62**, 107–136 (2006)
30. Rusu, D., Fortuna, B., Grobelink, M., Mladenic, D.: Semantic graphs derived from triplets with application in document summarization. Informatica, **33**, 357–362 (2009)
31. Sauper, C., Barzilay, R.: Automatically generating wikipedia articles: a structure-aware approach. In: Proceeding of the Joint Conference of the 47th Annual Meeting of the Association for Computational Linguistics and the 4th International Joint Conference on Natural Language Processing of the Asian Federation of Natural Language Processing (ACL-IJCNLP 2009), Singapore (2009)
32. Schlaefer, N., Ko, J., Betteridge, J., Sautter, G., Pathak, M., Nyberg, E.: Semantic extensions of the Ephyra QA system for TREC2007. In: Proceeding of the Text Retrieval Conference (TREC2007), Gaithersburg (2007)
33. Sekine, S.: On-demand information extraction. In: Proceeding of the Joint Conference of the International Committee on Computational Linguistics and the Association for Computational Linguistics (COLING-ACL 2006), Sydney (2006)
34. Vanderwende, L., Banko, M., Menezes, A.: Event-centric summary generation. In: Proceeding of the Document Understanding Conference (DUC 2004), Boston (2004)
35. Vikas, O., Meshram, A.K., Meena, G., Gupta, A.: Multiple document summarization using principal component analysis incorporating semantic vector space model. Comput. Linguist. Chin. Lang. Process. **13**(2), 141–156 (2008)
36. White, M., Korelsky, T., Cardie, C., Ng, V., Pierce, D., Wagstaff, K.: Multidocument summarization via information extraction. In: Proceeding of the Human Language Technologies (HLT 2001), Lisbon, pp. 263–269. (2001)
37. Yarowsky, D.: Word-sense disambiguation using statistical models of Rogets categories trained on large corpora. In: Proceeding of the 14th International Conference on Computational Linguistics (COLING 1992), Nantes (1992)

Part IV
Multi-Document Summarization

Chapter 10
Generating Update Summaries: Using an Unsupervized Clustering Algorithm to Cluster Sentences

Aurélien Bossard

Abstract This article presents a summarization system dedicated to update summarization. We first present the method on which this system is based, CBSEAS, and its adaptation to the update summarization task. Generating update summaries is a far more complicated task than generating "standard" summaries. We describe TAC 2009 "Update Task", used to evaluate the system. This international evaluation campaign allowed us to compare our system to other automatic summarization systems. The results obtained were mixed: our system ranked among the first quarter for informational content, but only above average for linguistic quality.

10.1 Introduction

During the past decade, automatic summarization, supported by evaluation campaigns and a large research community, has shown fast and deep improvements. Indeed, the research in this area is led by strong industrial needs: fast processing despite an ever increasing amount of data. As the field of automatic summarization widens, the applications become more varied. We can quote e-mail streams, blogs, scientific articles or newswire articles as application fields, and opinion-oriented, update and differential summaries as different kinds of summaries.

An update summary consists in summarizing some documents about a topic, under the assumption that the user has already read former documents about the same topic. The update summary has to summarize what is new to the user in the documents to be summarized. This automatic summarization task is strongly supported by industrial partners, as it meets real needs.

A. Bossard (✉)
Laboratoire d'Informatique de Paris-Nord (UMR 7030, CNRS et U. Paris 13), 99, av. J.-B. Clément, 93430 Villetaneuse, France
e-mail: aurelien.bossard@lipn.univ-paris13.fr

T. Poibeau et al. (eds.), *Multi-source, Multilingual Information Extraction and Summarization 11*, Theory and Applications of Natural Language Processing, DOI 10.1007/978-3-642-28569-1_10, © Springer-Verlag Berlin Heidelberg 2013

In this paper, we present our research in automatic update summarization: we first developed an automatic standard summarization system, and adapted it to update summarization. This system is based on sentence clustering, which efficiently improves the informational diversity in the summaries.

Automatic generation of update summaries has been proposed as a task of DUC[1] 2007, TAC[2] 2008 and 2009 evaluation campaigns. In order to evaluate our system, we participated in the "Update Task" of TAC 2008 and 2009. This task provided the evaluation of two different kinds of summary: standard request-guided summaries and update request-guided summaries, and allows us to estimate the quality of the summarization system and its adaptation to automatic generation of update summaries.

This article is based on a generic multi-document summarization system, CBSEAS—Clustering-Based Sentence Extractor for Automatic Summarization— [5], which uses unsupervized clustering to detect redundancy. Detecting redundancy is indeed crucial for multi-document summarization: it is hypothesized that the most repeated pieces of information are the most important, and several techniques are based on this assumption. Moreover, detecting the sentences that do not convey the same piece of information can help increase informational diversity. Apart from the standard summarization problem, the main issue in this article is update management: how can already known information be distinguished from new information?

This article shows how the different aspects of these two issues are processed. We also present the reassessment of CBSEAS system during TAC 2008 and 2009 evaluation campaigns. The evaluation of automatic summaries is still an open research area. Several evaluation metrics are used. We discuss how they differ in the method and in the aspects of the summaries they try to evaluate.

We first give a quick overview of the state of the art and then describe the generic summarization system, CBSEAS, and the adaptations made to manage the updating problem. Finally, we describe the evaluation and present the details of the results obtained on TAC 2008 and 2009 Update task.

10.2 State of the Art

In this section, we present an overview of existing methods for automatic summarization and update management.

10.2.1 Overview of Automatic Multi-document Summarization

Automatic summarization has been studied since the beginning of data processing. Automatic summarization process based on advanced linguistic theories has soon

---

[1] Document Understanding Conference: http://www-nlpir.nist.gov/projects/duc/index.html

[2] Text Analysis Conference: http://www.nist.gov/tac

proved to be too ambitious: indeed, it requires paraphrases recognition which involves the need of complex linguistic resources, and text generation processes which are still in an exploratory stage. Recently, some research such as those of Marcu [30] tried to analyze the rhetorical structure prior to the sentence selection process, but this method is still theoretically limited to specific areas.

Therefore, right from the 1950s [28], research in automatic summarization has focused on the excerpt of important sentences—creation of extracts—rather than on the generation of abstracts. The extracted sentences have to constitute a coherent text faithful to the ideas/information expressed in the original documents. Sentence extraction basically in scoring each sentence from the documents to be summarized, and to extract those which get the highest scores in the summary. The number of sentences or words in the summary can be set in advance, but can also be dynamically set using a compression ratio—for example 10% of the original documents.

Edmundson [14] defined textual clues which can be used to determine the importance of a sentence. In particular, he set a list of cue words, such as "hardly" or "impossible", and used term frequency, sentence position and the number of words occuring in the title. These clues are still used by recent systems, like Kupiec's [24].

Other systems focus on term frequency. Luhn [17] led the way of frequency-based sentence extraction systems. He proposed to build a list of important terms. The importance of a term depends on whether its frequency belongs to a predefined range or not. The more words belonging to this list a sentence presents, the more important it is. Radev [32] took advantage of the advances in text statistics by integrating the tf.idf metric to Luhn's method. The list of important terms, that Radev calls "centroid", is composed of the n terms with the highest tf.idf. The sentences are ranked according to their similarity to the centroid. The clue-based and term frequency-based methods are efficient when selecting the sentences reflecting the global content of the documents to be summed up. Such a sentence is called "central". However, these methods are not meant to generate good summaries according to informational diversity. Now, informational diversity is almost as important as centrality when evaluating a summary. Indeed, a summary should contain all the important pieces of information which should not be repeated.

In order to deal with diversity, the MMR [7] method—for Maximum Margin Relevance—selects iteratively the sentence which maximizes the score function shown in (10.1). This score function works with a user query, which means that the generated summaries are guided towards the informational need of a user. The MMR score function takes into account diversity by subtracting the similarity of the evaluated sentence to already selected sentences from its centrality score. This method has been widely used and adapted to different summarization tasks [8, 19, 33, 38].

$$MMR = \operatorname*{argmax}_{P_i \in D\setminus S} \left[\lambda sim_1(P_i, Q) - (1 - \lambda) \operatorname*{argmax}_{P_j \in S} sim_2(P_i, P_j) \right] \qquad (10.1)$$

where Q is the user query, D the sentences to summarize, S the already selected sentences, and λ the novelty factor.

Redundancy in multi-document summarization is a good clue to the importance of a piece of information. MMR does take redundancy into account, but only in order to filter sentences and not as an extraction criterion. Radev [15] took advantage of the recent advances in social network analysis to use information redundancy as the main criterion to judge sentence importance for automatic summarization. He builds a graph of the documents to be summarized, in which the nodes are the sentences, and the edges the sentences similarities. He then uses the *prestige* notion as in the social network area in order to extract the most important sentences. The nodes with the highest *prestige* are those which are strongly linked to other high *prestige* nodes.

The former methods consider the global content of the documents to summarize when evaluating sentence centrality. Yet the documents shall not be considered as a whole, but as multiple clusters of sentences grouped according to their informational content. MultiGen [1] identifies paraphrased sentences, and group them together, creating "themes". This is achieved by aligning sentence syntactic trees. For each theme, MultiGen proceeds a sentence fusion algorithm to generate a new sentence from the sentences of the theme. This method relies on language-dependent resources: syntactic analyzer and semantic knowledge. Thus some authors [3,20,22] tried to approximate the paraphrase identification, using only lexical data. The clusters are ranked upon their overlapping with the global content of the documents, then one sentence is extracted from every cluster.

We also assume that identifying themes in the documents is crucial for automatic summarization. However, in the former methods, errors in clustering may distort the cluster ranking. For this reason, CBSEAS generates summaries based on sentence clustering, and uses a combination of global centrality (the similarity to overall content) and local centrality (the similarity to cluster content) to rank sentences. The main issue of this article lies in the update management more than in the "standard" summary generation.

10.2.2   Overview of Update Summarization Systems

DUC 2007 and TAC 2008 "Upadte task" revealed that generating update summaries is a far more complex task than generating "standard" summaries [10]. We here present different strategies aiming at managing update summarization.

Some authors, such as Galanis and Malakasiotis [17] remove from the update documents all the sentences showing some similarity to a sentence from the initial documents beyond a predefined empirical threshold. Others chose to work on a global similarity between the two sets of documents [21]. The empirical sentence similarity threshold constitutes a bias that the authors have chosen to move to a more global view on the documents. Sentences are iteratively removed from the update document set until the similarity between the update cluster and the initial one falls under a threshold.

The method presented in [6] selects the sentences for the update summary using the MMR method described above. The weight of dissimilarity to the already selected sentences in the scoring function has been increased in order to ensure that the extracted sentences do not carry information that the user has already read.

Another method, exposed in [36], aims at evaluating the novelty of a word. The novelty factor (nf) of a word in a document published at a date t depends on its number of occurrences in the previous documents and its number of occurrences in the later documents, as shown in (10.2).

$$f n(w) = \frac{|n d_t|}{|p d_t| + |D|} \tag{10.2}$$

$nd_t = d : w \in d \wedge t_d\ t$
$pd_t = d : w \in d \wedge t_d \leq t$
$D = d : t_d\ t$

The novelty factor is used to measure sentence novelty. This method has proved to be efficient, be it with TAC 2008 or with TAC 2009. However, we want to evaluate a novel method based on the similarity between sentences, which doesn't need to empirically set a similarity threshold, and can be applied in a case where documents do not have timestamp.

10.3   CBSEAS, a Generic Approach for Automatic Multi-document Summarization

We specifically want to manage the multi-document aspect by considering redundancy as the main issue of multi-document summarization. Indeed, we consider the documents to summarize as made up by groups of sentences carrying the same information. In each of these clusters, one sentence can be considered as central. Extracting this sentence, and not another one, in every cluster can lead to summaries in which the risk of redundancy is minimized. The summaries generated with this method may carry a good informational diversity.

Our approach implements this method. The first step is to cluster sentences, and then select one sentence per cluster.

10.3.1   Sentence Clustering

We here describe the first part of our algorithm: sentence clustering. We wanted a flexible clustering algorithm in which we could easily adapt the clustering criterion. *Fast global k-means* seemed to be appropriate for that purpose. It takes a similarity or dissimilarity matrix as input. The model created after the sentence clustering can be used not only to extract sentences, but also for post-processing such as sentence ordering, which will be the subject of future publications.

10.3.1.1 Pre-processing

The input documents undergo preprocessings before they can be processed by CBSEAS. We here present the different preprocessings used by our system.

Sentence Segmentation

Some authors choose to work on small textual structures rather than on sentences. Therefore, they extract groups of syntactically related words, dividing sentences into clauses [30]. Extracting clauses rather than sentences leads to the problem of clauses identification—though automatic syntactic analysis has recently greatly improved—and clauses independence. Other authors have chosen to extract entire paragraphs in order to improve the linguistic coherence of the summaries. However, it increases the risk of extracting non pertinent sentences, as whole paragraphs are extracted.

We have made the choice to extract whole sentences in order to avoid generating ungrammatical summaries and extracting irrelevant sentences.

POS Tagging

The documents are morphosyntactically analyzed and the textual units tagged by tree-tagger[3] [35]. It enables the differentiation of morphosyntactic types during the sentence similarity computation, which should not be only considered as a simple similarity computation between sets of elements. In fact, natural language processing involves syntactic and semantic knowledge.

Named Entity Tagging

Named entity tagging enables the refinement of the sentence similarity computation. Indeed, the tagging enables to take into account a complex lexical group such as "President George W. Bush" as a unique lexical entity. Such a group should in fact be identified with a single named entity. Therefore it will be considered as a single term, not as four distinct terms. Named entities are tagged using GATE architecture and ANNIE system [9].

10.3.1.2 Sentence Similarity Computation

We put forward the hypothesis that sentence similarity must take into account the type of documents which CBSEAS has to summarize and the kind of summary

---

[3]Tree-tagger webpage: http://www.ims.uni-stuttgart.de/projekte/corplex/TreeTagger/

asked by the user. For example, the characteristics that will determine if two sentences are similar will differ whereas it is for opinion summarization of for market analysis summarization. In the first case, adjectives, adverbs and sentiment verbs are discriminant; in the second case, it will be currencies, amounts, and company names.

$$sim(s_1, s_2) = \frac{\dfrac{\sum\limits_{mt} weight(mt) \times fsim(s_1, s_2)}{fsim(s_1, s_2) + gsim(s_1, s_2)}}{\sum\limits_{mt} weight(mt)} \qquad (10.3)$$

$$fsim(s_1, s_2) = \sum_{n_1 \in s_1} \sum_{n_2 \in s_2} tsim(n_1, n_2) \times \frac{tfidf(n_1) + tfidf(n_2)}{2} \qquad (10.4)$$

$$gsim(s_1, s_2) = card((n_1 \in s_1, n_2 \in s_2) \mid tsim(n_1, n_2) < \delta) \qquad (10.5)$$

where mt are the morphological types, s_1 and s_2 the sentences, tsim the similarity between two terms using WordNet and the JCn similarity measure [23] and δ a similarity threshold.

We want to take this fact into account by using a parameterizable similarity measure, which can easily be adapted to the different tasks to which a summarization system can be confronted. We also want to take into account linguistic relationships between terms—e.g. synonymy, hyperonymy. For that purpose, we used the WordNet database [16] and the JCn similarity measure [23] which is based on the distance between terms in the WordNet taxonomy. Equations 10.3–10.5 present this measure.

Clustering Algorithm

Once the similarity matrix has been computed, CBSEAS automatically clusters the sentences, grouping together semantically close ones. This step is achieved using *fast global k-means* algorithm [25], an incremental variant of the well known *k-means* clustering algorithm [29]. *Fast global k-means* avoids the problematical choice of the *k* initial cluster centers. The incrementality of *fast global k-means* makes it also interesting so as to generate update summaries.

Fast global k-means first creates one cluster that contains all the elements. At each step of the algorithm, a new cluster is created, whose center is the farthest element to its cluster center. Each element is then placed in the cluster whose center is the closest to the element. The clusters centers are then reinitialized. The algorithm stops when it has created the number of clusters asked by the user.

Figure 10.1 presents a cluster generated by CBSEAS. The sentences come from eight different documents. Words shared by at least a half of this cluster's sentences are boldfaced in order to identify the reason of the clustering.

- p1 (doc1) : Chess legend **Bobby Fischer**, who faces prison if **Bobby** returns to the **United States**, can only avoid **deportation** from **Japan** if **Iceland** upgrades its granting of **residency** to full citizenship, a **Japanese** lawmaker said Wednesday.
- p2 (doc6) : **Fischer**, 62, has been detained in **Japan** awaiting **deportation** to the **United States**, where **Bobby** is wanted for violating economic sanctions against the former Yugoslavia by playing a highly publicized chess match there in 1992.
- p3 (doc7) : **Fischer**, 62, has been detained in **Japan** awaiting **deportation** to the **United States**, where **Bobby** is wanted for violating economic sanctions against the former Yugoslavia by playing a highly publicized chess match there in 1992.
- p4 (doc5) : But during Thursday's meeting the committee said it was awaiting more documents from **Japan** to be sure **Bobby** would be let go if given citizenship in **Iceland**.
- p5 (doc8) : Palsson said **Fischer**, 62, would be informed of the vote Tuesday morning in **Japan**, where he has been detained awaiting deportation to the **United States** on a warrant for violating economic sanctions against the former Yugoslavia by playing a chess match there in 1992.
- p6 (doc3) : **Iceland** recently issued **Fischer** a special passport and **residence** permit to help resolve the standoff over **Bobby**'s status.
- p7 (doc2) : **Iceland** had previously sent **Fischer** a passport which showed **residency** status but not full nationality, a document not deemed sufficient to save him from **deportation** to the **United States**.
- p8 (doc10) : **Iceland** had already agreed to issue a **resident**'s permit and special passport to **Fischer**, but that had not proved enough for the **Japanese** authorities.

Source : TAC 2009 "Update Task"

Fig. 10.1 Example of a cluster generated by CBSEAS

10.3.2 Sentence Selection

After the sentences clustering step, CBSEAS extracts one sentence per cluster. Let us remind the reader that the sentences thus extracted should minimize the information redundancy, thus providing a summary with a good informational diversity. The extraction criterion is as important as the clustering. In fact, the way the sentences are extracted influences the centrality of the summary.

Here are the three criteria used by CBSEAS to extract sentences: local centrality, global centrality, sentence length, and their encoding. The final score of a sentence is computed with a weighted sum of these three scores. The weights are set using a genetic algorithm which is detailed in [4].

10.3.2.1 Local Centrality

Local centrality is the relevance of a sentence to the content of its cluster. We want the extracted sentences to reflect best the information of their cluster. The idea behind the local centrality is the following one: the overall sentences of a cluster C express a set of pieces of information which we name I. The most central sentence in C is the one which contains I's most important pieces of information. We work under the assumption that information redundancy is correlated to information importance. The sentence which maximizes the sum of similarities to the other sentences, s_{max}, is the most central sentence. It receives 1 as local centrality score. The other sentences receive a local centrality score equal to their similarity to s_{max}. Figure 10.2 illustrates this measure.

List of the atomic pieces of information (AI) found in the Fig. 10.1 sentences:

AIs	Weight
Fischer faces deportation	5
Ficher chess player	4
Iceland issued Fischer a resident permit	4
Fischer violated economic sanctions against Yugoslavia	3
Iceland issued Fischer a passport	3
Fischer is 62	3
Fischer has been detained in Japan	3
Fischer faces prison	1
Fischer could avoid deportation	1
Iceland special passport can not avoid him deportation	1
Iceland wants to be sure Fischer would be let go if given citizenship	1

s2, s3 and s5 are the sentences carrying most of the AIs:

Sentences	s1	s2	s3	s4	s5	s6	s7	s8
Sum of AI weights	15	21	21	1	21	7	8	7

Sentences similarities:

	s1	s2	s3	s4	s5	s6	s7	s8	sum
s1	1.0	.171	.171	.156	.150	.1	.133	.133	2.014
s2	.171	1.0	1.0	.091	.821	.031	.094	.064	3.272
s3	.171	1.0	1.0	.091	.821	.031	.094	.064	3.272
s4	.156	.091	.091	1.0	.083	.075	.069	.077	1.642
s5	.15	.821	.821	.083	1.0	.027	.108	.055	3.065
s6	.1	.031	.031	.075	.027	1.0	.167	.315	1.746
s7	.133	.094	.094	.069	.108	.167	1.0	.115	1.78
s8	.133	.064	.053	.077	.055	.315	.115	1.0	1.812

Local centrality scores as defined in Sect. 10.3.2.1:

Sentence	s1	s2	s3	s4	s5	s6	s7	s8
Score	.171	1.0	1.0	.091	.821	.031	.094	.064

Fig. 10.2 Illustration of the local centrality concept

10.3.2.2 Global Centrality

The major problem of extracting sentences exclusively upon local centrality measure is that it does not take into account the global content nor a user query (if one). In order to generate precise summaries which meet the informational need of a user, we must add a global centrality score to the local centrality score. For that purpose, we identify two cases:

- The user has a query, so the summary must be connected to that query;
- The user does not have a query, so the summary must be relevant to the overall content of the documents.

In the first case, we use the similarity to the request as global centrality score. This sentence to query similarity is computed the same way as the similarity between sentences, defined in Sect. 10.3.1.2. In the second case, we use the *centroid* score as defined in [32].

10.3.2.3 Sentence Length

The length of the summaries is often limited to a certain amount of words. For this reason, we have chosen to give a length score to every sentence, in order to penalize too short or too long sentences. This score function is defined in (10.6).

$$score_{length} = \frac{1}{e^{(|length(sentence)-length_{required}|)}} \tag{10.6}$$

10.4   Generating Update Summaries

With the development of news feed websites, the update detection and summarization has become an important research problematic. Indeed, users tracking a topic do not want to have to read every newly published article, but only the information they have not already perused. Therefore, update summarization answers industrial needs as regards content access. Moreover, as standard summarization systems now achieve a good quality in terms of informational content, research can now concentrate on more complex tasks, such as those recently proposed by the DUC and TAC evaluation campaigns: opinion summarization, update summarization, or topic-based summarization. In this section, we present our method to manage update summarization.

10.4.1 Intuitions

CBSEAS—Clustering-Based Sentence Extractor for Automatic Summarization— clusters together semantically close sentences. In other terms, it creates different clusters for semantically distant sentences. Our clustering method can also be used to differenciate sentences carrying new pieces of information from sentences carrying already known pieces of information, and so for managing update. In fact, sentences carrying old pieces of information are semantically close to the sentences that a user has already read.

The main weakness of such a method resides in its lack of advanced dedicated linguistic processing. For example, the sentences "After hemming and hawing and bobbing and weaving, the board of directors of Fannie Mae finally jettisoned Franklin D. Raines, the mortgage finance giant's former chief executive, and Timothy Howard, its former chief financial officer." extracted from the

AQUAINT-2 corpus[4] is easily manually identifiable because of the use of "finally". However, if using linguistic tags can be helpful for detecting update sentences, it also limits the method to a unique language, whereas we want to apply our method to other languages later.

What is more, CBSEAS has proven to be efficient at grouping together semantically close sentences and differentiate semantically far ones. In fact, CBSEAS ranked itself at the third place for avoiding redundancy on TAC 2008 "Opinion Summarization Task" [5]. This is another reason for using our clustering method to differentiate update sentences from non-update ones.

10.4.2 Update Algorithm

Before trying to identify update sentences, we need to model the pieces of information that the user requesting the update summary has already read. We can then confront the new documents to this model in order to determine if sentences from these documents carry new pieces of information. Therefore the first step of our algorithm is to cluster the sentences from the documents already read by the user—which we call D_I—into k_I groups, as in Sect. 10.3.1 for the generation of a standard summary.

The model thus computed—M_I—is then used for the second step of our algorithm, which consists in determining if a sentence from the new documents—D_U—is to be grouped with the sentences from D_I, or to create a new cluster which will only contain update sentences. *Fast global k-means* algorithm, slightly modified, can be used to confront elements to a previously established model in order to determine if these elements can be an integral part of the model. We here describe the part of our algorithm dedicated to update.

First, our algorithm computes the similarities between sentences from D_U with the clusters centers of M_I and between all the sentences from D_U. Then it adds the new sentences to M_I, and iterates *fast global k-means* from the k_I iteration with the following constraints:

- The sentences from D_I can not be moved to another cluster; this is done to preserve the M_I model which encodes the old pieces of information. It also avoids to disturb the semantic range of the new clusters that bear novelty.
- The M_I clusters centers can not be recomputed; as the semantic range of a cluster directly depends on its center, this prevents the semantic range of M_I clusters from being changed by the integration of new elements from D_U.

The main problem of this algorithm, which is detailed in Fig. 10.3 lies on the choice of k_I—the number of M_I clusters—and that of k_U—the number of update clusters.

---

[4]The AQUAINT-2 collection is a subset of the LDC English Gigaword Third Edition composed of news articles from different press agencies.

Fig. 10.3 Update detection
algorithm

```
//M_I Clustering
for all p in P_I
do
        cluster(p) ← C_1
for end
for i in 1  k_I
do
        for n in 1  i
        do
                center(C_n) ← argmax( ∑      sim(p_j,p_m))
                             p_j∈C_n p_m∈C_n
        for end
        for all p in P_I
        do
                cluster(p) ← argmax (sim(center(C_m),p)
                            C_m,1<m<u
        for end
        if i < k_I alors
                cluster(argmin(sim(p,center(cluster(p))))) ← C_{i+1}
                        p∈D_I
        end if
for end
//Update detection
for all p in P_U
do
        cluster(p) ← argmax_{C_i,1<i<k_I}(sim(center(C_i),p))
for end
for i in k_I  k_I+k_U
do
        for n in k_I+1  i
        do
                center(C_n) ← argmax_{p_j∈C_n}(∑_{p_m∈C_n} sim(p_j,p_m))
        for end
        for all p dans P_U
        do
                cluster(p) ← argmax (sim(center(C_m),p))
                            C_m,1<m<i
        for end
        if i < k_U alors
                cluster(argmin_{p_m∈D_I}(sim(p_m,center(cluster(p))))) ← C_{i+1}
        end if
for end
```

We empirically decided to set k_U to the desired number of sentences for the update summary, and k_I to $\frac{|S_I|}{S_U} \times k_U$, where S_I and S_U are respectively the sentences from D_I and D_U. For our participation to TAC 2009 "Update Task", we have chosen to let a genetic algorithm decide the values of k_I and k_U. This algorithm, trained on TAC 2008 data, is presented in [4]. None of these solutions are ideal, as they suppose the existence of at least k_U new pieces of information carried by k_U different sentences. Other solutions could be conceived, such as determining whether adding update clusters improves or deteriorates the quality of the clustering. A clustering quality index such as Davies Bouldin's [12] could be used for that purpose.

Once the update clusters have been populated, the update summary is generated by extracting one sentence per update cluster, as in Sect. 10.3.2.

10.5   Evaluation: Participation to TAC "Update Task"

We evaluated our work under the "Update Task" of the TAC 2009 evaluation campaign conducted by the NIST.[5] We here present in details the task, the different metrics used for evaluation, and the results obtained by our update summarization system.

10.5.1   Detailed Description of the Task

The "Update Task" of TAC 2009 evaluation campaign requires to produce two different kinds of summaries: standard summaries and update summaries, both query-based.

The task consists of 44 topics given a short title, a query and two sets of documents: initial and update documents. The systems must generate two summaries for each of the document sets:

- **Standard summary**: a summary which synthesizes the ten initial documents;
- **Update summary**: a summary of the ten update documents "under the assumption that the user has already read the first ten documents. The purpose of the update summary is to inform the reader of new information about the topic" [10].

The summaries length is limited to 100 words, whatever the length of the original documents.

Each document set is composed of ten documents extracted from the AQUAINT-2 corpus. Theses documents are news articles written in English and coming from different sources: AFP, NYT, APW, LTW and Xinhua press agencies.

The queries are complex and are written in English. For example, the query of Topic D0902 is: "Describe the debate over use of emergency contraceptives, also called the morning-after pill, and whether or not it should be available without a prescription." Figure 10.4 presents a topic from TAC 2008 "Update Task". The 2008 "Update Task" was unchanged in 2009.

10.5.2   The Metrics Used for Evaluation

The NIST used three different methods for evaluating the participants runs. The first method is the widely used ROUGE[6] package [26]. The ROUGE metrics are

---

[5]NIST: National Institute of Standards and Technology
[6]ROUGE: Recall-Oriented Understudy for Gisting Evaluation

Topic D0848 : Airbus A380		
Describe developments in the production and launch of the Airbus A380		
Initial documents		
16/01/2005	AFP	The Airbus A380 : from drawing board to runway-ready in a decade
16/01/2005	AFP	A380 'superjumbo' will be profitable from 2008 : Airbus chief
16/01/2005	APW	Airbus prepares to unveil 1380 "superjumbo", world's biggest passenger jet
17/01/2005	LTW	Can Airports Accommodate the Giant Airbus A380 ?
19/01/2005	AFP	After fanfare, Airbus A380 now must prove it can fly
25/01/2005	AFP	Airbus mulls boosting A380 production capacity
10/04/2005	AFP	While US government moans, airports ready for Airbus giant
27/04/2005	AFP	Paris airport neighbors complain about noise from giant Airbus A380
27/04/2005	NYT	Giant Airbus 380 makes maiden flight
04/05/2005	AFP	Airbus A380 takes off on second test flight
Update documents		
01/06/2005	AFP	Airbus announces delay in delivering new superjumbo A380
03/06/2005	AFP	German wing of Airbus denies superjumbo A380 parts delay
05/10/2005	AFP	US aviation officials to study A380 turbulence
15/10/2005	AFP	Airbus says it cannot meet demand for A380 superjumbo
18/10/2005	APW	Second Airbus A380 makes maiden flight
13/11/2005	APW	Airbus executive says company will pay millions in compensation for late A380 deliveries
17/02/2006	APW	Airbus sees no delay to A380 after wing ruptured during test
22/02/2006	AFP	Airbus confident of A380 certification
26/03/2006	APW	33 people injured in evacuation frill for A380 super-jumbo
29/03/2006	APW	Airbus A380 superjumbo passes emergency evacuation test

Fig. 10.4 Example of a topic from the TAC 2009 "Update Task"

based on n-gram co-occurrences between the automatic summary and reference summaries established by experts. Their main advantage resides in their complete automation. However, the evaluation of summaries cannot be limited to n-grams of n-gram sequences comparison. Therefore, the NIST has chosen to use more precise evaluation methods which are not entirely automatic.

The second method used by the NIST is the Pyramid method, described in [31]. The authors define the notion of SCUs—Summarization Content Units—which are pieces of information that appear in the summaries. The Pyramid method first requires human judges to extract a list of SCUs from the reference summaries. The SCUs are then ranked according to their number of occurrences in the reference summaries, and can be seen as forming a Pyramid where the most important pieces of information are at the top and the least important at the base. The SCUs are also extracted from the evaluated summaries, and compared with the pyramid in order to obtain the Pyramid score.

The Pyramid score takes into account the linguistic quality of the summaries: if a sentence is ungrammatical, it doesn't carry any SCU. However, this evaluation method does not take into account the coherence between sentences nor the global structure of a summary. That is the reason why the NIST introduced completely manual evaluation measures. They are described in [11]. The manual measures evaluate both *overall responsiveness* and *readability*. *Overall responsiveness* reflects the degree to which a summary is responding to the informational need expressed in the topic statement, considering its informational content as well as linguistic quality. The readability score reflects the fluency and structure of a summary, regardless of content. It is based on grammaticality, non-redundancy, referential clarity, focus, structure and coherence. *Overall responsiveness* and *readability* were evaluated according to a five-point scale:

- **5**: very good
- **4**: good
- **3**: barely acceptable
- **2**: poor
- **1**: very poor.

It would have been interesting to have a look on the different aspects on which the readability measure was based, as it was the case for TAC 2008 "Opinion Summarization Task" evaluation. It would have allowed us to better understand the cause of our system's results.

10.5.3   Baselines

The NIST provided three baselines for the "Udpate Task" of TAC 2009. The first one (*Baseline 1*) consists in extracting the first sentences of the latest document until the limit of 100 words is reached. This baseline provides a lower bound of what can be achieved with an automatic summarizer.

The second baseline (*Baseline 2*) is built by randomly ordering the sentences of a reference summary. It gives an overview of the sentence ordering impact on linguistic quality and overall responsiveness.

The third baseline (*Baseline 3*) is made up of entire sentences manually extracted from the documents to summarize. The extraction method is detailed in [18]. The idea behind this baseline is to provide an upper bound of what can be achieved with a purely extractive summarizer, both in terms of content and linguistic quality.

10.5.4   Results and Discussion

In this section, we present the results of our update summarization system, compared with the other participants'.

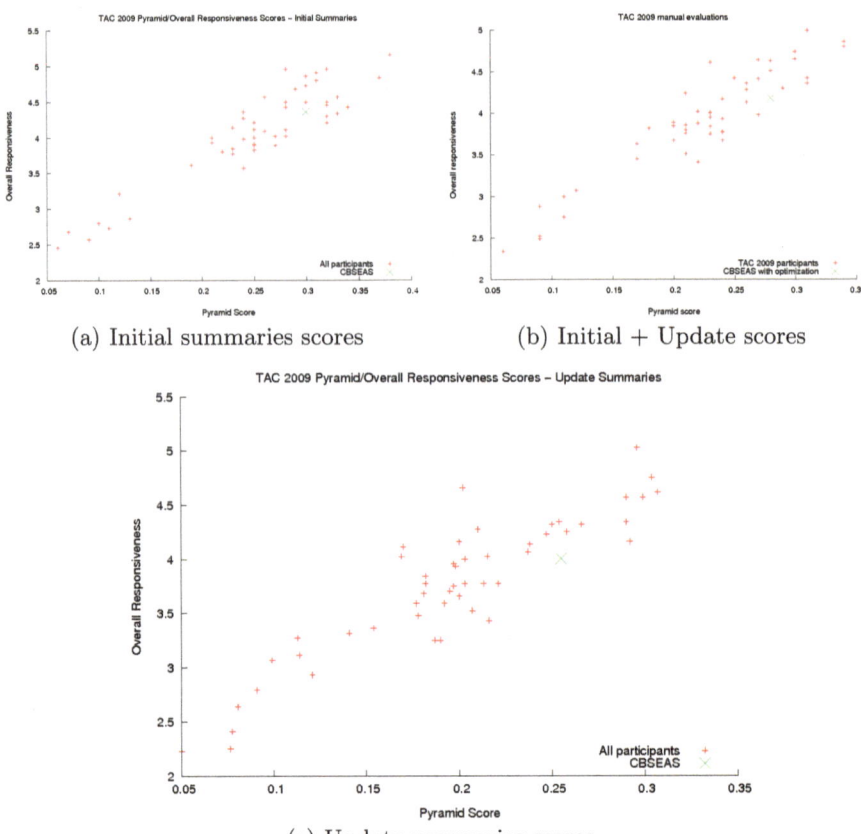

(a) Initial summaries scores   (b) Initial + Update scores

(c) Update summaries scores

Fig. 10.5 TAC 2009 "Update Task" results : Pyramid and overall responsiveness scores

Figure 10.5 shows the pyramid evaluation and the overall responsiveness score for all the participants. Our system performs well, ranking among the first third of participants for initial summaries, and among the ten best for update summaries. The "overall responsiveness" score is not as good. This is due to the poor linguistic quality of the summaries generated by our system. In fact, CBSEAS does not apply any post-processing, such as anaphora resolution or sentence ordering, which could improve the coherence of the summaries.

Figure 10.6 presents the two summaries generated by CBSEAS for the D0911 toopic. One can see that the last sentence is cut. This is due to the 100 words limit set by the NIST. CBSEAS does not automatically remove the entire sentence which exceeds this limit. This also negatively affects the linguistic score.

Figure 10.7 presents the different scores obtained by CBSEAS, and its position in relation to the other systems. One can see that the linguistic quality is the real weakness of our system. However, this display has brought to light the efficiency of our update management strategy. CBSEAS loses "only" 13% of its pyramid score

D0911 Bobby Fischer : initial summary
Describe efforts to secure asylum in Iceland for chess legend Bobby Fischer.
Chess legend Bobby Fischer was on Monday granted citizenship by the parliament of Iceland, a move which could allow him to avoid deportation from Japan to the United States where he is wanted for violating sanctions against the former Yugoslavia.
Chess legend Bobby Fischer, who faces prison if he returns to the United States, can only avoid deportation from Japan if Iceland upgrades its granting of residency to full citizenship. Iceland's parliament voted Monday to grant citizenship to fugitive U.S. chess star Bobby Fischer.
Lawmakers in Iceland are likely to grant citizenship to mercurial chess genius Bobby Fischer, a
Pyramid score: 0.622   Linguistic score: 6

D0911 Bobby Fischer : update summary
Describe efforts to secure asylum in Iceland for chess legend Bobby Fischer.
Iceland said Wednesday it hoped to give detained chess legend Bobby Fischer a passport before the weekend after granting him citizenship in a move that could allow him to avoid a US prison term.
An Icelandic supporter of Bobby Fischer said Tuesday he had paid a registration fee that would allow the American chess legend to settle in Iceland.
Chess legend Bobby Fischer could leave his Japanese detention cell by the weekend, his supporters said Tuesday, a day after Iceland's parliament voted to grant him citizenship.
Japan said Tuesday it may let detained chess legend Bobby Fischer leave for Iceland,
Pyramid score: 0.345   Linguistic score: 6

Fig. 10.6 Example of a pair of summaries generated by CBSEAS

when managing the update summarization, compared with the initial summaries score, while the overall participants show an average 21.5% decrease. Five systems on the 15 which overtake CBSEAS on initial summaries Pyramid scores undertake CBSEAS on update summaries pyramid scores. Generally speaking, update summarization is a difficult task, and one can notice that systems perform better on initial summaries than on update ones.

One could argue that the proposed evaluation is not complete: the initial and update summaries are evaluated independently. The update summary evaluation could have been pushed further, evaluating the presence of SCUs in the automatic summaries that can be found in the initial documents. The redundancy between the update summaries and the already known content is not evaluated for itself.

One interesting result is the linguistic score of the *Baseline 2*, presented in Fig. 10.8: with a score of 5.68, it is overtaken by two automatic summarizers. The best systems equal the *Baseline 3*—which consists of manually extracting sentences—in selecting the most important pieces of information (Pyramid score). However, these systems remain far away from this baseline (cf. Fig. 10.8) in units of linguistic quality and overall responsiveness. These facts prove the impact of sentence ordering on linguistic quality, and so on the user satisfaction—overall responsiveness—towards a summary.

The TAC 2009 evaluation campaign has shown that our system, although competitive for generating summaries, still needs to improve its linguistic quality.

Mean scores of initial and update summaries

	ROUGE-2	ROUGE-SU4	Pyr.	Ling.	Ov. resp.
CBSEAS rank	9/53	10/53	11/53	31/53	18/53
CBSEAS score	0.0919	0.1305	0.28	4.67	4.18
Meilleur score	0.0273	0.0583	0.06	3.40	2.34
Lowest score	0.1127	0.1452	0.34	5.78	4.99
Mean score	0.0786	0.1168	0.226	4.751	3.922

Initial summaries

	ROUGE-2	ROUGE-SU4	Pyr.	Ling.	Ov. resp.
CBSEAS rank	8/53	8/53	15/53	35/53	19/53
CBSEAS score	0.1027	0.1338	0.3	4.91	4.3
Lowest score	0.0282	0.0591	0.06	3.43	2.46
Best score	0.1216	0.1510	0.38	5.93	5.16
Mean score	0.0853	0.1214	0.252	4.762	4.075

Update summaries

	ROUGE-2	ROUGE-SU4	Pyr.	Ling.	Ov. resp.
CBSEAS rank	8/53	15/53	10/53	24/53	22/53
CBSEAS score	0.0811	0.1223	0.26	4.75	3.98
Lowest score	0.0264	0.0576	0.05	3.36	2.23
Best score	0.1039	0.1395	0.31	5.89	5.02
Mean score	0.0719	0.1122	0.198	4.742	3.769

Fig. 10.7 Detailed numeric results of TAC 2009 "Update Task"

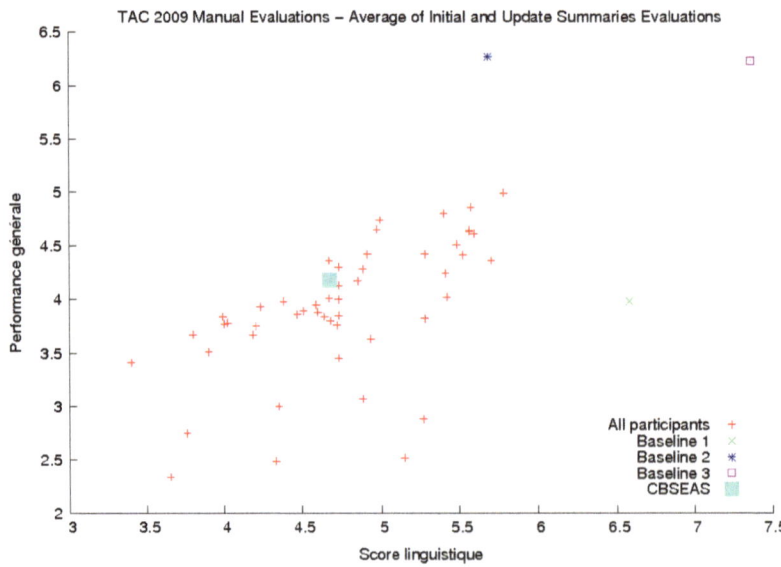

Fig. 10.8 TAC 2009 "Update Task": overall responsiveness and linguistic scores of the three baselines and all participants

Table 10.1 ROUGE scores of CBSEAS with different similarity measures on a french summarization corpus

	sim1	sim2	sim3	Boost
ROUGE1	0.39357	0.37026	0.39296	0.40551
ROUGE2	0.13577	0.11228	0.13667	0.14154
ROUGE-SU4	0.15122	0.13453	0.15024	0.15754

The summaries informational content is good, as it performs 83.3% of *Baseline 3*—which provides an upper bound of what can be achieved with an extractive summarizer—in units of Pyramid score. The update management is also satisfactory, as CBSEAS performs better than the vast majority of participants.

10.5.5 Is There a Place to Improve Informational Content?

A summarization system needs to perform well in two aspects: the pieces of information its summaries convey, and the presentation to the user (linguistic quality, structure...). We here describe how our system can be improved in terms of informational content.

10.5.5.1 Sentence Similarity

Sentence similarity affects the summary in two ways. First, it is the main criterion for sentence clustering. It is also used in order to rank sentences. We showed in recent experiments led with CBSEAS on a French corpus [13] that sentence similarity deeply impacts on summary quality. Table 10.1 presents ROUGE scores of summaries generated with different similarity measures:

- **sim1**: unigram $cosine_{tf.idf}$ similarity;
- **sim2**: bigram $cosine_{tf.idf}$ similarity;
- **sim3**: bigram skip-unit 4 $cosine_{tf.idf}$ similarity (there can be a space of 4 words maximum between the words of a bigram).
- **Boost**: harmonic mean of sim1 and sim2.

Refining sentence similarity computation can result in improvements in summaries quality.

10.5.5.2 Centrality Measure

This article is based on a CBSEAS version that uses two different centrality measures: one for global centrality and one for local centrality. Global centrality is computed using the *centroid* score as defined in [32]. Local centrality is the sum of the similarity to the elements belonging to the same cluster (similar to the *degree* method). Several methods now produce better summaries than these two ones, as shown by the recent TAC evaluation campaigns. If using *centroid* and *degree* methods allowed us to experiment a multi-document update summarization system, we should now focus on specifying new centrality measures to improve the quality of the summaries' informational content.

10.5.5.3 Update Management

CBSEAS ranked itself among the ten best systems for update management during TAC 2009 on Pyramid score. However, the analysis of TAC 2009 results reveals that *novelty factor*, introduced in [36], seems the most efficient way to generate update summaries. The combination of sentence clustering and novelty factor scoring to rank sentences inside update clusters should be investigated. However, such a method would remain specific to time stamped documents. Moreover, unlike the update summarization method exposed in this paper, it could not be used for differential summarization.[7]

10.5.6 How Improve Linguistic Quality?

Linguistic quality has been pointed out as the main weakness of our system. We here expose some ways to improve it.

10.5.6.1 Pronoun and Anaphora Resolution

The presence of unresolved anaphoras in a summary lowers its coherence. Figure 10.9 presents a summary with two unresolved pronouns. The first pronoun refers to Samaan in the summary, as in the original document. However, the second pronoun is misleading the reader: it apparently refers to Samaan, but it actually refers to Archbishop Basile Georges Camoussa in the original documents.

During TAC 2009, a second version of CBSEAS was proposed, which uses GATE and ANNIE system [9] to resolve pronouns. The results are not convincing: as one can see in Table 10.2, this second version (CBSEASb) scored lower than the version which doesn't use pronoun resolution (CBSEASa). We assume that the poor linguistic score is due to the replacement of every personal pronoun in the summaries, even the pronouns that directly follow their referents. However, the quality of the anaphora resolution is also in cause, as ROUGE and pyramid scores suffer from the addition of this module to CBSEAS. Better integration of anaphora resolution modules should be experimented.

---

[7]Differential summarization is a variant of update summarization. Its goal is to summarize the differences between two sets of documents, not what is new in a set compared to an earlier set.

D0808 : Christian minorities in Iraq
Describe the events related to Christian minorities in Iraq and their current status.
On Oct. 16, bomb attacks targeted five churches in Baghdad - which damaged buildings but caused no casualties. **Samaan** said Christians are vulnerable in predominantly Muslim Iraq. **He** said throughout the years, Iraq's Christians had kept to themselves trying to keep out of trouble. ~~He~~ was abducted and taken off in a car. Christians hold only one portfolio in the interim government of Iyad Allawi.

Fig. 10.9 Example of a summary with misleading unresolved pronouns

Table 10.2 Overall results of CBSEASa and CBSEASb on TAC 2009 "Update Task": average of initial and update summaries scores

Run	ROUGE-2	ROUGE-SU4	Pyr.	Ling.	Ov. Resp.
CBSEASa	0.0919	0.1305	0.28	4.67	4.18
CBSEASb	0.0823	0.1230	0.24	4.64	4.11

10.5.6.2 Sentence Ordering

Sentence ordering is crucial for automatic summarization system. If the sentences of a summary are not properly ordered, the reader is able to reorder them. However, the goal of a summary is to give immediate access to information. Several methods aim at ordering sentences:

- Document date and inner document sentence position based method [34];
- Lexical cohesion based method: two sentences will follow one another if they share the same vocabulary [37];
- Time model based methods [27];
- Inner document position and subtopic based method: sentences from the original documents are clustered. The clusters are ordered following the position of their sentences. The sentences are ordered following the clusters order [2].

The method presented in [2] has proved efficient and can easily be adapted to our system, as CBSEAS is based on sentence clustering. The adapation and integration of this method, and its impact on summaries quality should be investigated.

10.6 Conclusion

In this article, we presented CBSEAS, a generic summarization system, and a new algorithm designed to manage update summarization. Our system obtained competitive results on TAC 2009 "Update Task". The compared results of initial and update summaries show that our update management strategy is efficient. However, it could be pushed further by filtering the sentences from the update documents, using such a method as the one described in [36], which is based on words novelty factor. The results also put forward the quality of CBSEAS' sentence selection method. However, our system lacks in linguistic post-processing. This results in user satisfaction being barely better than the average of all TAC 2009 participants. If sentence ordering has been spotted as an important element for automatic summaries coherence and comprehension, the impact of other post-processings such as sentence compression, sentence ordering, or anaphora resolution should be evaluated in future works.

References

1. Barzilay, R., McKeown, K.R.: Sentence fusion for multidocument news summarization. Comput. Linguist. **31**(3), 297–328 (2005)
2. Barzilay, R., Elhadad, N., McKeown, K.: Inferring strategies for sentence ordering in multi-document news summarization. J. Artif. Intell. Res. (JAIR) **17**, 35–55 (2002)
3. Boros, E.P., Kantor, P.B., Neu, D.J.: A clustering based approach to creating multi-document summaries. In: Proceedings of the 4th Annual International ACM SIGIR Conference on Research and Development in Information Retrieval, New Orleans (2001)
4. Bossard, A., Rodrigues, C.: Combining a multi-document update summarization system – cbseas – with a genetic algorithm. Smart Innovation, Systems and Technologies. Springer (2011)
5. Bossard, A., Généreux, M., Poibeau, T.: Description of the lipn systems at tac2008: summarizing information and opinions. In: Notebook Papers and Results of TAC 2008, Gaithersburg (2008)
6. Boudin, F., Torres-Moreno, J.-M., El-Bèze, M.: A scalable MMR approach to sentence scoring for multi-document update summarization. In: Proceedings of the 2008 COLING Conference, Manchester, pp. 21–24 (2008)
7. Carbonell, J., Goldstein, J.: The use of mmr, diversity-based reranking for reordering documents and producing summaries. In: SIGIR '98: Proceedings of the 21st Annual International ACM SIGIR Conference, pp. 335–336. ACM, New York (1998)
8. Chowdary, C.R., Kumar, P.S.: Esum: an efficient system for query-specific multi-document summarization. In: Proceedings of the 31th European Conference on IR Research on Advances in Information Retrieval, ECIR '09, pp. 724–728. Springer, Berlin/Heidelberg (2009)
9. Cunningham, H., Maynard, D., Bontcheva, K., Tablan, V.: GATE: a framework and graphical development environment for robust NLP tools and applications. In: Proceedings of the 40th Anniversary Meeting of the Association for Computational Linguistics. Philadelphia (2002)
10. Dang, H.T., Owczarzak, K.: Overview of the TAC 2008 update summarization task. In: Notebook Papers and Results of TAC 2008, Gaithersburg, pp. 10–23 (2008)
11. Dang, H.T., Owczarzak, K.: Overview of the TAC 2009 update summarization task. In: Notebook Papers and Results of TAC 2009, Gaithersburg (2009)
12. Davies, D.L., Bouldin, D.W.: A cluster separation measure. IEEE Trans. Pattern Anal. Mach. Intell. **PAMI-1**(2), 224–227 (1979)
13. de Loupy, C., Guégan, M., Ayache, C., Seng, S., Moreno, J.M.T.: A french human reference corpus for multi-document summarization and sentence compression. In: Proceedings of LREC'10, Valletta (2010)
14. Edmundson, H.P., Wyllys, R.E.: Automatic abstracting and indexing—survey and recommendations. Commun. ACM **4**(5), 226–234 (1961)
15. Erkan, G., Radev, D.R.: Lexrank: graph-based centrality as salience in text summarization. J. Artif. Intell. Res. (JAIR) **22** (2004)
16. Fellbaum, C.: WordNet: An Electronic Lexical Database. MIT, Cambridge (1998)
17. Galanis, D., Malakasiotis, P.: Aueb at tac 2008. In: Notebook Papers and Results of TAC 2008, Gaithersburg (2008)

18. Genest, P.É., Lapalme, G., Yousfi-Monod, M.: Hextac: the creation of a manual extractive run. In: Notebook Papers and Results of TAC 2009, Gaithersburg (2009)
19. Goldstein, J., Mittal, V., Carbonell, J., Kantrowitz, M.: Multi-document summarization by sentence extraction. In: NAACL-ANLP 2000 Workshop on Automatic Summarization, vol. 4, pp. 40–48. Association for Computational Linguistics, Morristown (2000)
20. He, R., Liu, Y., Qin, B., Liu, T., Li, S.: Hitir's update summary at tac 2008: extractive content selection for language independence. In: Notebook Papers and Results of TAC 2008, Gaithersburg (2008)
21. He, T., Chen, J., Gui, Z., Li, F.: Ccnu at tac 2008: proceeding on using semantic method for automated summarization yield. In: Notebook Papers and Results of TAC 2008, Gaithersburg (2008)
22. Ji, P.: Multi-document summarization based on unsupervised clustering. In: Ng, H., Leong, M.K., Kan, M.Y., Ji, D. (eds.) Information Retrieval Technology. Lecture Notes in Computer Science, vol. 4182, pp. 560–566. Springer Berlin/Heidelberg (2006)
23. Jiang, J.J., Conrath, D.W.: Semantic similarity based on corpus statistics and lexical taxonomy. In: International Conference Research on Computational Linguistics (ROCLING X), Taiwan (1997)
24. Kupiec, J., Pedersen, J., Chen, F.: A trainable document summarizer. In: SIGIR '95: Proceedings of the 18th Annual International ACM SIGIR Conference on Research and Development in Information Retrieval, pp. 68–73. ACM, New York (1995). doi:http://doi.acm.org/10.1145/215206.215333
25. Likas, A., Vlassis, N., , Verbeek, J.: The global k-means clustering algorithm. Pattern Recognit. 36, 451–461 (2001)
26. Lin, C.Y.: Rouge: a package for automatic evaluation of summaries. In: Proceedings of the Workshop on Text Summarization Branches Out (WAS 2004), Barcelona (2004)
27. Lin, Z., Hoang, H.H., Qiu, L., Ye, S., Kan, M.Y.: NUS at TAC 2008: augmenting timestamped graphs with event information and selectively expanding opinion contexts. In: Proceedings of TAC 2008 Workshop on Automatic Summarization, Gaithersburg (2008)
28. Luhn, H.: The automatic creation of literature abstracts. IBM J. 2(2), 159–165 (1958)
29. MacQueen, J.: Some methods for classification and analysis of multivariate observations. In: Le Cam, L.M., Neyman, J. (eds.) Proceedings of the Fifth Berkeley Symposium on Mathematical Statistics and Probability, vol. 1, Statistics. University of California Press, Berkeley (1967)
30. Marcu, D.: Improving summarization through rhetorical parsing tuning (1998)
31. Nenkova, A., Passonneau, R.J., McKeown, K.: The pyramid method: incorporating human content selection variation in summarization evaluation. TSLP 4(2) (2007)
32. Radev, D., Winkel, A., Topper, M.: Multi document centroid-based text summarization. In: Proceedings of the ACL 2002 Demo Session, Philadelphia (2002)
33. Ribeiro, R., de Matos, D.M.: Extractive summarization of broadcast news: comparing strategies for european portuguese. In: Proceedings of the 10th International Conference on Text, Speech and Dialogue, TSD'07, pp. 115–122. Springer, Berlin/Heidelberg (2007)
34. Saggion, H., Gaizauskas, R.: Multi-document summarization by cluster/profile relevance and redundancy removal. In: Proceedings of the Document Understanding Conference 2004. NIST (2004)
35. Schmid, H.: Probabilistic part-of-speech tagging using decision trees. In: Proceedings of the International Conference on New Methods in Language Processing, Manchester (1994)
36. Varma, V., Bysani, P., Bharat, K.R.V., Kovelamudi, S., GSK, S., Kumar, K., Maganti, N.: Iit hyderabad at tac 2009. In: Notebook Papers and Results of TAC 2009, Gaithersburg (2009)
37. Wang, Y.W.: Sentence Ordering for Multi-Document Summarization in Response to Multiple queries. B.Sc, Northeastern University (2002)
38. Wang, B., Liu, B., Sun, C., Wang, X., Li, B.: Adaptive maximum marginal relevance based multi-email summarization. In: Proceedings of the International Conference on Artificial Intelligence and Computational Intelligence, AICI '09, pp. 417–424. Springer, Berlin/Heidelberg (2009)

Chapter 11
Multilingual Statistical News Summarization

Mijail Kabadjov, Josef Steinberger, and Ralf Steinberger

Abstract In this chapter we present a generic approach for summarizing clusters of multilingual news articles such as the ones produced by the Europe Media Monitor (EMM) system. Our approach uses robust statistical techniques as well as multilingual tools for named entity recognition and disambiguation to produce entity-centered summaries. We run experiments with the TAC 2008 and 2009 data sets (English corpora for summarization research), and we obtained very promising results; at TAC 2009 our runs attained top rank for linguistic quality and second best for overall responsiveness. We also run a small-scale evaluation on languages other than English, demonstrating thereby the multilinguality of our approach, but also providing interesting evidence that contradicts the pervasive assumption "if it works for English, it works for any language". Finally, we present an online system currently under development which will eventually incorporate all the elements of the summarization approach discussed hereby and we show sample output summaries in various languages.

11.1 Introduction

Automatic news summarization deals with the problem of producing a succinct informative gist for a set of news articles about the same topic. The aim of the task could be that the target language of the summary be the same as the input articles (standard single-/multi-document summarization) [23] or that the languages of summary/input articles be different (cross-language document summarization) [38]. Moreover, the task of handling several languages, with summary and input articles being in the same language, has been termed as multilingual summarization [16].

M. Kabadjov (✉) · J. Steinberger · R. Steinberger
EC Joint Research Centre, 21027, Ispra (VA), Italy
e-mail: mijail.kabadjov@jrc.ec.europa.eu; josef.steinberger@jrc.ec.europa.eu;
ralf.steinberger@jrc.ec.europa.eu

T. Poibeau et al. (eds.), *Multi-source, Multilingual Information Extraction and Summarization 11*, Theory and Applications of Natural Language Processing, DOI 10.1007/978-3-642-28569-1_11, © Springer-Verlag Berlin Heidelberg 2013

Summarization has been an active area of research for several decades [6, 17], but in particular over the past 17 years. The area was initially focused on single-document summarization [19], a fact reflected by the first US NIST's Document Understanding Conference (DUC) evaluation exercises [26]. Then, over the past decade the emphasis shifted to multi-document summarization exemplified by latter DUCs followed by the Text Analysis Conference (TAC).[1] However, it has been only until recently that interest in multilingual summarization has risen [13, 16] (see the introductory Chap. 1 for the current state of play in the area).

In this chapter we describe a statistical approach to multilingual news summarization based on the Latent Semantic Analysis (LSA) paradigm which builds on and extends work by Steinberger et al. [33]. We have been developing our approach as part of the Europe Media Monitor (EMM) project which we briefly introduce next.

The EMM news gathering engine collects over 100,000 news articles per day in about 50 languages from about 2,500 web news sources [1]. It feeds the news articles into four publicly accessible media monitoring applications (see http://press.jrc.it/overview.html), each with a different focus. The EMM applications cluster all these articles into major news stories, identify entities (locations, persons, organizations) [28,29], send out breaking news alerts to subscribed users, monitor the development of a story over time, link news clusters across languages [34], and more. Currently, however, it does not provide succinct and comprehensive summaries for the news clusters. This is clearly a desirable feature since these clusters may contain hundreds of news articles which would otherwise be impossible to read in full within a short time frame and yet, this is often the case of decision makers within the European Union which make use of the EMM system on a daily or even hourly basis.

Another reason for needing to put a summarization module in place within the EMM system is to bypass the need for having to go to the source articles for further examination of a given news story. Currently, the EMM system serves as a news aggregator and general trend visualization tool, but for more information on a particular story one has to go to the source news article. Providing a high quality summary on site, thus, would substantially improve the usability of the EMM system.

A distinctive characteristic of EMM that significantly adds to the overall complexity of the system is the high multilinguality of the raw data that the system must handle. Thus, one of the main challenges of an effort to develop a news summarizer for EMM is that such a summarizer must be necessarily multilingual, an issue that has not been addressed much in the literature on text summarization.

We intrinsically evaluate our summarization approach on two data sets: for English on TAC 2008 and 2009 data and for English, French, Russian, Arabic, Spanish, German and Czech on parallel data from project-syndicate[2] [37]. This latter evaluation is by no means to the depth of the evaluation carried out by Saggion et al. [30], which is to our knowledge the most thorough multilingual

---

[1]http://www.nist.gov/tac

[2]http://www.project-syndicate.org/

summarization evaluation study done thus far, but it is to a greater breadth than theirs (i.e., their study covers only English, French and Spanish).

We also present NewsGist, an online summarization system currently under development which has been fully integrated into EMM (i.e., summarizes the EMM live news clusters). We show and discuss a few interesting and representative sample summaries in several languages produced by NewsGist.

The remainder of the chapter is organized as follows: in Sect. 11.2 we discuss the related work; in Sect. 11.3 we present the Latent Semantic Analysis model for summarization which we use as our starting point; in Sect. 11.4 we describe the system for multilingual entity disambiguation we used to enhance the LSA representation; in Sect. 11.5 we discuss extensions to LSA with semantic information; in Sect. 11.6 we present and discuss results over the TAC data; in Sect. 11.7 we present and discuss results over the project-syndicate data; in Sect. 11.8 we present an online multilingual summarization system purpose-built for EMM, called NewsGist; and in the last section we conclude the chapter with pointers to future work.

11.2 Related Work

Work on Text Summarization has been quite varied and abundant. A basic processing model for Text Summarization, proposed by Sparck-Jones [11] comprises three main stages: source text interpretation (I) to construct a source representation (e.g., lexical chains, semantic graphs, discourse models), source representation transformation (T) to form a summary representation (e.g., Singular Value Decomposition, SVD), and summary text generation (G). More practically-motivated approaches that use shallow linguistic analysis and only partially cover this processing model, as well as more ambitious ones attempting all three stages using deep semantic analysis have been proposed in the literature.

There are approaches based on shallow linguistic analysis such as word frequencies [17], cue phrases (e.g., "in conclusion", "in summary") and location (e.g., title, section headings) [6]; there are machine learning approaches that combine a number of surface features [14] and/or more elaborate features exploiting discourse structure [36] to train classifiers using specialized corpora formed by pairs of documents and their hand-written summaries; there are also more sophisticated approaches, but still working at the surface level, exploiting cohesive relations like co-reference [4] and lexical cohesion [2] to identify salience or purely lexical approaches trying to identify 'implicit topics' by conflating together words using methods inspired by Latent Semantic Analysis LSA [7]; using yet deeper linguistic analysis, there are approaches purely based on discourse structure (e.g., RST) [20] and others combining discourse structure with surface features [10] or lexical with higher level semantic information such as anaphora [33]; and finally there are knowledge-rich approaches, where the source undergoes a substantial semantic analysis during the process of filling in a predefined template [22] or the source data is available in a more structured way (i.e., events have been identified already) [21].

Over the past decades the area has gradually moved from single-document summarization, to multi-document summarization and recently on to more subtle summarization goals such as genre-focused [35], cross-language summarization [38] and multilingual summarization [13, 16].

Summarization evaluation, a closely related issue, is a particularly challenging problem [25], and even more so when more complexity is introduced by covering more than one language [30]. Saggion et al. [30] carry out a thorough study with three languages: English, French and Spanish and address the very interesting problem of bypassing the need for model summaries. They used the Jensen-Shannon divergence for ranking summarization systems and found that for English on certain tasks the measure correlates strongly with Pyramid and Responsiveness scores, but on other tasks, such as biographical information and entity opinion summarization, the correlations were rather weak. For French and Spanish they found that Jensen-Shannon divergence correlates well with ROUGE [15] scores.

Simultaneously, [37] proposed a method for multilingual summarization evaluation that takes advantage of sentence-level-aligned parallel data such as that of the project-syndicate.[3] Among other things, they found that a language-independent summarization system selected on average only 35% of the same sentences for a given pair of languages for their set of seven languages (English, French, Russian, Arabic, Spanish, German and Czech) covering an ample set of language family branches as are Germanic, Romance, Slavic and Semitic. This latter observation suggests that the pervasive assumption that "if it works for English, it works for any language" may not actually hold in the case of summarization and hence, for the future such claims should be attested empirically.

In this work we address the problem of multilingual news summarization building on and extending previous work by Kabadjov et al. [13] and Steinberger et al. [33] (we discuss this work in the next section). We adopt two evaluation methodologies: firstly, the one proposed at the Text Analysis Conference (TAC) and for that purpose we make use of the TAC data sets; secondly, the evaluation methodology for multilingual summarization evaluation put forward by Turchi et al. [37] with the corresponding data set.

11.3 LSA-Based Summarization

The LSA approach to summarization first builds a term-by-sentence matrix from the source, then applies Singular Value Decomposition (SVD) and finally uses the resulting matrices to identify and extract the most salient sentences. SVD finds the latent (orthogonal) dimensions, which in simple terms correspond to the different topics discussed in the source.

---

[3]http://www.project-syndicate.org/

More formally, it first builds matrix $A = [A_1, A_2, \ldots, A_n]$, where each column $A_j = [a_{1j}, a_{2j}, \ldots, a_{nj}]^T$ represents the weighted term-frequency vector of sentence j in a given set of documents. Each element in this vector is defined as:

$$a_{ij} = L(f_{ij}) \cdot G(f_{ij}), \tag{11.1}$$

where f_{ij} denotes the frequency with which term i occurs in sentence j, $L(f_{ij})$ is the local weight for term i in sentence j, and $G(f_{ij})$ is the global weight for term i in the whole set of documents. The weighting scheme we found to work best is using a binary local weight and an entropy-based global weight:

$$L(f_{ij}) = \begin{cases} 1 & if \quad f_{ij} \geq 1 \\ 0 & otherwise \end{cases} \tag{11.2}$$

$$G(f_{ij}) = 1 - \sum_j \frac{p_{ij} \log(p_{ij})}{\log(n)} \tag{11.3}$$

where $p_{ij} = f_{ij}/g_i$, g_i is the total number of times that term i occurs in the whole set of documents and n is the number of sentences in the set. If there are m terms and n sentences in the set, an $m \times n$ matrix A will be obtained.

After that step Singular Value Decomposition (SVD) is applied to the above matrix as $A = U \Sigma V^T$. U is the term matrix, whose columns are called left singular vectors Σ is a diagonal matrix, whose diagonal elements are non-negative singular values sorted by magnitude in descending order, and finally, V^T is the sentence matrix, whose rows are called right singular vectors.

The last step of the whole process is the selection of prominent sentences. We employ the method put forward in [31], whereby sentences are extracted iteratively based on the sentence vectors' lengths in matrix $B = \Sigma^2 \cdot V^T$ reduced to r dimensions (see below). Elevating Σ to the power of 2 draws on the finding by Ding [5] which shows that the statistical significance of each latent dimension is roughly the square of its singular value. Thus, instead of selecting sentences containing the largest index value in matrix V^T as in the original method [7], the idea is to select sentences with highest combined weight across all topics.

It is highly desirable to perform automatically the dimensionality reduction, that is, to compute the value of r. [33] proposed an intuitive method for achieving that based on the sought summary ratio (ssr) as a percentage which is taken as input to the summarizer. Then, $r = ssr/100 * n$, where n is the total number of sentences and ssr is the sought summary ratio (e.g., 15%).

11.4 Multilingual Entity Recognition and Disambiguation

The purely statistical tool described in the previous section helps to select the most important sentences by covering the most important LSA dimensions. The better performance achieved at the TAC 2009 competition compared to that of TAC

2008 led us to believe that adding named entity (NE) information and semantic generalizations to the LSA input is beneficial. The system used in both years was based on the LSA summarizer, but in addition to the term-by-sentence source matrix used in 2008, we added NE information and MeSH thesaurus term recognition information to the matrix in 2009. The task of named entity recognition and disambiguation was discussed in detail in Chaps. 4, 5, 6, and 8 of this book (for current state of affairs in the area of information extraction see the introductory Chap. 1).

The improved performance may be linked to several reasons. Firstly, by giving sentences containing NEs more weight, the system is presumably likely to favor more self-contained sentences that read well on their own. Indeed, the system was evaluated (by human evaluators) the best system of all 43 submitted runs regarding linguistic quality. Secondly, both to identify the most important information aspects (LSA dimensions) and to avoid redundant information, it is useful to capture para-phrases describing the same information. This includes mentions of the same entities spelled differently, but also generic and specific references for the same concept (i.e., hypernym/hyponym relations such as 'lung disease' and 'tuberculosis').

We thus make use of named entity recognition and disambiguation tools that have so far been developed for 20 languages [29]. The tools recognize names in the text, distinguish whether they are references to persons or locations (e.g., the person 'Paris Hilton' vs. the location 'Paris') and which of several homographic places is being referred to (e.g., there are 15 different locations each called 'Paris', 'Berlin' and 'Roma').

The following subsections provide details on the automatically extracted infor-mation aspects person and organization (Sect. 11.4.1), location (Sect. 11.4.2) and MeSH thesaurus entries (Sect. 11.4.3).

11.4.1   Recognition and Disambiguation of Person and Organization Names

The software tool is based on hand-written, language-independent recognition pat-terns that make reference to language-specific word lists contained in separate files. These files contain long lists of words or regular expressions that are typically found next to names, including titles (e.g., 'Minister'), words indicating nationality (e.g., 'German'), age (e.g., '32-year old'), occupation (e.g., 'playboy'), significant verbal phrases (e.g., 'has declared'), and more. We refer to these patterns generically as trigger words. Name stop words are used to exclude identifying frequent uppercase words (such as 'Monday') as part of the name. These trigger expressions can be concatenated and they can also be partially nested so as to allow the recognition of the names in the strings 'Tony Blair, the 58-year-old former British Prime Minister' and 'the current Secretary-General of the United Nations, Ban Ki-moon'. Long lists of international first names are also used to recognize strings of uppercase words that contain a known first name (e.g., Ivan `Uppercase`). Organization names are recognized through patterns that make use of typical organization name parts such as 'ministry', 'club' or 'union', modifiers such as 'international', words that may link these parts such as 'of the', etc.

While the rules themselves are hand-crafted and the word lists are manually checked, the trigger word lists are partially learned automatically, e.g., by searching for words and phrases occurring next to known named entities identified through lookup, or next to new entities found with an initial seed set of trigger words.

In our multi-annual and multilingual media monitoring activity, we discovered that names can often be spelled in many different ways, not only across scripts (Latin, Arabic, Cyrillic, etc.) and across languages, but also within the same language. These spelling differences are due to varying transliteration standards for different target languages (e.g., German 'Ustinow', French 'Oustinov' and English 'Ustinov'), different character sets (e.g., the Polish name 'Wałęsa' is written as 'Walesa' in languages not using the two diacritics), frequent typos (e.g., 'Condoleeza' or 'Condolezza' instead of the correct form 'Condoleezza'), etc. In order to ground these and more variants to the same entity, a name variant mapping tool was developed that can recognize and map variants. The tool applies a number of transliteration and normalization rules to all newly found names in order to produce an abstract and purely pragmatically motivated canonical form. If this canonical form is the same as that of any known name that has been recognized in the past, and if the calculated distance of the two name variants is within a certain similarity threshold, then the two name variants get automatically stored as known variants of the same named entity. The transliteration rules are standard transliteration rules for the Arabic/Farsi, Cyrillic, Greek and Hindi scripts. The approximately 30 normalization rules—all motivated by empirical observations of frequent partial spelling variations—remove diacritics, single double consonants, replace 'ou' by 'u', replace word-final '-ow' by '-ov', etc. Finally, all vowels are removed to produce this so-called canonical form. The edit distance is calculated on both the lower-cased (and transliterated) form and on the normalized form (before vowel removal). The two distances are combined into an overall similarity value, which is then compared to the cut-off point, above which the two variants are automatically added to the same entity. Names that have been recognized in at least five different news clusters are considered to be known names. All known names will be looked up using a finite state tool so that they are recognized even without trigger words.

Name inflections and other variations are dealt with by pre-generating, for each known name, a number of morphological variants. Language-specific hand-crafted rules generate inflected names by adding or replacing the suffixes of known names. For instance, for the Slavic language Slovene, the forms Blaira, Blairo, Blairu, Blairom and five other inflected forms of the name 'Blair' will be generated automatically. Name variant generation is also carried out for names with the Arabic word particle 'al' and for words with a hyphen, as both are sometimes dropped. Name variant generation is thus not only relevant for highly inflected languages.

Table 11.1 Aggregative
source representation

	unit$_1$	unit$_2$	unit$_3$	...
ngram$_1$				
ngram$_2$	Lexical info			
...				
entity$_1$				
entity$_2$	Entity info			
...				
hypernym$_1$				
hypernym$_2$	Hypernymy info			
...				

These automatically generated name variants are looked up using the same finite state tool, which then returns both the inflected and the normal version [29]. The disambiguated and normalized names identified in the media reports are fed to the term-by-sentence matrix shown in Table 11.1 and are thus considered in the sentence selection process.

11.4.2 Recognition and Disambiguation of Geographical Locations

Historically (e.g., MUC-7 [9]), place name recognition consists of identifying references to locations in text and to disambiguate them from homographic person names or from other homographic words. For instance, there are places called 'Javier' (Spain) and 'Solana' (Philippines), and there are places called 'And' (Iran), 'To' (Ghana) and 'Be' (India). Within the EMM framework, we need to go beyond this MUC task by furthermore disambiguating between various homographic place names in order to identify precise latitude-longitude information and to put a dot on a map. Place name homography is very common: for instance, there are 205 places with the name 'San Antonio'.

For the recognition, disambiguation and grounding of place names, a gazetteer is indispensable. We thus use a multilingual gazetteer that combines entries from many different sources. The geo-tagging task is thus a lookup task to identify place name candidates, followed by disambiguation steps. The person-location homography is solved by giving preference to the person reading as person recognition is more reliable. For the selection of the most likely place name among several homographic place names, we use a number of heuristics: The geo-disambiguation rules in EMM are language-independent, as they make use of features such as the information whether an entity has previously been tagged as being a person or organization name, gazetteer-provided information on the size class of the locations (capital, major city, city, town, etc.), information on the country of origin of the news source, information on other potential geographic references in the same text, as well as kilometric distance between the ambiguous place and other, non-ambiguous places in the same text. Apart from the gazetteer, the only language-specific information is a geo-stop word list, which can be created within a few hours of work. Such a geo-stop word list is used to exclude false positives for locations like 'And' and 'By' (locations in Iran and in Sweden, respectively), which are homographic with high frequency words of a language. Just like with the NER rules described in the previous section, language-specific information is kept out of the disambiguation rules, so that any new language can be plugged into the system if a gazetteer for this language is available [28]. In our experiments we make use of that EMM component to augment the term-by-sentence matrix (see Table 11.1) with disambiguated and normalized location information.

11.4.3 Tagging Texts with MeSH Thesaurus Entries

The Medical Subject Headings thesaurus (MeSH, see http://www.nlm.nih.gov/mesh/) is a medical thesaurus initially created for human indexing of documents. It is hierarchically structured and its approximately 30,000 terms are related via hyponymy (and hypernymy) and synonymy relations. While it was developed for the medical domain, it contains many generic concepts regarding animals, food, chemicals and drugs, social interaction, psychology, and more. We have access to a third-party tool[4] that automatically tags texts with MeSH terms. We use this tool and thesaurus because it is available in various natural languages. The tool typically identifies tens of MeSH terms per news article.

The tool thus produces lists of hierarchically organized MeSH identifiers. Similarly to the person, organization and location identifiers described in the previous sub-sections, these MeSH identifiers are appended to the LSA term-by-sentence matrix (see Table 11.1), but additionally to the MeSH identifier, a weighted list of its hypernym identifiers (i.e., more generic terms) is added. The purpose of this is to allow LSA to make the link between specific and generic terms even if these do not share any surface similarity. For instance, the terms 'voting', 'lobbying', 'liberalism' and 'politics' are closely related in the MeSH tree structure.

11.5 Enriching LSA with Semantic Information

The Vector Space Model is one of the most widely used models in Information Retrieval and in general in Statistical NLP. According to the model, a textual unit can be represented as a vector; terms occurring in the unit define the dimensions of the space and their (weighted) frequencies the coordinates within that space. Then,

---

[4]The multilingual MeSH term recognition software was developed by Health-on-the-Net HON, see http://www.hon.ch/

by varying the operationalization of a textual unit (typically a whole document) and a term (commonly a word token) one ends up with quite different knowledge representations, which may be more or less suitable for different tasks.

Steinberger et al. [33] generalized the notion of term to entail, in addition to words, mentions of discourse entities, thus enhancing the original LSA-based summarization of Gong and Liu [7], who use a word-by-sentence representation.

In modeling local coherence as defined by the Centering theory [8], Barzilay and Lapata [3] put forward simultaneously a similar document representation, called 'entity grid' which was essentially an entity-by-sentence matrix. Though, as opposed to Steinberger et al. [32] they did not attempt to combine it with a purely lexical representation.

Combining several sources of knowledge in the vector space model, among which key words and entities, was independently proposed by Steinberger et al. [34] while working on language-independent news cluster representation for cross-lingual news cluster linking. For that representation, they developed multilingual tools for geo-tagging and entity disambiguation [29].[5]

In this chapter, we propose to build on the aforementioned strands of work and to construct an aggregative semantically-enriched source representation which combines several sources of knowledge. These are lexical information, information about entities and about hypernymy relationships such as those found in IS-A taxonomies (e.g., WordNet, MeSH[6]). The intuition is that words alone are weaker anchors due to morphological variance and synonymy, while the extracted entity information used has been disambiguated and normalized. For instance, person names are frequently written differently across documents [29], but after name variant recognition they can all be represented in the same way.

In our preliminary experiments, we followed the approach adopted by Steinberger et al. [33] and generalized the notion of 'term' to entail not only word tokens, but also references to real entities. Thus, we likewise extended the term-by-sentence matrix A used as input to the LSA with information about disambiguated entities (see Table 11.1). However, as opposed to [33] we did not use a coreference resolver.[7] Instead, we used a more general multilingual named entity disambiguator and geo-tagger (cf. previous section).

Augmenting the initial matrix with information about disambiguated entities naturally provides not only stronger inter-sentential cohesion (i.e., the LSA clusters sentences from different documents that make reference to the same entities),

---

[5]The use of the multilingual tools in higher-level applications can be seen at http://emm. newsexplorer.eu/

[6]The Medical Subject Headings (MeSH) thesaurus is prepared by the US National Library of Medicine for indexing, cataloging, and searching for biomedical and health-related information and documents. Although, it was initially meant for biomedical and health-related documents, since it represent a large IS-A taxonomy it can be used in more general tasks (http://www.nlm.nih. gov/mesh/meshhome.html)

[7]Steinberger et al. [33] worked on monolingual single-document summarization.

but also provides multilingual capabilities inherited by the multilingual entity disambiguation. Thus, this approach to summarization is not only multi-document, but also multilingual.[8]

11.6 Experiments with TAC Data (and Participation Therein)

11.6.1 TAC 2008

Using a standard English corpus for Summarization research developed by the US National Institute for Standards and Technology (NIST) for the 2008 Text Analysis Conference (hereafter TAC-08), we obtained promising, though, not statistically significant improvements over a lexical-only baseline ranked in the top 15–24% across all evaluation metrics at the 2008 TAC competition. For our preliminary experiments we used the popular ROUGE metric to evaluate the performance. The results are presented in Tables 11.2–11.4.

On the standard multi-document summarization task (see Table 11.2), we include the official scores at TAC-08 of a lexical-only summarizer that we used as a baseline for our experiments (cf. first row of the table) as well as an improved version

Table 11.2 Multi-document Summarization results

Approach	R_1	R_2	R_{SU4}
lexical only TAC-08	0.355	0.088	0.123
lexical only	0.359	0.087	0.125
lexical + entities	0.367	0.093	0.13
Best TAC-08 system		0.111	0.142

Table 11.3 Update summarization results

Approach	R_1	R_2	R_{SU4}
TAC-08	0.348	0.081	0.12
lexical only	0.363	0.091	0.13
lexical + entities	0.364	0.92	0.132
Best TAC-08 system		0.101	0.136

Table 11.4 Overall performance results

Approach	R_1	R_2	R_{SU4}
TAC-08	0.352	0.084	0.122
lexical only	0.361	0.089	0.128
lexical + entities	0.366	0.093	0.131

---

[8]The multilingual named entity disambiguator and geo-tagger developed at the JRC have already been used for cross-lingual linking of multilingual news clusters produced by the EMM system [34].

of it referred to as 'lexical only' (cf. second row). The third row of Table 11.2 corresponds to the results obtained by the LSA extension proposed in Sect. 11.5. The last row shows the results obtained by the best system at TAC-08 and is included only for reference purposes.

From Table 11.2 we can see that the performance of the 'lexical+entities' version of the system is higher than the 'lexical only' version, our baseline, but we note the improvement is not statistically significant after running a t test.[9]

It is worth noting that the EMM entity disambiguation module we used in this experiment has been optimized for precision, since in the EMM's context the vast amounts of data (i.e., over 80 K processed articles per day) make up for the compromise on recall. However, in the TAC-08 context there is substantial room for improvement of the recall of entity mentions within a document by bringing in an intra-document coreference resolution system, such as GuiTAR [12], and aggregating the output with that of the entity disambiguator. In the light of this we believe the attained results are promising.

On the update summarization task (see Table 11.3),[10] we present the same evaluation dimensions as for the standard summarization task. The picture is similar to the previous case, though, the improvement this time is clearly insignificant. We believe this is possibly due to this second task being much more specific than the standard summarization task and hence needing either more elaborate fine-tuning of the model or a much bigger corpus for a larger-scale evaluation.

Table 11.4 presents a combined picture of both tasks.

11.6.2   TAC 2009

For the TAC 2009 submissions, we had access to the multilingual tools for geo-tagging and entity disambiguation developed by R. Steinberger et al., and also to the MeSH taxonomy. In what follows, we briefly discuss the results of our participation in TAC 2009 summarization task.

The first part of the summarization task was to write a short (100 words) summary of a set of newswire articles (an update summary), under the assumption that the user has already read a given set of earlier articles (used for the creation of an initial summary).

Fifty-two automatically created sets of summaries were submitted by 27 participating groups and compared against three baselines. Baseline 1 (*run1*) returns all the

---

[9]Note that the statistical test we used to attest significance was ran against the improved version of the lexical-only summarizer and not the official TAC-08 scores, since we considered it was the fairer comparison.

[10]The purpose of the update summarization task is to produce a summary of only the novel information contained in a newer set of news articles with respect to an older set, both covering the same news story.

Table 11.5 TAC'09 results of the update summarization task—initial summaries

Run ID	Overall responsiveness	Linguistic quality	Pyramid score	ROUGE-2	ROUGE-SU4	BE
2	6.364	5.477	0.646	0.331	0.344	0.248
3	6.341	7.477	0.358	0.106	0.138	0.053
40	**5.159**	5.636	**0.383**	**0.121**	**0.151**	**0.064**
24	4.955	5.682	0.316	0.098	0.133	0.056
19	4.955	**5.932**	0.277	0.094	0.129	0.052
11	4.795	5.773	0.314	0.096	0.130	0.054
1	3.636	6.705	0.175	0.063	0.099	0.029

Table 11.6 TAC'09 results of the update summarization task—update summaries

Run ID	Overall responsiveness	Linguistic quality	Pyramid score	ROUGE-2	ROUGE-SU4	BE
2	6.182	5.886	0.690	0.319	0.337	0.250
3	6.114	7.250	0.329	0.097	0.136	0.057
40	4.568	5.500	0.290	**0.104**	**0.140**	0.062
24	**5.023**	**5.886**	**0.296**	0.096	0.135	**0.064**
19	4.318	5.182	0.266	0.077	0.116	0.045
11	4.227	5.182	0.247	0.083	0.121	0.047
1	4.318	6.455	0.160	0.051	0.091	0.024

leading sentences (up to 100 words) in the most recent document. This baseline provides a lower bound on what can be achieved with a simple fully automatic extractive summarizer. Baseline 2 (*run2*) returns a copy of one of the model summaries for the docset, but with the sentences randomly ordered. It provides a way of testing the effect of poor linguistic quality on the overall responsiveness of an otherwise good abstractive summary. Baseline 3 (*run3*) returns a summary consisting of sentences that have been manually selected from the docset. It provides an approximate upper bound on what can be achieved with a purely extractive summarizer.

The NIST assessors assigned a content score using Columbia University's Pyramid method [24], a readability score and an overall responsiveness score (combining both of the previous ones) to each of the automatic and human summaries. The score is an integer between 1 (very poor) and 10 (very good). Standard automatic scores ROUGE-2, ROUGE-SU4, and BE were calculated as well.

We submitted two runs. For our first priority run (*run19*) we used all types of features: lexical (we used unigrams and bigrams), entity (the output of the entity disambiguation systems) and MeSH terms. In our second run (*run11*) we used all types of features except the MeSH-based ones, to evaluate the contribution of taxonomic information.

Next, we discuss first the results of initial summaries (Table 11.5) followed by the results of update summaries (Table 11.6). The top two rows show the scores of the two upper-bound baselines (*run2* and *run3*) and the last row corresponds to the lower-bound baseline (*run1*). Below the upper bounds we show the results of the best systems evaluated by overall responsiveness (*run40* scored highest in the case of initial summaries and *run24* in the case of update summaries). Results of our runs can be found below (*run19* and *run11*).

Our two runs received very good scores for initial summaries. Our *run19* was the best run overall in linguistic quality, and *run11* was second. *Run19* also received the second highest score in overall responsiveness. (*run11* was seventh in this case.) *run11* did better according to the Pyramid score, 11th; *run19* was 19th.

In the case of update summaries, our *run19* was evaluated again as better than *run11*—9th in overall responsiveness/14th in linguistic quality/8th in Pyramid, compared to 13th/14th/13th. The fact that *run19* scored higher than *run11* in all human-based scores except for Pyramid with initial summaries suggests that using taxonomic information has a positive impact on summary quality, although the differences between the runs are not statistically significant. Automatic measures did not seem to correlate well with human-based measures this year.

11.7 Experiments with Project-Syndicate Data

In this section we carry out a multilingual summarization evaluation of our summarization approach on seven different languages: English, French, Russian, Arabic, Spanish, German and Czech. We follow the evaluation method proposed by Turchi et al. [37] and make use of their data set from project-syndicate.

11.7.1 Project-Syndicate Data

Turchi's et al. [37] approach to multilingual extractive summarization evaluation is grounded on two key ideas. Firstly, they proposed to make use of sentence-aligned parallel corpora to create simultaneously model summaries for seven languages and having human annotators select summary-worthy sentences only for one 'pivot' language (English), and secondly, to use different degrees of inter-annotator agreement to vary the gold standard for the evaluation. Their corpus consisted of four clusters: Israeli-Palestinian conflict, Malaria, Genetics and Science-and-Society. Each cluster contains five documents downloaded from project-syndicate. The average size of documents in terms of number of sentences is in the couple of hundreds range.

After experimenting with several evaluation methods they concluded that the most suitable one is the one taking into account sentences selected by at least two annotators, which they named the binary model.

11.7.2 Experimental Results

We ran our summarizer in two configurations: one only using lexical features (i.e., unigrams and bigrams)—which also serves as a baseline run—and another

Table 11.7 Summarization results

Summarizers	Summary Size (number of sentences)					
	1	3	5	10	15	20
English						
lexical + entities	1.0	0.67	0.6	0.6	0.5	0.43
lexical	0.0	0.67	0.6	0.6	0.47	0.45
French						
lexical + entities	0.5	0.67	0.6	0.55	0.47	0.43
lexical	0.0	0.5	0.6	0.45	0.47	0.4
German						
lexical + entities	1.0	0.83	0.7	0.55	0.47	0.35
lexical	0.5	0.5	0.7	0.55	0.43	0.38
Spanish						
lexical + entities	1.0	0.83	0.7	0.45	0.37	0.4
lexical	0.5	0.67	0.5	0.5	0.37	0.43
Russian						
lexical + entities	1.0	0.67	0.6	0.65	0.53	0.6
lexical	1.0	0.67	0.6	0.5	0.57	0.6
Arabic						
lexical + entities	0.0	0.5	0.7	0.55	0.47	0.5
lexical	0.5	0.67	0.5	0.6	0.53	0.53
Czech						
lexical + entities	0.0	0.67	0.6	0.5	0.43	0.48
lexical	0.5	0.67	0.7	0.7	0.53	0.48
Overall						
lexical + entities	0.64	0.69	0.64	0.55	0.46	0.45
lexical	0.43	0.62	0.6	0.56	0.48	0.46
Lead	–	–	0.3	0.25	0.26	0.25
Random	0.22	0.22	0.22	0.22	0.22	0.22

one using lexical and entity features. The latter run also subsumes a multilingual coreference resolver which improves the recall of within document entity mentions identified by the entity recognition and disambiguation system described earlier (Sect. 11.4).

The experimental results are presented in Table 11.7. Each summary score is computed by first calculating the intersection of sentences selected by the summarizer with those selected by at least two annotators divided by the number of sentences in the system summary.

The first thing we observe is that overall (see bottom part of Table 11.7) for target summaries of size three sentences or smaller incorporating entities works better than the baseline 'lexical' case and both perform better than two baseline summarizers: one selecting the first sentence of each document in the cluster (labeled 'Lead' in Table 11.7) and another one selecting random sentences (labeled

'Random'). One possible reason for that is that by adopting a more semantically-aware representation the summarization machinery is able to produce succinct summaries of better quality than the 'lexical'-only method, but as soon as the summarization compression rate is relaxed the benefit of including entities becomes less visible (and even in some cases yields worse results).

The variation in summarization performance across languages can be in part explained by the inconsistent performance of the entity recognition and disambiguation system (augmented with coreference) due to lack of or noisy resources for the languages. For instance, for languages like English and German we have good entity disambiguation performance which also translates into decent summarization performance, whereas for Czech the performance is notably lower.

One of the things that [37] showed was that just running a language-independent summarizer, such as the LSA summarizer that we used, on different languages does not translate into consistent output from the summarizer. That is, the same summarizer does not necessarily select the same sentences across languages.

In order to assess how this might be affected by adopting a more semantically-aware approach, we compute the average percentage of common sentences selected by the summarizer for each language with each other language and finally computing an overall average. So effectively this score measures the consistency of the output of the summarizer across languages. For the 'lexical' summarizer the output consistency score is 0.34, whereas for the 'lexical+entities' summarizer is 0.35, or in other words very slim improvement. However, if we leave out the languages for which the underpinning lexical resources are less well developed, in this case ar, ru and cz, for the rest (en, fr, de and es) we obtained an output consistency score for the 'lexical' run of 0.372, whereas for 'lexical + entities' 0.395. This suggests that, indeed, a more semantically-aware representation of source documents improves the consistency of the summarizer's output across languages. Or in other words, by incorporating entities into the process, we are able to build a more language-independent source representation than the one produced by the baseline 'lexical' run.

To summarize, our multilingual summarization evaluation showed that producing short informative summaries (from one to three sentences) is better addressed by bringing in entities than without them, but not necessarily so for longer summaries. Additionally, poor entity resolution and disambiguation performance due to underlying less-polished lexical resources seems to translate into summarization performance degradation. Finally, bringing in entities seems to improve output consistency of the summarizer across languages.

11.8 Online Live Summaries by NewsGist

In this section we describe the summarization system currently under development for EMM, which we have named NewsGist. We first

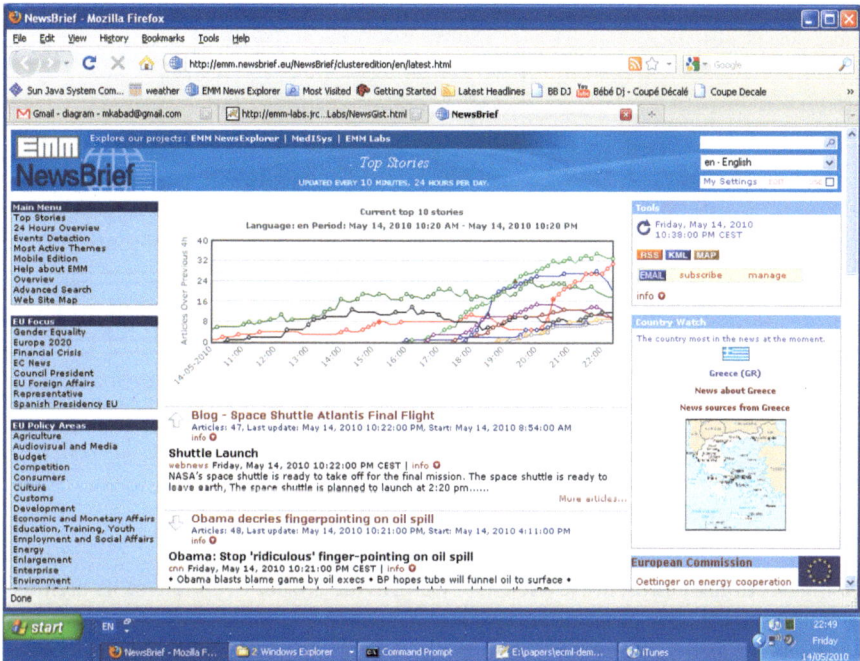

Fig. 11.1 NewsBrief, EMM's near real-time news analysis application

11.8.1 Europe Media Monitor

The Europe Media Monitor is a web-based multilingual news aggregation system that collects over 100k news articles per day in about 50 languages from more than 2,500 web news sources. The system employs text mining techniques to provide a picture of the present situation in the World (as conveyed in the media). Every ten minutes it automatically clusters all the collected news articles and displays the ten largest clusters per language by plotting them on a time-by-size graph. It also provides all the necessary hyperlinks to navigate through the clusters and to go to the source for a detailed exploration. In addition, it applies some deeper information analysis techniques, as for example, to automatically detect violent events, derive reported social networks and analyze media impact (Fig. 11.1).

The public website (http://emm.newsbrief.eu) provides a user interface to all this information. This public website is visited on a regular basis by some 30,000 human users, and gets some 1.2M hits per day.[11]

---

[11]For more details on EMM see [1].

Fig. 11.2 EMM's main
processing phases

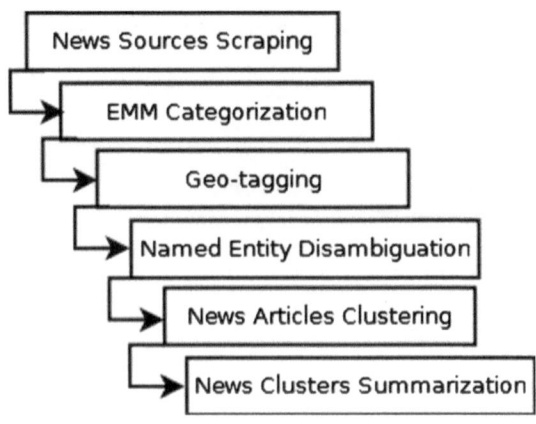

11.8.2 NewsGist

In this section we present NewsGist.[12] We provide a brief overview of its architecture and show screenshots for several languages.

As mentioned earlier, NewsGist was developed as part of EMM which is built on a pipeline architecture, where an input text document undergoes several processing phases during which the source is augmented with several layers of metadata such as named entities recognized in the text and semantic categories triggered by the text. The data interchange format between processing phases is RSS, a light-weight type of XML typically used by online news providers.

Thus, the input to NewGist is an RSS file enriched with information acquired by previous processing phases. Most importantly, by the time the RSS file reaches NewsGist, it already contains the outcome of the clustering of news articles and as output NewsGist produces a summary for each distinct news cluster (see Fig. 11.2).

The core system is pretty compact and is implemented as a Java servlet, running on top of Apache's Tomcat web server.[13]

The three-stage processing model for summarization discussed in the previous section is captured in an abstract class AbstractSummariser which defines three abstract methods to be implemented by classes extending this class: interpret(), transform() and generate(). Additionally, it provides a basic implementation for the method summarise() which takes a generic input and passes it through the three abstract methods, implementations for which are to be provided by inheriting classes.

---

[12]Online demo of the system is available at http://emm-labs.jrc.it/EMMLabs/NewsGist.html

[13]http://tomcat.apache.org/

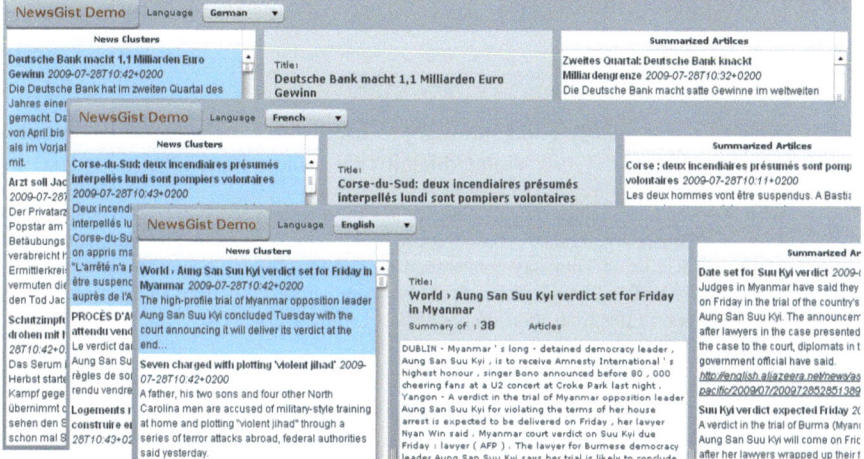

Fig. 11.3 NewsGist's overlaid screenshots for English, French and German

An implementer class of `AbstractSummariser` is `NewsGistLSAImple-menter`. This class in association with class `TermBySentenceMatrix` contain the core functionality for building the initial term-by-sentence matrix, invoking the singular value decomposition and finally selecting the most salient sentences to be included in the end summary. Language-specific tokenization and sentence splitting is provided by CORLEONE [27]. Reading and writing of RSS files is provided by EMM utility libraries. Matrix operations, and in particular singular value decomposition, is provided by the matrix-toolkits-java libraries[14] and also, alternatively, by the Java Matrix Package (JAMA).[15]

Entities and geo-locations are recognized in the EMM's processing phases prior to summarization and thus are already available in the input RSS files for NewsGist.

In Fig. 11.3 we show overlaid screenshots of NewsGist's online demo (http://emm-labs.jrc.it/EMMLabs/NewsGist.html) for three languages of the European Union: English, French and German.

11.8.3   Sample Summaries in Various Languages

In this section we show and briefly discuss a few reasonably good summaries from the online system, NewsGist, followed by a few problematic summaries.

---

[14]http://code.google.com/p/matrix-toolkits-java/

[15]http://math.nist.gov/javanumerics/jama/

11.8.3.1 Acceptable Summaries

English

An example of a reasonably good summary in English for a cluster of 107 articles titled "Ireland's Abortion Laws Violated Rights Of Cancer Patient":

(11.4) Irish abortion ban violates rights: court. Ireland must reform abortion laws. The judgment from the Strasbourg, France - based court backed the right of a. The European Court of Human Rights on Thursday condemned Ireland for obliging a woman suffering from cancer and who feared a pregnancy would worsen her health to have an abortion abroad. European court rules Irish abortion ban violates women's rights.

Spanish

A similar example to the above in Spanish for a cluster of 12 articles titled "Cumbre centroamericana culmina con anuncio de dilogo Nicaragua-Costa Rica" (Summit in Central America concludes with an announcement for Nicaragua-Costa Rica dialogue):

(11.5) Nicaragua y Costa Rica aceptan diálogo. Costa Rica pide salida de tropas nicaragüenses. El proceso de ... El Gobierno de Costa Rica manifestó hoy que "aprecia" los esfuerzos realizados por Guatemala en el marco de la Cumbre del Sistema de Integración Centroamericana (SICA) que se celebra en Belice, para propiciar un diálogo con Nicaragua que ayude a resolver el litigio fronterizo que mantienen ambos países. Pugna Nicaragua – Costa Rica acapara agenda de cumbre centroamericana.

Russian

And finally an example in Russian for a cluster of 63 articles titled "Регионалы останутся в парламенте на всю ночь" (Locals remain in Parliament for the whole night):

(11.6) Депутаты Верховной Рады избили оппозиционера. Депутат от БЮТ, избитый в ВР, находиця в тяжелом состоянии. Врачи скорой помощи также ушли из зала. Мы должны обеспечить на утро нормальную работу парламента, – отметил он. В Верховную Раду прибыла милиция.

11.8.3.2 More Problematic Summaries

English

A redundant summary of a cluster of seven articles, "45 arrested in major Garda crackdown on drug dealing":

(11.7) Gardai smash Dublin drugs network. 45 arrested over drugs in 48 hours in Dublin. 45 arrested in major Garda crackdown on drug dealing. Gardai smash Dublin drugs network.

> Gardai believe they have smashed one of Dublin ' s main drugs networks after 45 people were arrested following a six – month covert operation.

Spanish

Likewise the following is a redundant summary in Spanish for a cluster of 105 articles, "Julian Assange ya está en libertad" (Julian Assange walks away free):

(11.8)　Tribunal británico ratifica libertad bajo fianza a Julian Assange. Fundador de WikiLeaks, Assange, sale en libertad bajo fianza. El Tribunal Superior de Londres ha rechazado el recurso de la Fiscalía sueca en contra de la libertad del fundador de Wikileaks. Julian Assange, en libertad tras abonar la fianza. El creador de Wikileaks queda en libertad condicional tras pagar una fianza después de que el Tribunal Superior de Londres desestimara un recurso de Suecia que reclama su extradición.

French

Another example in French for a cluster of 45 articles, "Europe—Un sommet pour mettre en place un mécanisme de protection de l'euro" (Europe—A summit to put in place a mechanism for protection of the Euro):

(11.9)　Les dirigeants de l 'Union européenne sont tombés d'accord jeudi soir pour créer un Fonds de secours permanent en faveur des pays de la zone euro en difficulté. Le traité de Lisbonne sera modifié en ce sens. Accord de l'UE pour changer son traité et créer un Fonds de secours. Les dirigeants des pays de l'UE sont tombés d'accord jeudi soir pour modifier le traité de Lisbonne afin de permettre la création d'un Fonds de secours permanent en faveur des pays de la zone euro, en cas de crise grave, a indiqué à l'AFP une source diplomatique. Zone euro: l'Europe prête à mobiliser tous les moyens nécessaires .

11.9　Conclusion

We presented a generic statistical approach to summarization based on the Latent Semantic Analysis (LSA) paradigm. The key extension to previous LSA-based work is the use of a multilingual and multi-document entity disambiguation system when building the source representation. The entity disambiguation system is able to identify and ground references to geographical locations in news articles in 20 languages to the corresponding entries in gazetteers as well as to link references to persons and organizations to respective records in a continuously updated database of known names built over the past 8 years. We showed promising results from experiments with English using the Text Analysis Conferences corpora (TAC 2008 and 2009). Additionally, we presented a multilingual evaluation over seven languages English, French, Russian, Arabic, Spanish, German and Czech on parallel data from project-syndicate. This latter evaluation, although not to the level of depth of the evaluation carried out by Saggion et al. [30], covers a larger set of languages.

We also described an online multilingual summarization system currently under development for the Europe Media Monitor (EMM) and discussed a few representative sample summaries in several languages.

In future work we believe that there is much more to be done on multilingual summarization evaluation and approaches such as that by Saggion et al. [30] devised to bypass the need for model summaries, which are very costly and time consuming to produce, hold a lot of promise. However, also there is more to be done on bridging the gap between research and practical applications—in our case our online system NewsGist, where time performance is of utmost importance, is still based only on core functionality of the LSA approach presented in previous sections and we are gradually incorporating more and more functionality without making prohibitive performance compromises.

Acknowledgements We would like to thank the EMM team for providing a stable and robust news gathering infrastructure.

References

1. Atkinson, M., der Goot, E.V.: Near real time information mining in multilingual news. In: Proceedings of the 18th International World Wide Web Conference (WWW 2009), Madrid, pp. 1153–1154 (2009)
2. Barzilay, R., Elhadad, M.: Using lexical chains for text summarization. In: Mani, I. (ed.) Proceedings of the Workshop on Intelligent and Scalable Text Summarization at the Annual Joint Meeting of the ACL/EACL, Madrid (1997)
3. Barzilay, R., Lapata, M.: Modeling local coherence: an entity-based approach. In: Proceedings of the 43rd Annual Meeting of the Association for Computational Linguistics, Ann Arbor (2005)
4. Boguraev, B., Kennedy, C.: Salience-based content characterisation of text documents. In: Mani, I. (ed.) Proceedings of the Workshop on Intelligent and Scalable Text Summarization at the Annual Joint Meeting of the ACL/EACL, Madrid (1997)
5. Ding, C.H.Q.: A probabilistic model for latent semantic indexing. J. Am. Soc. Inf. Sci. Technol. **56**(6), 597–608 (2005)
6. Edmundson, H.: New methods in automatic extracting. J. Assoc. Comput. Mach. **16**(2), 264–285 (1969)
7. Gong, Y., Liu, X.: Generic text summarization using relevance measure and latent semantic analysis. In: Proceedings of ACM SIGIR, New Orleans (2002)
8. Grosz, B., Aravind, J., Scott, W.: Centering: a framework for modelling the local coherence of discourse. Comput. Linguist. **21**(2), 203–225 (1995)
9. Hirschman, L.: MUC-7 coreference task definition, version 3.0. In: Chinchor, N. (ed.) Proceedings of the 7th Message Understanding Conference, Virginia. NIST (1998). Available online at http://www-nlpir.nist.gov/related_projects/muc/proceedings/muc_7_toc.html
10. Hovy, E., Lin, C.: Automated text summarization in summarist. In: Mani, I. (ed.) Proceedings of the Workshop on Intelligent and Scalable Text Summarization at the Annual Joint Meeting of the ACL/EACL, Madrid (1997)
11. Jones, K.S.: Automatic summarising: factors and directions. In: Mani, I., Maybury, M. (eds.) Advances in Automatic Text Summarization. MIT, Cambridge (1999)
12. Kabadjov, M.A.: A comprehensive evaluation of anaphora resolution and discourse-new recognition. Ph.D. thesis, Department of Computer Science, University of Essex (2007)

13. Kabadjov, M.A., Steinberger, J., Pouliquen, B., Steinberger, R., Poesio, M.: Multilingual statistical news summarisation: preliminary experiments with english. In: Proceedings of the Workshop on Intelligent Analysis and Processing of Web News Content at the IEEE/WIC/ACM International Conferences on Web Intelligence and Intelligent Agent Technology (WI-IAT), Milan (2009)
14. Kupiec, J., Pedersen, J., Chen, F.: A trainable document summarizer. In: Proceedings of the 18th Annual International ACM SIGIR Conference on Research and Development in Information Retrieval, Seattle, pp. 68–73 (1995)
15. Lin, C.Y.: ROUGE: a package for automatic evaluation of summaries. In: Proceedings of the Workshop on Text Summarization Branches Out, Barcelona (2004)
16. Litvak, M., Last, M., Friedman, M.: A new approach to improving multilingual summarization using a genetic algorithm. In: Proceedings of the 48th Annual Meeting of the Association for Computational Linguistics, Uppsala, pp. 927–936. Association for Computational Linguistics (2010)
17. Luhn, H.: The automatic creation of literature abstracts. IBM J. Res. Dev. 2(2), 159–165 (1958)
18. Mani, I. (ed.): Proceedings of the Workshop on Intelligent and Scalable Text Summarization at the Annual Joint Meeting of the ACL/EACL, Madrid (1997)
19. Mani, I., Maybury, M. (eds.): Advances in Automatic Text Summarization. MIT, Cambridge (1999)
20. Marcu, D.: From discourse structures to text summaries. In: Mani, I. (ed.) Proceedings of the Workshop on Intelligent and Scalable Text Summarization at the Annual Joint Meeting of the ACL/EACL, Madrid (1997)
21. Maybury, M.: Generating summaries from event data. In: Mani, I., Maybury, M. (eds.) Advances in Automatic Text Summarization. MIT, Cambridge (1999)
22. McKeown, K., Radev, D.: Generating summaries of multiple news articles. In: Proceedings of the 18th Annual International ACM SIGIR Conference on Research and Development in Information Retrieval, Seattle, pp. 74–82 (1995)
23. Nenkova, A., Louis, A.: Can you summarize this? identifying correlates of input difficulty for generic multi-document summarization. In: Proceedings of the 46th Annual Meeting of the Association for Computational Linguistics, Columbus, pp. 825–833. Association for Computational Linguistics (2008)
24. Nenkova, A., Passonneau, R.: Evaluating content selection in summarization: the pyramid method. In: Proceedings of the Meeting of the North American Chapter of the Association for Computational Linguistics (NAACL), Boston (2004)
25. Nenkova, A., Passonneau, R., McKeown, K.: The pyramid method: incorporating human content selection variation in summarization evaluation. ACM Trans. Speech Lang. Process. 4(2), 4 (2007)
26. Over, P., Dang, H., Harman, D.: DUC in context. Inf. Process. Manag. 43(6), 1506–1520 (2007). Special Issue on Text Summarisation (Donna Harman, ed.)
27. Piskorski, J.: CORLEONE – core linguistic entity online extraction. Tech. Rep. EN 23393, Joint Research Centre of the European Commission (2008)
28. Pouliquen, B., Kimler, M., Steinberger, R., Ignat, C., Oellinger, T., Blackler, K., Fuart, F., Zaghouani, W., Widiger, A., Forslund, A.C., Best, C.: Geocoding multilingual texts: recognition, disambiguation and visualisation. In: Proceedings of the 5th International Conference on Language Resources and Evaluation (LREC 2006), Genoa, pp. 53–58 (2006)
29. Pouliquen, B., Steinberger, R.: Automatic construction of multilingual name dictionaries. In: Goutte, C., Cancedda, N., Dymetman, M., Foster, G. (eds.) Learning Machine Translation. NIPS series. MIT, Cambridge (2009)
30. Saggion, H., Torres-Moreno, J.M., da Cunha, I., SanJuan, E., Velazquez-Morales, P.: Multilingual summarization evaluation without human models. In: Proceedings of the International Conference on Computational Linguistics, Beijing, pp. 1059–1067 (2010)
31. Steinberger, J., Ježek, K.: Update summarization based on novel topic distribution. In: Proceedings of the 9th ACM DocEng, Munich (2009)

32. Steinberger, J., Kabadjov, M.A., Poesio, M., Sanchez-Graillet, O.: Improving LSA-based summarization with anaphora resolution. In: Proceedings of the Conference on Empirical Methods in Natural Language Processing (EMNLP), Vancouver (2005)
33. Steinberger, J., Poesio, M., Kabadjov, M.A., Ježek, K.: Two uses of anaphora resolution in summarization. Inf. Process. Manag. **43**(6), 1663–1680 (2007). Special Issue on Text Summarisation (Donna Harman, ed.)
34. Steinberger, R., Pouliquen, B., Ignat, C.: Using language-independent rules to achieve high multilinguality in text mining. In: Fogelman-Soulié, F., Perrotta, D., Piskorski, J., Steinberger, R. (eds.) Mining Massive Data Sets for Security. IOS-Press, Amsterdam/Holland (2009)
35. Stewart, J.G.: Genre oriented summarization. Ph.D. thesis, Language Technologies Institute, School of Computer Science, Carnegie Mellon University (2008)
36. Teufel, S., Moens, M.: Sentence extraction as a classification task. In: Mani, I. (ed.) Proceedings of the Workshop on Intelligent and Scalable Text Summarization at the Annual Joint Meeting of the ACL/EACL, Madrid (1997)
37. Turchi, M., Steinberger, J., Kabadjov, M., Steinberger, R.: Using parallel corpora for multilingual (multi-document) summarisation evaluation. In: Proceedings of CLEF-10, Padua, pp. 52–63. Springer, Berlin (2010)
38. Wan, X., Li, H., Xiao, J.: Cross-language document summarization based on machine translation quality prediction. In: Proceedings of the 48th Annual Meeting of the Association for Computational Linguistics, Uppsala, pp. 917–926. Association for Computational Linguistics (2010)

Chapter 12
A Bottom-Up Approach to Sentence Ordering for Multi-Document Summarization

Danushka Bollegala, Naoaki Okazaki, and Mitsuru Ishizuka

Abstract In Chap. 1, multi-document summarization is introduced as a potential solution to the information explosion problem. A major challenge in creating a summary from information extracted from multiple sources is to decide the order in which those information must be presented in the summary. Incorrect ordering of information selected from multiple sources would lead to misunderstandings. In this chapter, we discuss the challenges involved when ordering information selected from multiple sources and present several approaches to overcome those challenges. We also introduce several semi-automatic evaluation measures to empirically evaluate an ordering of sentences created by an algorithm.

12.1 Multi-Document Summarization and the Problem

The rapid growth of the World Wide Web has resulted in a large amount of electronically available textual information. We use Web search engines to retrieve the relevant information regarding a particular query. However, often Web search

D. Bollegala (✉)
Graduate School of Information Science and Technology, The University of Tokyo,
7-3-1, Hongo, Bunkyo-ku, Tokyo, 113-8656, Japan
e-mail: danushka@iba.t.u-tokyo.ac.jp

N. Okazaki
Department of System Information Sciences, Graduate School of Information Sciences, Tohoku
University, 6-3-09 Aramakiaza-Aoba, Aoba-ku, Sendai, 980-8579, Japan
e-mail: okazaki@ecei.tohoku.ac.jp

M. Ishizuka
Department of Information and Communication Engineering, Graduate School of Information
Science and Technology, The University of Tokyo, 7-3-1, Hongo, Bunkyo-ku, Tokyo, 113-8656,
Japan
e-mail: ishizuka@i.u-tokyo.ac.jp

T. Poibeau et al. (eds.), *Multi-source, Multilingual Information Extraction and Summarization 11*, Theory and Applications of Natural Language Processing, DOI 10.1007/978-3-642-28569-1__12, © Springer-Verlag Berlin Heidelberg 2013

1. Such storms have maximum sustained winds greater than 155 mph and can cause catastrophic damage.
2. Earlier Wednesday, Gilbert was classified as a Category 5 storm, the strongest and deadliest type of hurricanes.
3. Tropical Storm Gilbert formed in the eastern Caribbean and strengthened into a hurricane Saturday night.

Fig. 12.1 Randomly ordered sentences in a summary

queries return more than one relevant search results. A user must read all those Web documents and obtain the necessary information. A text summarization system can reduce the time and effort required by a user to read a set of documents by automatically generating a short and informative summary of all the information that exist in the set of documents. The problem of generating a single coherent summary from a given set of documents that describes a particular event is referred to as Multi-document Summarization. The related problem of generating a single coherent summary from a *single document* is named as single document summarization.

Multi-document summarization [5,9,25] tackles the information overload problem by providing a condensed and coherent version of a set of documents. Among a number of sub-tasks involved in multi-document summarization including sentence extraction, topic detection, sentence ordering, information extraction, and sentence generation most multi-document summarization systems have been based on an extraction method, which identifies important textual segments (e.g. sentences or paragraphs) in source documents. It is important for such multi-document summarization systems to determine a coherent arrangement for the textual segments extracted from multi-documents, in order to reconstruct the text structure for summarization.

A summary with improperly ordered sentences confuses the reader and degrades the quality/reliability of the summary itself. Barzilay et al. [2] has provided empirical evidence to show that the proper order of extracted sentences significantly improves their readability. Lapata [15] experimentally shows that the time taken to read a summary strongly correlates with the arrangement of sentences in the summary.

For example, consider the three sentences shown in Fig. 12.1, selected from a reference summary in Document Understanding Conference (DUC) 2003 dataset. The first and second sentences are extracted from the same source document, whereas the third sentence is extracted from a different document. Although all three sentences are informative and talk about the storm, *Gilbert*, the sentence ordering shown in Fig. 12.1 is inadequate. For example, the phrase, *such storms*, in sentence 1, in fact refers to *Category 5 storms*, described in sentence 2. A better arrangement of sentences in this example would be 3-2-1.

In single document summarization, where a summary is created using only one document, it is natural to arrange the extracted information in the same order as in the original document. In contrast, for multi-document summarization, we need to establish a strategy to arrange sentences extracted from different documents.

Therefore, the problem of sentence ordering is more critical for multi-document summarization systems compared to single document summarization systems. In this chapter, we will be mainly focusing on the sentence ordering problem in multi-document summarization.

Ordering extracted sentences into a coherent summary is a non-trivial task. Rhetorical relations [20] such as *cause-effect* relation and *elaboration* relation exist between sentences in a coherent text. If we can somehow determine the rhetorical relations that exist among a given set of sentences, then we can use those relations to infer a coherent ordering of the set of sentences. For example, if a sentence A is the effect of the cause mentioned in a sentence B, then we might want to order the sentence A after the sentence B in a summary that contains both sentences A and B. Unfortunately, the problem of automatically detecting the rhetorical structure of an arbitrary text is a difficult and unsolved problem. The performance reported by the state-of-the-art rhetorical structure analysis systems is not sufficient to be used in a sentence ordering system.

The task of constructing a coherent summary from an unordered set of sentences has several unique properties that make it a difficult problem. Source documents for a summary may have been written by different authors, have different writing styles, or written on different dates, and based on the different background knowledge. Often a multi-document summarization system is presented with a set of articles that discuss about a particular news event. Those news articles are selected from different newspapers. Although the articles themselves are related and discuss a particular event, those articles are written by different authors. Therefore, the collection of texts that the multi-document summarization system receives are not always coherent with regard to the authorship. We cannot expect a set of extracted sentences from such a diverse set of documents to be coherent on their own.

The problem of information ordering is not limited to automatic text summarization, and concerns natural language generation applications. A typical (NLG) [26] system consists of six components: content determination, discourse planning, sentence aggregation, lexicalization, referring expression generation, and orthographic realization. Among those, information ordering is particularly important in discourse planning, and sentence aggregation [7,8,12]. In concept-to-text generation [26], given a concept (e.g. a keyword, a topic, or a collection of data), the objective is to produce a natural language text about the given concept. For example, consider the case where generating game summaries, given a database containing statistics of American football. A sentence ordering algorithm can support a natural language generation system by helping to order the sentences in a coherent manner. The techniques that we present in this Chapter are specifically designed for multi-document news summarization.

Existing methods for sentence ordering are divided into two approaches: making use of chronological information [2, 17, 21, 22], and learning the natural order of sentences from large corpora [1, 11, 14]. A newspaper usually disseminates descriptions of novel events that have occurred since the last publication. For this reason, the chronological ordering of sentences is an effective heuristic for multi-document summarization [17,21]. Barzilay et al. [2] proposed an improved version of chronological ordering by first grouping sentences into sub-topics discussed in the source documents, then arranging the sentences in each group chronologically.

Okazaki et al. [22] proposed an algorithm to improve the chronological ordering by resolving the presuppositional information of extracted sentences. They assume that each sentence in newspaper articles is written on the basis that presuppositional information should be transferred to the reader before the sentence is interpreted. The proposed algorithm first arranges sentences in a chronological order, and then estimates the presuppositional information for each sentence by using the content of the sentences placed before each sentence in its original article. The evaluation results show that the proposed algorithm improves the chronological ordering significantly.

Lapata [14] presented a probabilistic model for text structuring and its application in sentence ordering. Her method computes the transition probability from one sentence to the next in two sentences, from a corpus based on the Cartesian product using the following features: verbs (precedent relationships of verbs in the corpus), nouns (entity-based coherence by keeping track of the nouns), and dependencies (structure of sentences). Lapata [15] also proposed the use of Kendall's rank correlation coefficient (Kendall's τ) for the automatic evaluation that quantifies the differences between orderings produced by an algorithm and by a human. Although she has not compared her method with chronological ordering, it could be applied to generic domains, not relying on the chronological clue specific to newspaper articles.

Barzilay and Lee [1] proposed *content models* to deal with the topic transition in domain specific text. The content models are implemented by Hidden Markov Models (HMMs), in which the hidden states correspond to topics in the domain of interest (e.g. earthquake magnitude or previous earthquake occurrences), and state transitions capture possible information-presentation orderings. The evaluation results showed that their method outperformed Lapata's approach by a wide margin. They did not compare their method with chronological ordering as an application of multi-document summarization.

Ji and Pulman [11] proposed a sentence ordering algorithm using a semi-supervised sentence classification and historical ordering strategy. Their algorithm includes three steps: the construction of sentence networks, sentence classification, and sentence ordering. First, they represent a summary as a network of sentences. Nodes in this network represent sentences in a summary, and edges represent transition probabilities between two nodes (sentences). Next, the sentences in the source documents are classified into the nodes in this network. The probability $p(c_k|s_i)$, of a sentence s_i in a source document belonging to a node c_k in the network, is defined as the probability of observing s_k as a sample from a Markov random walk in the sentence network. Finally, the extracted sentences are ordered to the weights of the edges. They compare the sentence ordering produced by their method against manually ordered summaries using Kendall's τ. Unfortunately, they do not compare their results against the chronological ordering of sentences, which has been shown to be an effective sentence ordering strategy in multi-document news summaries.

In Sect. 12.2, we present four heuristic criteria to capture the association of sentences in the context of multi-document summarization for newspaper articles. These criteria are then integrated into one criterion by a supervised learning approach in Sect. 12.3. We describe a bottom-up approach in arranging sentences, which repeatedly concatenates textual segments until the overall segment with all sentences is arranged.

12.2   Heuristic Approaches for Sentence Ordering

We define notation $a \succ b$ to represent that sentence a precedes sentence b. We use the term *segment*, to describe a sequence of ordered sentences. When segment A consists of sentences $a_1, a_2, \ldots, a_m$ in this order, we denote it as:

$$A = (a_1 \succ a_2 \succ \ldots \succ a_m). \tag{12.1}$$

The two segments A and B can be ordered as either B after A, or A after B. We define the notation $A \succ B$ to show that segment A precedes segment B. In the following subsections we introduce four heuristics (which we refer to as sentence ordering criteria) to determine the ordering between two sentences.

12.2.1   Chronology Criterion

Chronology criterion reflects the chronological ordering [17, 21], by which sentences are arranged in the chronological order of publication timestamps. A newspaper usually deals with novel events that have occurred since the last publication. Consequently, the chronological ordering of sentences has shown to be particularly effective in multi-document news summarization. Publication timestamps are used to decide the chronological order among sentences extracted from different documents. However, if no timestamp is assigned to documents, or if several documents have the identical timestamp, the chronological ordering does not provide a clue for sentence ordering. Inferring temporal relations among events [18, 19] using implicit time references (such as tense system) [16], and explicit time references (such as temporal adverbials) [10], might provide an alternative clue for chronological ordering. However, inferring temporal relations across a diverse set of multiple documents is a difficult task. Consequently, by assuming the availability of temporal information in the form of timestamps, we define the strength of association in arranging segments B after A, measured by a chronology criterion $f_{\text{chro}}(A \succ B)$ in the following formula:

$$f_{\text{chro}}(A \succ B) = \begin{cases} 1 & T(a_m) < T(b_1) \\ 1 & [D(a_m) = D(b_1)] \wedge [N(a_m) < N(b_1)] \\ 0.5 & [T(a_m) = T(b_1)] \wedge [D(a_m) \neq D(b_1)] \\ 0 & otherwise \end{cases}. \tag{12.2}$$

D. Bollegala et al.

(a) The earthquake crushed cars, damaged hundreds of houses and terrified people for hundreds of kilometers around.
(b) A major earthquake measuring 7.7 on the Richter scale rocked north Chile Wednesday.
(c) Authorities said two women, one aged 88 and the other 54, died when they were crushed under collapsing walls.

Fig. 12.2 Three sentences from a summary about an earthquake

Here, a_m represents the last sentence in segment A, b_1 represents the first sentence in segment B, $T(s)$ is the publication date of the sentence s, $D(s)$ is the unique identifier of the document to which sentence s belongs, and $N(s)$ denotes the line number of sentence s in the original document. The chronological order of segment B arranging after A is determined by comparing the last sentence in the segment A and the first sentence in the segment B.

The chronology criterion assesses the appropriateness of arranging segment B after A if sentence a_m is published earlier than sentence b_1, or if sentence a_m appears before b_1 in the same article. For sentences extracted from the same source document, preferring the original order in the source document has proven to be effective for single document summarization [2]. The second condition in the chronological criterion defined in formula (12.2) imposes this constraint. If sentence a_m and b_1 are published on the same day, but appear in different articles, the criterion assumes the order to be undefined. If none of the above conditions are satisfied, the criterion estimates that segment B will precede A. By assigning a score of zero for this condition in formula (12.2), the chronological criterion guarantees that sentence orderings which contradicts with the definition of chronological ordering are not produced.

12.2.2 Topical-Closeness Criterion

A set of documents discussing a particular event usually contains information related to multiple topics. For example, a set of newspaper articles related to an earthquake typically contains information about the magnitude of the earthquake, its location, casualties, and rescue efforts. Grouping sentences by topics has shown to improve the readability of a summary [1, 2]. For example, consider the three sentences shown in Fig. 12.2, selected from a summary of an earthquake in Chile. Sentences (a) and (c) in Fig. 12.2 present details about the damage by the earthquake, whereas sentence (b) conveys information related to the magnitude and location of the earthquake. In this example, sentences (a) and (c) can be considered as topically related. Consequently, when the three sentences are ordered as show in Fig. 12.2, we observe abrupt shifts of topics from sentence (a) to (b), and from (b) to (c). A better arrangement of the sentences that prevents such disfluencies is (b)-(a)-(c).

> (a) Honduran death estimates grew from 32 to 231 in the first days, to 6,076 with 4,621 missing.
> (b) Honduras braced as category 5 Hurricane Mitch approached.
> (c) The EU approved 6.4 million in aid to Mitch's victims.

Fig. 12.3 Precedence relations in a summary

The topical-closeness criterion deals with the association of two segments, based on their topical similarity. The criterion reflects the ordering strategy proposed by Barzilay et al. [2], which groups sentences referring to the same topic. To measure the topical closeness of two sentences, we represent each sentence by using a vector. First, we remove *stop words* (i.e. functional words such as *and, or, the*, etc.) from a sentence and lemmatize verbs and nouns. Second, we create a vector in which each element corresponds to the words (or lemmas in the case of verbs and nouns) in the sentence. Values of elements in the vector are either 1 (for words that appear in the sentence) or 0 (for words that do not appear in the sentence).[1]

We define the topical closeness of two segments A and B as follows,

$$f_{\text{topic}}(A \succ B) = \frac{1}{|B|} \sum_{b \in B} \max_{a \in A} \text{sim}(a, b). \qquad (12.3)$$

Here, $\text{sim}(a, b)$ denotes the similarity of sentences a and b, calculated by the cosine similarity of two vectors corresponding to the sentences. For sentence $b \in B$, $\max_{a \in A} \text{sim}(a, b)$ yields the similarity between sentences b and $a \in A$, which is the most similar to b. The topical-closeness criterion $f_{\text{topic}}(A \succ B)$ assigns a higher value when the topic referred to by segment B is the same as by segment A.

12.2.3 Precedence Criterion

In extractive multi-document summarization, only the important sentences that convey the main points discussed in source documents are selected to be included in the summary. However, a selected sentence can presuppose information from other sentences that were not selected by the sentence extraction algorithm. For example, consider the three sentences shown in Fig. 12.3, selected from a summary on hurricane Mitch. Sentence (a) describes the after-effects of the hurricane, whereas sentence (b) introduces the hurricane. To understand the reason for the deaths mentioned in sentence (a), one must first read sentence (b). Consequently, it is appropriate to arrange the three sentences in Fig. 12.3 in the order (b)-(a)-(c).

---

[1] Using the frequencies of words instead of the binary (0, 1) values as vector elements, did not have a positive impact in our experiments. We think this is because, compared to a document, a sentence typically has a lesser number of words, and a word does not appear many times in a single sentence.

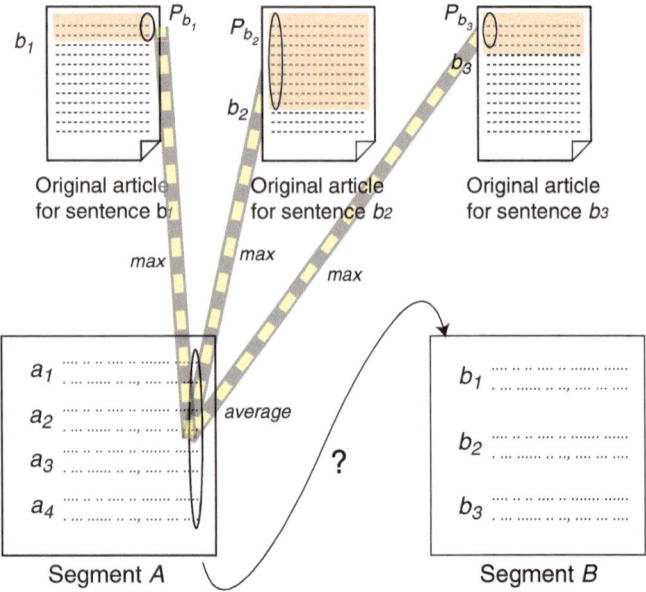

Fig. 12.4 Precedence criterion

In general, it is difficult to perform such an in-depth logical inference on a given set of sentences. Instead, we use source documents to estimate precedence relations. For example, assuming that in the source document where sentence (a) was extracted, there exist a sentence that is similar to sentence (b), we can conclude that sentence (b) should precede sentence (a) in the summary.

To formally define the precedence criterion, let us consider the case illustrated in Fig. 12.4, where we arrange segment A before B. Each sentence in segment B has the presuppositional information such as background information or introductory facts that should be conveyed to a reader in advance. Given sentence $b \in B$, such presuppositional information may be presented by the sentences appearing before the sentence b in the original article. However, we cannot guarantee whether a sentence-extraction method for multi-document summarization chooses any sentences before b for a summary, because the extraction method usually determines a set of sentences within the constraint of summary length that maximizes information coverage and excludes redundant information. The *precedence criterion* measures the substitutability of the presuppositional information of segment B (e.g., the sentences appearing before sentence b) as segment A. This criterion is a formalization of the sentence-ordering algorithm proposed by Okazaki et al. [22].

We define the precedence criterion in the following formula,

$$f_{\mathrm{pre}}(A \succ B) = \frac{1}{|B|} \sum_{b \in B} \max_{a \in A, p \in P_b} \mathrm{sim}(a, p). \qquad (12.4)$$

Here, $f_{pre}(A \succ B)$ is the strength of association for ordering segment B after A, measured using the precedence criterion. P_b is a set of sentences appearing before sentence b in the original article from which b was extracted. If b is the first sentence in its source document, then P_b is the empty set. For each sentence p in set P_b we compute the cosine similarity $sim(a, b)$ between p and sentences a in segment A. Cosine similarity between sentences are computed exactly as described in the topical-closeness criterion. We find the maximum similarity between p and any sentence from segment A. Finally, we average the similarity scores by dividing from the number of sentences in segment B. Figure 12.4 shows an example of calculating the precedence criterion for arranging segment B after A. We approximate the pre-suppositional information for sentence b by sentences P_b, i.e., sentences appearing before the sentence b in the original article. Calculating the maximum similarity in the possible pairs of sentences in P_b and A, Formula (12.4) is interpreted as the average similarity of the precedent sentences $\forall P_b (b \in B)$ to the segment A.

12.2.4   Succession Criterion

In extractive multi-document summarization, sentences that describe a particular event are extracted from a set of source articles. Usually, there exist a logical sequence among the information conveyed in the extracted sentences. For example, in Fig. 12.2, sentence (a) describes the results of the earthquake described in sentence (b). It is natural to order a sentence that describes the result or an effect of a certain cause after a sentence that describes the cause. Therefore, in Fig. 12.2, sentence (a) should be ordered after sentence (b) to create a coherent summary. We use the information conveyed in source articles to propose *succession criterion* to capture the coverage of information for sentence ordering in multi-document summarization.

The general problem of deciding whether a sentence a can be *inferred* from another sentence b is difficult. Given two text segments, the textual entailment recognition task[2][6] attempts to decide whether the meaning of one text can be inferred from another text. Despite the recent progress in research on textual entailment [4, 30], automatically detecting textual entailment of sentences remains a challenging task. Therefore, we use the information conveyed in source articles to propose *succession criterion* to capture the coverage of information for sentence ordering in multi-document summarization.

The succession criterion assesses the coverage of the succeeding information for segment A by arranging segment B after A:

$$f_{\text{succ}}(A \succ B) = \frac{1}{|A|} \sum_{a \in A} \max_{s \in S_a, b \in B} sim(s, b). \tag{12.5}$$

---

[2]http://www.pascal-network.org

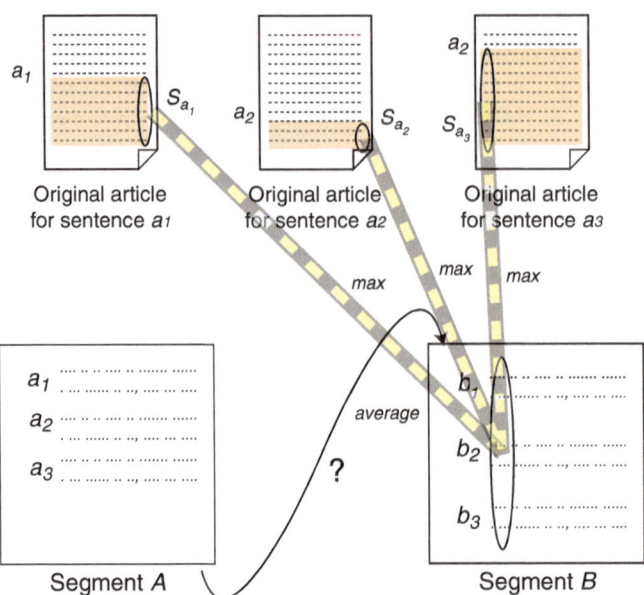

Fig. 12.5 Succession criterion

Here, for each sentence a in segment A, S_a denotes the set of sentences appearing after sentence a in the original article (i.e. article from which a was extracted). For each sentence s in set S_a, we compute the cosine similarity $sim(s, b)$, between sentences s and sentences b in segment B. Cosine similarity is computed exactly as described in the topical-closeness criterion. Figure 12.5 shows an example of calculating the succession criterion to arrange segments B after A. We approximate the information that should follow segment A by the sentences in segments S_a. We then compare each segment S_a with segment B to measure how much segment B covers this information. The succession criterion measures the substitutability of the succeeding information, (e.g., the sentences appearing after the sentence $a \in A$) such as segment B.

12.3 Integrating Different Heuristics Using Machine Learning

We described four criteria for measuring the strength and direction of association between two segments of texts. However, it is still unclear which criteria and conditions increase the performance. A human may use a combination of criteria to produce a summary. Bollegala et al. [3] use summaries created by humans as training data to find the optimum combination of the proposed criteria so that the

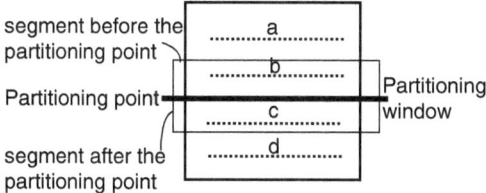

Fig. 12.6 Partitioning a human-ordered extract into pairs of segments

$$+1 : [f_{\text{chro}}(A \succ B), f_{\text{topic}}(A \succ B), f_{\text{pre}}(A \succ B), f_{\text{succ}}(A \succ B)]$$
$$-1 : [f_{\text{chro}}(B \succ A), f_{\text{topic}}(B \succ A), f_{\text{pre}}(B \succ A), f_{\text{succ}}(B \succ A)]$$

Fig. 12.7 Two vectors in a training data generated from two ordered segments $A \succ B$

combined function fits to the human-made summary. They integrate the four criteria: chronology, topical-closeness, precedence, and succession, to define the function $f(A \succ B)$ to represent the association direction and strength of the two segments A and B (Formula (12.6)). More specifically, given the two segments A and B, function $f(A \succ B)$ yields the integrated association strength based on four values, $f_{\text{chro}}(A \succ B)$, $f_{\text{topic}}(A \succ B)$, $f_{\text{pre}}(A \succ B)$, and $f_{\text{succ}}(A \succ B)$. Formalizing the integration task as a binary classification problem, they employ a Support Vector Machine (SVM) to model the function.

The first step of this method is to partition a human-ordered extract into pairs each of which consists of two non-overlapping segments. Let us explain the partitioning process taking four human-ordered sentences, $a \succ b \succ c \succ d$ shown in Fig. 12.6. Firstly, we place the partitioning point just after the first sentence a. Focusing on sentences a and b at the boundary of the partition point, we extract the pair $\{(a), (b)\}$ of two segments (a) and (b). Enumerating all possible pairs of two segments appearing just before/after the partitioning point, we obtain the following pairs, $\{(a), (b)\}$, $\{(a), (b \succ c)\}$ and $\{(a), (b \succ c \succ d)\}$. Similarly, segment pairs, $\{(b), (c)\}$, $\{(a \succ b), (c)\}$, $\{(b), (c \succ d)\}$, $\{(a \succ b), (c \succ d)\}$, are obtained from the partitioning point between sentences b and c. Collecting the segment pairs from the partitioning point between sentences c and d (i.e. $\{(c), (d)\}$, $\{(b \succ c), (d)\}$ and $\{(a \succ b \succ c), (d)\}$), in total, ten pairs were extracted from the four sentences shown in Fig. 12.6. In general, this process yields $n(n^2 - 1)/6$ pairs from ordered n sentences. From each pair of segments, we generate one positive and one negative training instance as follows.

Given a pair of two segments A and B, arranged in an order $A \succ B$, we obtain a positive training instance (labeled as $+1$) by computing a four dimensional vector (Fig. 12.7) with the following elements: $f_{\text{chro}}(A \succ B)$, $f_{\text{topic}}(A \succ B)$, $f_{\text{pre}}(A \succ B)$, and $f_{\text{succ}}(A \succ B)$. Similarly, we obtain a negative training instance (labeled as -1) corresponding to $B \succ A$. We use a manually ordered set of summaries and assume an ordering $A \succ B$ as a positive sentence ordering (for training purposes) if in a manually ordered summary the sentence A precedes the sentence B. For such two sentences A and B, we consider the ordering $B \succ A$ as a negative sentence ordering (for training purposes). Accumulating these instances as training data, we construct a binary classifier modeled by a Support Vector Machine. The SVM classifier yields the association direction of two segments (e.g. $A \succ B$ or $B \succ A$) with the class information (i.e., $+1$ or -1).

We assign the association strength of two segments by using the posterior probability that the instance belongs to a positive $(+1)$ class. When an instance is classified into a negative (-1) class, we set the association strength as zero (see the definition of Formula (12.6)). Because SVM is a large-margin classifier, the output of an SVM is the distance from the decision hyperplane. However, the distance from a hyperplane is not a valid posterior probability. We use sigmoid functions to convert the distance into a posterior probability (see [24] for a detailed discussion on this topic).

Let us consider a bottom-up approach in arranging the sentences. Starting with a set of segments initialized with a sentence for each, we concatenate two segments, with the strongest association (discussed later) of all possible segment pairs, into one segment. Repeating the concatenating will eventually yield a segment with all sentences arranged. The algorithm is considered as a variation of agglomerative hierarchical clustering, with the ordering information retained at each concatenating process.

We introduce a function $f(A \succ B)$ to represent the direction and strength of the association of two segments, A and B,

$$f(A \succ B) = \begin{cases} p & (if\ A\ precedes\ B) \\ 0 & (if\ B\ precedes\ A) \end{cases}, \tag{12.6}$$

where p $(0 \le p \le 1)$ denotes the association strength of the segments A and B. The association strengths of the two segments with different directions, e.g., $f(A \succ B)$ and $f(B \succ A)$, are not always identical by our definition,

$$f(A \succ B) \ne f(B \succ A). \tag{12.7}$$

Figure 12.8 shows the process of arranging four sentences a, b, c, and d. We first initialize four segments with a sentence for each,

$$A = (a), B = (b), C = (c), D = (d).$$

Supposing that $f(B \succ A)$ has the highest value of all possible pairs, e.g., $f(A \succ B)$, $f(C \succ D)$, etc, we concatenate B and A to obtain a new segment,

$$E = (b \succ a).$$

Fig. 12.8 Arranging four sentences A, B, C, and D with a bottom-up approach

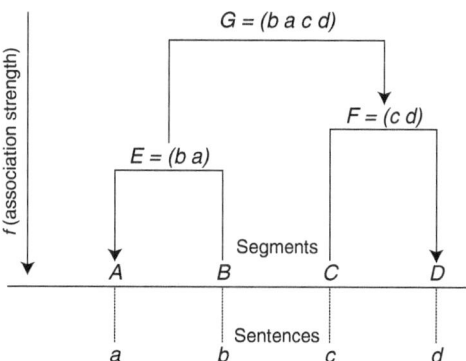

Algorithm 2 Sentence ordering algorithm

1: $P \leftarrow \{(s_1), (s_2), \ldots\}$
2: **while** $|P| > 1$ **do**
3:     $(p_a, p_b) \leftarrow \arg\max_{p_i, p_j \in P, \; p_i \neq p_j} f(p_i \succ p_j)$
4:     **for** $s \in p_b$ **do**
5:         $p_a \leftarrow p_a \oplus s$
6:     **end for**
7:     $P \leftarrow P \setminus \{p_b\}$
8: **end while**
9: **return** P

Then, we search for the segment pair with the strongest association. Supposing that $f(C \succ D)$ has the highest value, we concatenate C and D to obtain a new segment,

$$F = (c \succ d).$$

Finally, comparing $f(E \succ F)$ and $f(F \succ E)$, we obtain the final sentence ordering,

$$G = (b \succ a \succ c \succ d).$$

Algorithm 2 presents the pseudo code of the sentence ordering algorithm. Algorithm 2 takes a set of extracted sentences S as input, and returns a single segment of ordered sentences. First, for each sentence s_i in S, we create a segment p_i that contains s_i only. Subsequently, we find the two segments, p_a and p_b in the set of segments P, that have the maximum strength of association (Line No. 3). The *for* loop in lines 4–6 then appends the sentences in segment p_b to the end of segment p_a. The operator $\oplus$ in Line No. 5 denotes this appending operation. We then remove the segment p_b from P (Line No. 7). This process is repeated until we are left with a single segment in P. In Algorithm 2, we use the notation $|P|$ to denote the number of elements (i.e. segments) in P.

12.4   Semi-automatic Evaluation of Information Ordering

Automatically evaluating the correctness of information ordering is a difficult problem because of several reasons. First, there can be more than one correct ordering for a set of given sentences that creates a coherent summary. Therefore, an evaluation measure must be able to take into consideration those multiple possible orderings. This challenge is common among numerous natural language generation systems such as machine translation (MT) systems. Often there exist more than one possible translation for a given text. Second, the notion of an acceptable ordering of sentences differs from one person to another. The fluency in the particularly language in which summary is created as well as the level of background knowledge the reader has regarding the topic discussed in the summary can affect the judgement of a reader regarding the ordering of information. For example, a native speaker might find it comparatively easier to determine the ordering among a given set sentences than a non-native speaker in the language in which the summary is composed. Moreover, if the reader is already aware of the complete story, then he or she might be able to still understand a summary presented in out of order. Third, an evaluation measure must only take into consideration the ordering of information and not the content. If both ordering and content change among different summaries, then it might be difficult separate the quality of ordering without being influenced by the content. This challenge is often circumvent by sentence ordering systems because in the sentence ordering task we assume that the set of sentences to be ordered is given to the system and we only must determine the relative ordering among those sentences. Therefore, the summaries generated by different sentence ordering algorithms for the same set of sentences will have the same content.

One solution that is administered by researchers studying the problem of sentence ordering is to perform subjective evaluations. Specifically, a group of people (referred to as human judges or annotators) are given a list of guidelines to determine the quality of a summary. For example, four-level grades such as the ones shown below are frequently used to subjectively evaluate an ordering of sentences produced by an algorithm.

Perfect A *perfect* summary is a text that we cannot improve any further by re-
    ordering.
Acceptable An *acceptable* summary is one that makes sense, and is unnecessary to
    revise even though there is some room for improvement in terms of its readability.
Poor A *poor* summary is one that loses the thread of the story at some places, and
    requires minor amendments to bring it up to an acceptable level.
Unacceptable An *unacceptable* summary is one that leaves much to be improved
    and requires overall restructuring rather than partial revision.

Figure 12.9 shows a summary that obtained a *perfect* grade. The ordering 1 − 4 − 5 − 6 − 7 − 8 − 2 − 3 − 9 − 10 was assigned an *acceptable* grade, whereas 4 − 5 − 6 − 7 − 1 − 2 − 3 − 8 − 9 − 10 was given a *poor* grade. A random ordering of the ten sentences 4 − 7 − 2 − 10 − 8 − 3 − 1 − 5 − 6 − 9 received an *unacceptable* grade.

1. Hurricane Gilbert, one of the strongest storms ever, slammed into the Yucatan Peninsula Wednesday and leveled thatched homes, tore off roofs, uprooted trees and cut off the Caribbean resorts of Cancun and Cozumel.
2. Tropical Storm Gilbert formed in the eastern Caribbean and strengthened into a hurricane Saturday night.
3. Gilbert reached Jamaica after skirting southern Puerto Rico, Haiti and the Dominican Republic.
4. The Mexican National Weather Service reported winds gusting as high as 218 mph earlier Wednesday with sustained winds of 179 mph.
5. More than 120,000 people on the northeast Yucatan coast were evacuated, the Yucatan state government said.
6. Shelters had little or no food, water or blankets and power was out.
7. The storm killed 19 people in Jamaica and five in the Dominican Republic before moving west to Mexico.
8. Prime Minister Edward Seaga of Jamaica said Wednesday the storm destroyed an estimated 100,000 of Jamaica's 500,000 homes when it throttled the island Monday.
9. The National Hurricane Center said a hurricane watch was in effect on the Texas coast from Brownsville to Port Arthur and along the coast of northeast Mexico from Tampico north.
10. The National Hurricane Center said Gilbert was the most intense storm on record in terms of barometric pressure.

Fig. 12.9 An example of a *perfect* grade summary

In addition to providing guidelines, it is also helpful to provide some examples that belong to each grade. Although how well we define the grades, the gradings assigned by different human judges can vary. Inter-judge agreement measures such as the Kappa statistic and Kendall's coefficient of concordance (Kendall's W) are popularly used to measure the grades assigned by different human judges for orderings of sentences produced by a particular algorithm. It is both time consuming and costly to obtain human judgements for sentence orderings. The human judges must not only read the summaries but also read the source articles at times. Crowd sourcing tools such as the Amazon's Mechanical Turk[3] are gaining popularity as cheap alternatives to elicit human judgements for natural language generation systems [27]. However, the quality of the evaluations conducted by non-expert human judges is often unpredictable and one must perform careful pre-processing to exclude noisy gradings before they can be accepted. Despite the costs involved in performing subjective evaluations a major drawback in using human judgements is that we cannot reproduce subjective evaluations. Each time we improve a sentence ordering algorithm we must re-perform the subjective evaluations. The time taken to complete a subjective evaluation can be significantly long. This can also be a problem when iteratively improving a sentence ordering system.

As a solution to this problem several semi-automatic evaluation measures have been proposed for the purpose of evaluating an ordering of sentences. Before we discuss the semi-automatic evaluation measures in detail, it is useful to note the

---

[3]www.mturk.com

$$T_{eval} = (e \succ a \succ b \succ c \succ d)$$
$$T_{ref} = (a \succ b \succ c \succ d \succ e)$$

Fig. 12.10 An example of an ordering under evaluation T_{eval} and its reference T_{ref}

commonalities among those measures. All the semi-automatic evaluation measures we discuss compare an ordering of sentences produced by an algorithm against one or more orderings of those sentences created by human annotators. Therefore, the evaluation measure are not fully automatic but requires humans to construct correct orderings before-hand. However, the value of the semi-automatic evaluation measures is that we can easily reproduce the evaluation for any number of orderings produced by different algorithms for the same set of sentences. This is important because we can directly compare different algorithms for their ability to order a given set of sentences as done by human annotators. However, it must be emphasized that any work studying the problem of information ordering must not be satisfied by performing only semi-automatic evaluation but also consider performing subjective evaluations as a final step.

As mentioned earlier, even though subjective grading consumes much time and effort, we cannot reproduce the evaluation afterwards. Semi-automatic evaluation measures are particularly useful when evaluations must be performed quickly and repeatedly to tune an algorithm. Previous studies [1, 14, 15] employed rank correlation coefficients including Spearman's rank correlation coefficient, and Kendall's rank correlation coefficient (Kendall's τ), to compare a sentence ordering produced by a system against a manual ordering. In this section, we briefly survey the existing semi-automatic evaluation measures (specifically, Kendall τ and Spearman rank correlation coefficient), and introduce *average continuity* as an alternative evaluation measure for sentence orderings.

Let $S = s_1 \dots s_N$ be a set of N items to be ranked. Let π and σ denote two distinct orderings of S. Then, Kendall's rank correlation coefficient [13] (also known as Kendall's τ) is defined as follows,

$$\tau = \frac{4C(\pi, \sigma)}{N(N-1)} - 1. \tag{12.8}$$

Here, $C(\pi, \sigma)$ is the number of concordant pairs between π and σ (i.e. the number of sentence pairs that have the same relative positions in both π and σ). For example, in Fig. 12.10 between T_{eval} and T_{ref}, there are six concordant sentence pairs: (a, b), (a, c), (a, d), (b, c), (b, d), and (c, d). These six concordant pairs yield a Kendall's τ of 0.2. Kendall's τ is in the range $[-1, 1]$. It takes the value 1 if the two sets of orderings are identical, and -1 if one is the reverse of the other.

Likewise, Spearman's rank correlation coefficient (r_s) between orderings π and σ is defined as follows,

$$r_s = 1 - \frac{6}{N(N+1)(N-1)} \sum_{i=1}^{N} (\pi(i) - \sigma(i))^2. \tag{12.9}$$

Here, $\pi(i)$ and $\sigma(i)$ respectively denote the ith ranked item in π and σ. Spearman's rank correlation coefficient for the example shown in Fig. 12.10 is 0. Spearman's rank correlation, r_s, ranges from $[-1, 1]$. Similarly to Kendall's τ, the r_s value of 1 is obtained for two identical orderings, and the r_s computed between an ordering and its revers is -1.

In addition to Spearman's and Kendall's rank correlation coefficients, we propose an *average continuity* metric, which extends the idea of the continuity metric [22], to continuous k sentences.

A text with sentences arranged in proper order does not interrupt the process of a human reading from one sentence to the next. Consequently, the quality of a sentence ordering produced by a system can be estimated by the number of continuous sentence segments that it shares with the reference sentence ordering. For example, in Fig. 12.10 the sentence ordering produced by the system under evaluation (T_{eval}) has a segment of four sentences ($a \succ b \succ c \succ d$), which appears exactly in that order in the reference ordering (T_{ref}). Therefore, a human can read this segment without any disfluencies and will find to be coherent.

This is equivalent to measuring a precision of continuous sentences in an ordering against the reference ordering. We define P_n to measure the precision of n continuous sentences in an ordering to be evaluated as,

$$P_n = \frac{m}{N - n + 1}. \tag{12.10}$$

Here, N is the number of sentences in the reference ordering, n is the length of continuous sentences on which we are evaluating, and m is the number of continuous sentences that appear in both the evaluation and reference orderings. In Fig. 12.10, we have two sequences of three continuous sentences (i.e., ($a \succ b \succ c$) and ($b \succ c \succ d$)). Consequently, the precision of three continuous sentences P_3 is calculated as:

$$P_3 = \frac{2}{5 - 3 + 1} = 0.67. \tag{12.11}$$

The Average Continuity (AC) is defined as the logarithmic average of P_n over 2 to k:

$$AC = \exp\left(\frac{1}{k - 1}\sum_{n=2}^{k} \log(P_n + \alpha)\right). \tag{12.12}$$

Here, k is a parameter to control the range of the logarithmic average, and α is a fixed small value. It prevents the term inside the logarithm from becoming zero in case P_n is zero. We set $k = 4$ (i.e. more than five continuous sentences are not included for evaluation), and $\alpha = 0.001$. The average continuity is in the range $[0, 1]$. It becomes 0 when the evaluation and reference orderings share no continuous sentences and 1 when the two orderings are identical.

Let us compute the average continuity between the two orderings T_{eval} and T_{ref} as shown in Fig. 12.10. First, we observe that there are three segments of two consecutive sentences (i.e. in Formula (12.10), for $n = 2$, $m = 3$) between T_{eval} and T_{ref}. Namely, $(a \succ b)$, $(b \succ c)$, and $(c \succ d)$. Therefore, P_2 computed using Formula (12.10) is 0.75 (i.e. $3/(5 - 2 + 1)$). Similarly, we observe there are two segments of three consecutive sentences between T_{eval} and T_{ref}. They are $(a \succ b \succ c)$ and $(b \succ c \succ d)$. Therefore, P_3 is 0.67 (i.e. $2/(5 - 3 + 1)$). Finally, we observe that there exist a single segment of four consecutive sentences (i.e. $a \succ b \succ c \succ d$) between T_{eval} and T_{ref}. Therefore, P_4 as computed using Formula (12.10) is 0.5 (i.e. $1/(5 - 4 + 1)$). There are no segments with five or more consecutive sentences. We then use Formula (12.12) to compute average continuity between T_{eval} and T_{ref} as follows,

$$ AC = \exp\left(\frac{1}{4 - 1} \times (\log(P_2 + \alpha) + \log(P_3 + \alpha) + \log(P_4 + \alpha)) \right). $$

Substituting values for P_2, P_3, and P_4 computed as above, and setting $\alpha = 0.001$, we obtain an average continuity of 0.63 between T_{eval} and T_{ref}.

The underlying idea of Formula (12.12) is similar to that of BLEU metric [23], which was developed for the semi-automatic evaluation of machine-translation (MT) systems. BLEU compares the overlap between a translation produced by an MT system and a translation created by a human using n-grams (of words). If the two translations (by the MT system and by the human) share a large number of n-grams, then the MT system receives a high BLEU score. BLEU first computes the precision for n-grams with different lengths and then computes the geometric mean of all precision values using a formula similar to Formula (12.12) used to compute average continuity. Between two orderings of the same set of sentences, the set of words that appear in both orderings is identical; only the ordering of sentences is different. Therefore, average continuity measures the overlap between a summary produced by a system and a human-made ordering of those sentences using segments of continuous sentences instead of n-grams of words.

12.4.1 Using Multiple Reference Orderings to Evaluate Sentence Orderings

There can be more than one way to order a set of sentences to create a coherent summary. Therefore, to evaluate a sentence ordering produced by an algorithm, we must compare it with multiple reference orderings produced by different human annotators. However, all evaluation measures described in Sect. 12.4 compare a system ordering against one reference ordering. In this section, we modify the evaluation measures introduced in Sect. 12.4 to handle more than one reference ordering.

When comparing a system ordering against multiple reference orderings using the Kendall's τ, we consider a sentence pair (a, b) in the system ordering to be

$$T_{eval} = (e \succ a \succ b \succ c \succ d)$$

$$T_A = (a \succ b \succ c \succ d \succ e)$$

$$T_B = (b \succ c \succ d \succ e \succ a)$$

Fig. 12.11 Comparing multiple reference orderings with a system ordering

concordant if at least one of the reference orderings has sentence a before sentence b. For example, let us consider the case shown in Fig. 12.11 in which we compare a system ordering, T_{eval}, against two reference orderings T_A and T_B. If we only use T_A as the reference, then pair (e, a) in T_{eval} will not be a concordant pair. However, if we compare T_{eval} against T_B, we find that e is followed by a. Therefore, we conclude that (e, a) is a concordant pair. Kendall's τ between a system ordering π and a set $\{\sigma_1, \ldots, \sigma_t\}$ of t multiple reference orderings is defined as,

$$\tau = \frac{4C(\pi, \sigma_1, \ldots, \sigma_t)}{N(N-1)} - 1$$

$$\text{where, } C(\pi, \sigma_1, \ldots, \sigma_t) = \sum_{i=1}^{N} \sum_{j=i+1}^{N} \vee_{k=1}^{t} (\pi(i) < \sigma_k(j)). \qquad (12.13)$$

Here, N is the number of sentences being ordered, and $\vee_{k=1}^{t}$ is a disjunctive function that returns the value 1 if at least one of the $(\pi(i) < \sigma_k(j))$ inequalities is satisfied. It returns the value 0 if none of the inequalities are satisfied.

We modify the Spearman's rank correlation coefficient (Formula (12.9)) by replacing the term under summation by using the minimum distance between the corresponding ranks of a system ordering, and any one of the reference orderings. The revised formula is given by,

$$r_s = 1 - \frac{6}{N(N+1)(N-1)} \sum_{i=1}^{N} \min_{1 \le j \le t} (\pi(i) - \sigma_j(i))^2. \qquad (12.14)$$

We consider continuous sentence sequences between a system ordering and multiple reference orderings, to extend the average continuity (Formula (12.12)). Specifically, when computing m (the number of continuous sentences that appear in both a system and reference ordering) in Formula (12.10), we compare the continuous sequences of sentences in the system ordering with all reference orderings. If at least one of the reference orderings contains the sequence under consideration, then it is accepted. For example, in Fig. 12.11, the sequence $e \succ a$ in T_{eval} appears in T_B. It is counted as a continuous sequence of length 2, when computing P_2 along with $a \succ b$, $b \succ c$, and $c \succ d$. Once P_n values are computed as describe above, we use Formula (12.12) to compute the average continuity.

Table 12.1 Correlation between two sets of human-ordered extracts

Metric	Mean	Std. Dev	Min	Max
Spearman	0.739	0.304	−0.2	1
Kendall	0.694	0.290	0	1
Average continuity	0.401	0.404	0.001	1

12.4.2 Results of Semi-automatic Evaluation

Having described semi-automatic evaluation measures in Sect. 12.4, we are now in a position to evaluate the different sentence ordering heuristics presented in Sect. 12.2 and the machine learning approach presented in Sect. 12.3. As evaluation data, we use the third Text Summarization Challenge (TSC-3) corpus.[4] Text Summarization Challenge is a multiple document summarization task organized by the "National Institute of Informatics Test Collection for IR Systems" (NTCIR) project.[5] TSC-3 dataset was introduced in the 4th NTCIR workshop held in June 2–4, 2004. The TSC-3 dataset contains multi-document summaries for 30 news events. The events are selected by the organizers of the TSC task. For each topic, a set of Japanese newspaper articles are selected using some query words. Newspaper articles are selected from Mainichi Shinbun and Yomiuri Shinbun, two popular Japanese newspapers. All newspaper articles in the dataset have their date of publication annotated. Moreover, once an article is published, it is not revised or modified. Therefore, all sentences in an article bares the time stamp of the article.

For each topic, the organizers of the TSC task provide a manually extracted set of sentences. On average, a manually extracted set of sentences for a topic contains 15 sentences. The participants of the workshop are required to run their multi-document summarization systems on newspaper articles selected for each of the 30 topics and submit the results to the workshop organizers. The output of each participating system is compared against the manually extracted set of sentences for each of the topics using precision, recall and F-measure. Essentially, the task evaluated in TSC is sentence extraction for multi-document summarization. In order to construct the training data for the machine learning approach, two human subjects are required to arrange the extracts. The two human subjects worked independently and arranged sentences extracted for each topic. They were provided with the source documents from which the sentences were extracted. They read the source documents before ordering sentences in order to gain background knowledge on the topic. From this manual ordering process, we obtained 30(topics) × 2(humans) = 60 sets of ordered extracts. Table 12.1 shows the agreement of the ordered extracts between the two subjects. The correlation is measured by three metrics: Spearman's rank correlation,

---

[4]http://lr-www.pi.titech.ac.jp/tsc/tsc3-en.html

[5]http://research.nii.ac.jp/ntcir/index-en.html

Table 12.2 Comparison with human-made ordering

Method	Spearman	Kendall	Average continuity
Random ordering	−0.127	−0.069	0.011
Topical-closeness criterion	0.414	0.400	0.197
Precedence criterion	0.415	0.428	0.293
Succession criterion	0.473	0.476	0.291
Chronology criterion	0.583	0.587	0.356
Machine learning (integration)	0.603	0.612	0.459

Kendall's rank correlation, and average continuity. The mean correlation values (0.74 for Spearman's rank correlation and 0.69 for Kendall's rank correlation) indicate a strong agreement in sentence orderings made by the two subjects. In 8 out of the 30 extracts, sentence orderings created by the two human subjects were identical.

To train and test the machine learning approach, leave-one-out method is used. Specifically, the leave-out-out method arranges an extract by using an SVM model trained from the rest of the 29 extracts. Repeating this process 30 times with a different topic for each iteration, we generated a set of 30 orderings for evaluation.

Table 12.2 reports the resemblance of orderings produced by six algorithms to the two human-made ones, using the modified versions of Spearman's r_s, Kendall's τ, and average continuity described in Sect. 12.4.1. We experimentally determine the optimum parameter values (i.e. C = 32 and gamma = 0.5) for the radial basis functions (RBF) kernel in SVM. RBF kernel with those parameter values are used for all experiments. libSVM[6] was used as the SVM implementation.

As seen from Table 12.2, the machine learning approach outperforms the rest in all evaluation metrics. Among the different heuristics compared, chronological criterion appeared to play the most major role. Succession criterion performs slightly better than precedence criterion when compared using the Kendall coefficient and the Spearman coefficient. However, the two methods are indistinguishable using average continuity. Topical-closeness criterion reports the lowest performance among the four heuristic criteria. Random ordering gained almost zero in all three evaluation metrics. The one-way analysis of variance (ANOVA) verified the effects of different algorithms for sentence orderings with all metrics ($p < 0.01$). Moreover, we conduct the Tukey Honest Significant Differences (HSD) [28] test to compare the differences among these algorithms. The Tukey HSD test revealed that machine learning approach is significantly better than the rest.

---

[6]http://www.csie.ntu.edu.tw/~cjlin/libsvm/

12.5 Summary and Future Work

In this Chapter, we introduced the problem of information ordering in the context of multi-document summarization systems. An in depth review of numerous methods proposed to find a coherent ordering of sentences to create a summary were presented. We also looked at how we can combine the different heuristic approaches proposed in previous work on multi-document summarization using a machine learning technique. In brief, the machine learning approach learns a binary classification model that determines the order in which two sentences must be placed in a summary. A complete discussion of this machine learning method is beyond the scope of this book. We would like to direct the interested reader to [3].

There are still numerous open-issues in the field of information ordering. First, how to efficiently search the potentially combinatorial space of possible sentence orderings without using greedy algorithms remains an unsolved problem. Given n number of sentences, there are factorial n (i.e. $n!$) number of possible orderings among those sentences. Although there can be more than one acceptable orderings for a given set of sentences, the number of such orderings is much smaller than the possible number of orderings. Therefore, the problem of searching for the best ordering of for a given set of sentences is a combinatorial search problem which is not adequately studied. Most existing sentence ordering algorithms (including the machine learning-based method introduced in this chapter) resort to simple greedy search algorithms that do not perform an exhaustive search over this combinatorial ordering space. Recently, there has been some progress in listwise rank learning methods in the machine learning community that exploits permutation probability to learn a total ordering for a given set of items [29].

Second, there is a growing need for semi-automatic evaluation measures for sentence ordering. The measures discussed in this chapter such as Kendall's correlation measure have shown a high degree of correlation between subjective gradings assigned by human judges. Therefore, semi-automatic measures discussed in this chapter can be reliably use to evaluate a sentence ordering system. Despite these empirical evidence that support semi-automatic evaluation measures, they are not perfect alternatives for performing subjective evaluations. Given the fact that there can be more than one correct ordering for a given set of sentences, we must further study the possibilities of using multiple reference orderings in semi-automatic evaluation of sentence ordering. The approach presented in Sect. 12.4.1 would assume a pairwise ordering of two sentences as correct if it appears in any one of the human-made orderings. This simple rule is only tested for the case where we have only two sets of reference orderings (created by two independent human annotators). It remains unknown how well it would perform when we have more than two sets of references. For example, under this rule, a randomly produced sentence ordering would obtain higher evaluation scores when the number of references increase. To incorporate multiple references in semi-automatic evaluation measures, we must overcome this bias.

Ordering is only one of the factors that determines the readability of a summarization system. Readability of a text in general can be improved using many other techniques such as paraphrasing, sentence re-structuring (i.e. passive vs. active voice), expansion of acronyms, substituting zero pronouns and resolving co-references. Each of those techniques have been studied individually in the natural language community and numerous algorithms have been proposed. However, the exact combination of those techniques to improve the readability of a summary is not well studied. The future research in information ordering is expected to bring exciting new developments.

References

1. Barzilay, R., Lee, L.: Catching the drift: probabilistic content models, with applications to generation and summarization. In: HLT-NAACL 2004: Proceedings of the Main Conference, Boston, pp. 113–120 (2004)
2. Barzilay, R., Elhadad, N., McKeown, K.: Inferring strategies for sentence ordering in multi-document news summarization. J. Artif. Intell. Res. **17**, 35–55 (2002)
3. Bollegala, D., Okazaki, N., Ishizuka, M.: A bottom-up approach to sentence ordering for multi-document summarization. Inf. Process. Manag. **46**(1), 89–109 (2010)
4. Bos, J., Maekert, K.: Recognising textual entailment with logical inference. In: Proceedings of the conference on Human Language Technology and Empirical Methods in Natural Language Processing (HLT-EMNLP 2005), Vancouver, pp. 628–635 (2005)
5. Carbonell, J., Goldstein, J.: The use of mmr, diversity-based reranking for reordering documents and producing summaries. In: Proceedings of the 21st Annual International ACM SIGIR Conference on Research and Development in Information Retreival, Melbourne, pp. 335–336 (1998)
6. Dagan, I., Glickman, O.: Probabilistic textual entailment: generic applied modeling of language variability. In: Proceedings of PASCAL Workshop on Learning Methods for Text Understanding and Mining, Grenoble (2004)
7. Duboue, P., McKeown, K.: Empirically estimating order constraints for content planning in generation. In: Proceedings of the 39th Annual Meeting of the Association for Computational Linguistics (ACL'01), Toulouse, pp. 172–179 (2001)
8. Duboue, P., McKeown, K.: Content planner construction via evolutionary algorithms and a corpus-based fitness function. In: Proceedings of the Second International Natural Language Generation Conference (INLG'02), New York, pp. 89–96 (2002)
9. Elhadad, N., McKeown, K.: Towards generating patient specific summaries of medical articles. In: Proceedings of the NAACL 2001 Workshop on Automatic Summarization, Pittsburgh (2001)
10. Filatova, E., Hovy, E.: Assining time-stamps to event-clauses. In: Proceedings of the 2001 ACL Workshop on Temporal and Spatial Information Processing, Toulouse (2001)
11. Ji, P.D., Pulman, S.: Sentence ordering with manifold-based classification in multi-document summarization. In: Proceedings of Empherical Methods in Natural Language Processing, Sydney, pp. 526–533 (2006)
12. Karamanis, N., Manurung, H.M.: Stochastic text structuring using the principle of continuity. In: Proceedings of the Second International Natural Language Generation Conference (INLG'02). Columbia University, New York, pp. 81–88 (2002)
13. Kendall, M.G.: A new measure of rank correlation. Biometrika **30**, 81–93 (1938)
14. Lapata, M.: Probabilistic text structuring: experiments with sentence ordering. In: Proceedings of the Annual Meeting of ACL 2003, Sapporo, pp. 545–552 (2003)
15. Lapata, M.: Automatic evaluation of information ordering. Comput. Linguist. **32**(4), 471–484 (2006)

16. Lapata, M., Lascarides, A.: Learning sentence-internal temporal relations. J. Artif. Intell. Res. **27**, 85–117 (2006)
17. Lin, C., Hovy, E.: Neats:a multidocument summarizer. In: Proceedings of the Document Understanding Workshop (DUC) (2001)
18. Mani, I., Wilson, G.: Robust temporal processing of news. In: Proceedings of the 38th Annual Meeting of ACL (ACL 2000), Hong Kong, pp. 69–76 (2000)
19. Mani, I., Schiffman, B., Zhang, J.: Inferring temporal ordering of events in news. In: Proceedings of North American Chapter of the ACL on Human Language Technology (HLT-NAACL 2003), Edmonton, pp. 55–57 (2003)
20. Mann, W., Thompson, S.: Rhetorical structure theory: toward a functional theory of text organization. Text **8**(3), 243–281 (1988)
21. McKeown, K., Klavans, J., Hatzivassiloglou, V., Barzilay, R., Eskin, E.: Towards multidocument summarization by reformulation: progress and prospects. In: AAAI/IAAI, Orlando, pp. 453–460 (1999)
22. Okazaki, N., Matsuo, Y., Ishizuka, M.: Improving chronological sentence ordering by precedence relation. In: Proceedings of 20th International Conference on Computational Linguistics (COLING 04), Geneva, pp. 750–756 (2004)
23. Papineni, K., Roukos, S., Ward, T., Zhu, W.J.: Bleu: a method for automatic evaluation of machine translation. In: Proceedings of the 40th Annual Meeting of the Association for Computational Linguistics (ACL), Philadelphia, pp. 311–318 (2002)
24. Platt, J.: Probabilistic outputs for support vector machines and comparison to regularized likelihood methods. In: Smola, J., et al. (eds.) Advances in Large Margin Classifiers, pp. 61–74. MIT, Cambridge (2000)
25. Radev, D.R., McKeown, K.: Generating natural language summaries from multiple on-line sources. Comput. Linguist. **24**(3), 469–500 (1999)
26. Reiter, E., Dale, R.: Building Natural Language Generation Systems. Cambridge University Press, Cambridge/New York (2000)
27. Snow, R., O'Connor, B., Jurafsky, D., Ng, A.Y.: Cheap and fast – but is it good? evaluating non-expert annotations for natural language tasks. In: EMNLP'08, Honolulu (2008)
28. Tukey, J.W.: Exploratory Data Analysis. Addison-Wesley, Reading (1977)
29. Xia, F., Liu, T.Y., Wang, J., Zhang, W., Li, H.: Listwise approach to learning to rank: theory and algorithm. In: ICML 2008, Helsinki, pp. 1192–1199 (2008)
30. Zanzotto, F.M., Moschitti, A.: Automatic learning of textual entailments with cross-pair similarities. In: Proceedings of the 21st International Conference on Computational Linguistics and the 44th Annual Meeting of the ACL, Sydney, pp. 401–408 (2006)

Chapter 13
Improving Speech-to-Text Summarization by Using Additional Information Sources

Ricardo Ribeiro and David Martins de Matos

Abstract Speech-to-text summarization systems usually take as input the output of an automatic speech recognition (ASR) system that is affected by issues like speech recognition errors, disfluencies, or difficulties in the accurate identification of sentence boundaries. We describe the inclusion of related, solid background information to cope with the difficulties of summarizing spoken language and the use of multi-document summarization techniques in single document speech-to-text summarization. In this work, we explore the possibilities offered by phonetic information to select the background information and conduct a perceptual evaluation to better assess the relevance of the inclusion of that information. Results show that summaries generated using this approach are considerably better than those produced by an up-to-date latent semantic analysis (LSA) summarization method and suggest that humans prefer summaries restricted to the information conveyed in the input source.

13.1 Introduction

News has been the subject of summarization for a long time, demonstrating the importance of both the subject and the process. Systems like NewsInEssence [48], Newsblaster [36], or even Google News substantiate this relevance that is also supported by the spoken language scenario, where most speech summarization

R. Ribeiro (✉)
L^2F – INESC ID/ISCTE – Instituto Universitário de Lisboa, Rua Alves Redol, 9, 1000-029 Lisboa, Portugal
e-mail: ricardo.ribeiro@inesc-id.pt

D. Martins de Matos
L^2F – INESC ID/IST, Rua Alves Redol, 9, 1000-029 Lisboa, Portugal
e-mail: david.matos@inesc-id.pt

T. Poibeau et al. (eds.), *Multi-source, Multilingual Information Extraction and Summarization 11*, Theory and Applications of Natural Language Processing, DOI 10.1007/978-3-642-28569-1_13, © Springer-Verlag Berlin Heidelberg 2013

systems concentrate on broadcast news [8, 27, 28, 34, 49]. Nevertheless, although the pioneering efforts on summarization go back to the work of Luhn [32], in 1958, and Edmundson [10], in 1969, it is only after the renaissance of summarization as a research area of great activity—following up on the Dagstuhl Seminar [13] in 1995—that the first multi-document news summarization system, SUMMONS [35], makes its breakthrough [48, 52]. Regarding speech summarization, the state of affairs is more problematic: news summarization systems appeared later and still focus only on single document summarization. In fact, while text summarization has attained some degree of success [20, 37, 52] due to the considerable body of work, speech summarization still requires further research, both in speech and text analysis, in order to overcome the specific challenges of the task [15, 37]. Issues like speech recognition errors, disfluencies, and difficulties in accurately identifying sentence boundaries must be taken into account when summarizing spoken language. And, for example, if on the one hand, recognition errors seem not to have a considerable impact on the summarization task [42, 43], on the other hand, spoken language summarization systems often explore ways of minimizing that impact [19,22,54]. In addition to recognition errors, it is well-known that speech segmentation influences general spoken document processing [45] and speech summarization [30, 34], in particular. The same happens with disfluencies: see, for instance, the work of Charniak and Johnson [6].

We argue that by including related solid background information from a different source less prone to this kind of errors (e.g., a textual source) in the summarization process, we are able to reduce the influence of recognition errors on the resulting summary. To support this argument, we developed a new approach to speech-to-text summarization that combines information from multiple information sources to produce a summary driven by the spoken language document to be summarized. The idea mimics the natural human behavior, in which information acquired from different sources is used to build a better understanding of a given topic. Endres-Niggemeyer [11, 12] describes, from a cognitive sciences perspective, the human understanding/summarization process as a knowledge-based activity, where prior knowledge, namely domain/factual knowledge, is used to build a mental model of an information source. Wan et al. [53] and Ribeiro and de Matos [50, 51] explore that idea in computational terms. Furthermore, we build on the conjecture that this background information is often used by humans to overcome perception difficulties. In that sense, one of our goals is also to understand what is expected in a summary: a comprehensive, shorter, text that addresses the same subject of the input source to be summarized (possibly introducing new information); or a text restricted to the information conveyed in the input source.

In our work, we explore the use of phonetic domain information to overcome speech recognition errors and disfluencies. Instead of using the traditional output of the ASR module, we use the phonetic transliteration of the output and compare it to the phonetic transliteration of solid background information. This enables the use of text, related to the input source, free from the common speech recognition issues, in further processing.

Broadcast news are used as a case study and news stories from online newspapers provide the background information. Media monitoring systems, used to transcribe and disseminate news, provide an adequate framework to test the proposed method.

The next section describes the related work; Sect. 13.3 presents a characterization of the speech-to-text summarization problem and how we propose to address it; Sect. 13.4 describes our use of phonetic domain information, given the previously defined context; Sect. 13.5 reports the case study, including the experimental set up and results; finally, conclusions are drawn in Sect. 13.6.

13.2 Related Work

Spoken language summarization is often depicted as a much harder task than text summarization [15, 37]. In fact, the previously enumerated problems that make speech summarization such a difficult task constrain the application of text summarization techniques to speech summarization (although in the presence of planned speech, as it partly happens in the broadcast news domain, that portability is more feasible [9]). On the other hand, speech offers possibilities like the use of prosody and speaker identification to ascertain relevant content.

Furui [15] identifies three main approaches to speech summarization: sentence extraction-based methods, sentence compaction-based methods, and combinations of both. Sentence extractive methods comprehend, essentially, methods like LSA [18, 42], Maximal Marginal Relevance [30, 56], and feature-based methods [33, 43, 55]. Feature-based methods combine several types of features: current work uses lexical, acoustic/prosodic, structural, and discourse features to summarize documents from domains like broadcast news, meetings, or lectures. Even so, spoken language summarization is still quite distant from text summarization in what concerns the use of discourse features, and shallow approaches are what can be found in state-of-the-art work such as the one presented by Maskey and Hirschberg [33] or Murray et al. [43]. Recent work advances on the use of discourse information by capturing the rhetorical structure of lecture speech (which has a rigid linguistic structure) from acoustic and linguistic features [55]. Sentence compaction methods are based on word removal from the transcription, with recognition confidence scores playing a major role [19, 29]. A combination of these two types of methods was developed by Kikuchi et al. [22], where summarization is performed in two steps: first, sentence extraction is done through feature combination; second, compaction is done by scoring the words in each sentence and then a dynamic programming technique is applied to select the words that will remain in the sentence to be included in the summary.

From the perspective of the techniques used in summarization process, we can considered approaches based on the detection of the most salient content by means of statistical (or significance) measures, usually considered unsupervised; approaches based on classification, which formulate the sentence selection process as a binary classification problem, usually considered supervised; and, the approaches based on the sentence structure or location information [8, 27].

Regarding the use of additional related information to summarize a single document, we can find work both on written and spoken language. Chatain et al. [7] base topic-adapted speech summarization on a sentence scoring function in which one of the components is an n-gram linguistic model that is computed from given data. However, in the two experiments performed, one using talks and the other using broadcast news, only the one using talks used a topic-adapted linguistic model and the data used for the adaptation consisted of the papers in the conference proceedings of the talk to be summarized. Chen et al. [8] propose a unified probabilistic generative framework for speech summarization ($P(S|D) = P(D|S)P(S)$, where S and D are random variables representing sentences and documents, respectively). Sentence ranking is done by combining $P(D|S)$ (relevance in context) and $P(S)$ (relevance by itself). Several approaches to compute $P(D|S)$ in which additional data is used are experimented with: language model using a general text news collection; relevance model [25] using a set of documents returned by an information retrieval (IR) system, using the sentence as query; Sentence Topical Mixture Model training based on a text news collection. Sentence prior probability computation is also based on the set of documents returned by the IR system. CollabSum [53] explores document proximity in single document text summarization: by means of a clustering algorithm, documents related to the document to be summarized are grouped in a cluster and used to build a graph that reflects the relationships between all sentences in the cluster; then informativeness is computed for each sentence and redundancy removed to generate the summary.

13.3 Problem Characterization

Here we introduce an unambiguous conceptual characterization of our view of the problem, clarifying the objectives and the abstractions of our model. This serves as a basis for a better comprehension of the case study presented in Sect. 13.5, providing, at the same time, a framework that helps to understand how our method can be applied in other contexts. Some of the introduced concepts are intentionally underspecified, since we want to stress the need for those concepts, while leaving their concretization to specific contexts of application.

Summarization can be seen as a reductive transformation Φ that, given an input source I, produces a summary S:

$$S = \Phi(I), \tag{13.1}$$

such that $len(S) < len(I)$ and $inf(S)$ is as close as possible of $inf(I)$; where $len()$ is the length of the given input and $inf()$ is the information conveyed by its argument. While $len()$ can be objectively defined, for example, in terms of the number of segments (words [28] or sentences [33, 49]), $inf()$ is an ideal function that returns the informative content of an input source, that is approximated by every summarization model Φ.

The problem is that in order to compute S, we are not using I, but $\tilde{I}$, a noisy representation of I. Thus, we are computing $\tilde{S}$, which is a summary affected by the noise present in $\tilde{I}$:

$$\tilde{S} = \Phi(\tilde{I}). \tag{13.2}$$

This means that

$$inf\,(\tilde{S}) \subset inf\,(S) \subset inf\,(I), \text{ whereas} \tag{13.3}$$

$$len\,(\tilde{S}) \approx len\,(S) < len\,(I). \tag{13.4}$$

Our argument is that using a similar reductive transformation Ψ, where solid related background information B is also given as input, it is possible to compute a summary $\hat{S}$:

$$\hat{S} = \Psi(\tilde{I}, B), \text{ such that} \tag{13.5}$$

$$inf\,(\tilde{S}) \subset (inf\,(\hat{S}) \cap inf(S)) \subset inf(I), \text{ with} \tag{13.6}$$

$$len\,(\hat{S}) \approx len\,(\tilde{S}) \approx len(S) < len(I). \tag{13.7}$$

As seen in Sect. 13.2, the most common method to perform these transformations is by selecting sentences (or extracts) from the corresponding input sources.

Thus, let the input source representation $\tilde{I}$ be composed by a sequence of extracts,

$$\tilde{I} = e_1, e_2, \ldots, e_n \tag{13.8}$$

and the background information be defined as a sequence of sentences

$$B = s_1, s_2, \ldots, s_m. \tag{13.9}$$

The proposed method consists of selecting sentences s_i from the background information B such that

$$sim(s_i, e_j) < \varepsilon \wedge 0 \leq i \leq m \wedge 0 \leq j \leq n, \tag{13.10}$$

with $sim()$ being a similarity function and ε an adequate threshold. The difficulty lies in defining both the function and the threshold.

13.4 Working in the Phonetic Domain

Our approach to the method proposed in (13.8)–(13.10) minimizes the effects of recognition errors through the selection, from previously determined background knowledge, of sentences close to the sentence-like units (SUs) in the news story

Table 13.1 Phonetic features

Feature	Values
Type	Vowel, consonant
Vowel length	Short, long, diphthong, schwa
Vowel height	High, mid, low
Vowel frontness	Front mid back
Lip rounding	Yes, no
Consonant type	Stop, fricative, affricative, nasal, liquid
Place of articulation	Labial, alveolar, palatal, labio-dental, dental, velar
Consonant voicing	Yes, no

transcription. Although bearing similarities (SUs resemble the concept of written text sentence in speech), the concept of SU is different from the concept of sentence, since, even though semantically complete, SUs can be smaller than a sentence [31]. In order to select sentences, while diminishing recognition problems, we compute the similarity between sentences and SUs at the phonetic level. The estimation of the threshold ε (13.10) is based on the distance, measured in the phonetic domain, between the output of the ASR and its hand-corrected version. This means that manual transcriptions are required to make the proposed procedure work. We describe the data collections used in our case study (Sect. 13.5) in Sect. 13.5.2.

The selection of sentences from the background information is based on the alignment cost of the phonetic transcriptions of SUs from the input source and sentences from the background information. Sentences from the background information with alignment costs below the estimated threshold are selected to be used in summary generation.

13.4.1 Similarity Between Segments

There are several ways to compute phonetic similarity. Kessler [21] states that phonetic distance can be seen as, among other things, differences between acoustic properties of the speech stream, differences in the articulatory positions during production, or as the perceptual distance between isolated sounds. Choosing a way to calculate phonetic distance is a complex process.

The phone similarity function used in this process is based on a model of phone production, where the phone features correspond to the articulatory positions during production: the greater the matching between phone features, the smaller the distance between phones. The phone features used are described in Table 13.1.

The computation of the similarity between SUs or SUs and sentences is based on the alignment of the phonetic transcriptions of the given segments. The generation of the possible alignments and the selection of the best alignment is done through the use of Weighted Finite-State Transducers (WFSTs) [41, 46].

13.4.2 Threshold Estimation Process

To estimate the threshold to be used in the sentence selection process, we use the algorithm presented in Fig. 13.1. The procedure consists of comparing automatic transcriptions and their hand-corrected versions: the output is the average difference between the submitted inputs.

Our intuition is that the phonetic distance between the automatic transcription and its hand-corrected version should be similar to the phonetic distance between the automatic transcription and adequate background information. In fact, the automatic transcription of a given spoken document and its hand-corrected version report the same information. The main difference lies in the noise present in the automatic version of that report. We conjecture that the average distance between those two versions of the same information report is similar to average distance between segments from the automatic transcription of a given spoken document and segments from a document from a different source addressing the same topic. Even though this heuristic may appear naive, we believe it is adequate as a rough approach, considering the target material (broadcast news).

13.5 A Case Study Using Broadcast News

In this work, broadcast news are used as a case study. The experiments made explore the use of news stories from online newspapers as additional information sources. Media monitoring systems, used to transcribe and disseminate news, such as the SSNT system, provide an adequate framework to test the proposed method.

13.5.1 Media Monitoring System

SSNT [3] is a system for selective dissemination of multimedia contents, working primarily with Portuguese broadcast news services. The system, depicted in Fig. 13.2, is based on an ASR module, that generates the transcriptions used by the topic segmentation and topic indexing, and summarization modules. User profiles enable the system to deliver e-mails containing relevant news stories. These messages contain the name of the news service, a generated title, a summary, a link to the corresponding video segment, and a classification according to a thesaurus used by the broadcasting company.

Preceding the speech recognition module, a jingle detection (JD) module detects the beginning of the show and the middle commercials, and an audio preprocessing module (APP), based on multi-layer perceptrons (MLPs), classifies the audio in accordance with several criteria: speech/non-speech, speaker segmentation and clustering, gender, and background conditions.

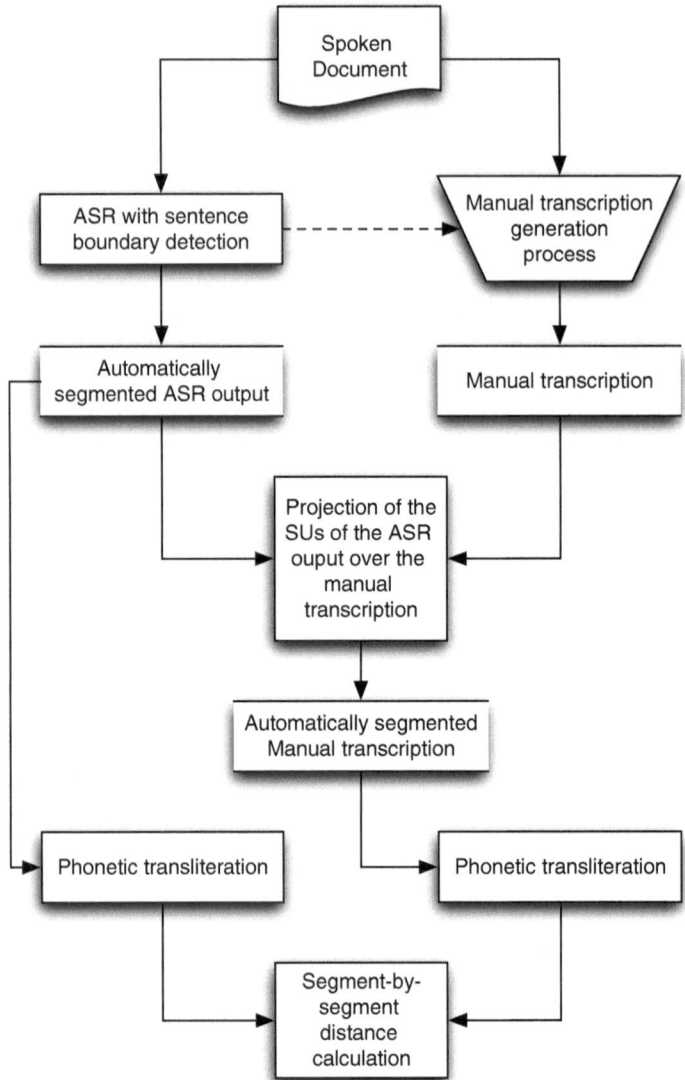

Fig. 13.1 Threshold estimation process: side open boxes represent data; regular boxes represent processes; the trapezoid represents a manual process; the *top* shape represents the documents to be processed; the *dashed line arrow* represents possible use

The ASR module, based on a hybrid speech recognition system that combines Hidden Markov Models with MLPs, with an average word error rate (WER) of 24% [3], greatly influences the performance of the subsequent modules.

The topic segmentation and topic indexing module was developed by Amaral and Trancoso [1]. Topic segmentation is based on clustering and groups transcribed segments into stories. The algorithm relies on a heuristic derived from the structure

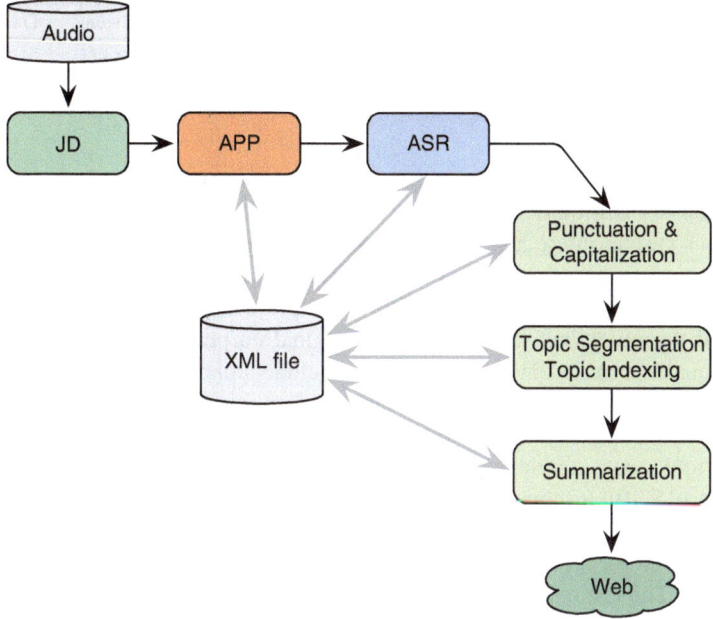

Fig. 13.2 SSNT architecture

of the news services: each story starts with a segment spoken by the anchor. This module achieved an *F-measure* of 68% [3]. The main problem identified by the authors was boundary deletion: a problem which also impacts the summarization task. Topic indexing is based on a hierarchically organized thematic thesaurus provided by the broadcasting company. The hierarchy has 22 thematic areas on the first level, for which the module achieved a correctness of 91.4% [2, 3].

Batista et al. [4] inserted a module for recovering punctuation marks, based on maximum entropy models, after the ASR module. The punctuation marks addressed were the "full stop" and "comma", which provide the sentence units necessary for use in the summarization module. This module achieved an *F-measure* of 56% and SER (Slot Error Rate, the measure commonly used to evaluate this kind of task) of 0.74.

Currently, the summarization module produces a summary composed by the first *n* sentences, as detected by the previous module, of each news story and a title (the first sentence). Our work focuses on improving this module.

13.5.2 Corpora

Two corpora were used in this experiment: a broadcast news corpus, the subject of our summarization efforts; and a written newspaper corpus, used to select the background information.

Table 13.2 Broadcast news corpus composition

Corpus	Stories	SUs	Tokens	Duration (h)
Train	121	2,064	40,121	5
Test	26	428	8,430	1

The broadcast news corpus is composed by excerpts of six Portuguese news programs. In addition to the automatically generated transcriptions, created using the SSNT system, there are hand-corrected versions, that constitute the reference. This corpus was developed as a complement to the Speech Recognition corpus created in the context of the ALERT European project [38, 39], and its development followed the guidelines defined for the original corpus. Both the automatic and the manual transcription of this corpus include punctuation and capitalization information, as well as story segmentation information. In the automatic version, the information was automatically generated by the SSNT system. The manual version took about ten times real time to be produced in our laboratory by trained annotators [5, 40]. Its composition (number of stories, number of sentence-like units (SUs), number of tokens, and duration) is detailed in Table 13.2. To estimate the threshold used for the selection of the background information, five news programs were used. The last one was used for evaluation.

The written newspaper corpus consists of the online version a Portuguese newspaper, downloaded daily from the Internet. In this experiment, three editions of the newspaper were used, corresponding to the day and the two previous days of the news program to be summarized. The corpus is composed by 135 articles, 1,418 sentence-like units, and 43,102 tokens.

13.5.3 The Summarization Process

The summarization process we implemented is characterized by the use of LSA to compute the relevance of the extracts (sentences/sentence-like units) of the given input source. Our choice was mainly motivated by the following rationale: the performance over automatically generated speech transcriptions and the corresponding manual transcriptions reported by previous work [42]; the room for improvement demonstrated by the results obtained by making small changes over the simple application of the method [49], being at the same time a natural framework for the integration of additional information; and, a critical analysis [47] that demonstrates that text summarization models applied to speech transcriptions still have an up-to-date performance.

LSA is based on the Singular Value Decomposition [16] of the term-sentence frequency $m \times n$ matrix, M. U is an $m \times n$ matrix of left singular vectors; Σ is the $n \times n$ diagonal matrix of singular values; and, V is the $n \times n$ matrix of right singular vectors (only possible if $m \geq n$):

$$M = U \Sigma V^{T}.$$

The idea behind the method is that the decomposition captures the underlying topics of the document by means of co-occurrence of terms (the latent semantic analysis), and identifies the best representative sentence-like units of each topic. Summary creation can be done by picking the best representatives of the most relevant topics according to a defined strategy.

For this summarization process, we implemented a module following the original ideas of Gong and Liu [17] and the ones of Murray et al. [42] for solving dimensionality problems, and using, for matrix operations, the GNU Scientific Library.[1]

13.5.4 Experimental Results

Our main objective was to understand if it is possible to select relevant information from background information that could improve the quality of speech-to-text summaries. To assess the validity of this hypothesis, five different processes of generating a summary were considered. To better analyze the influence of the background information, all automatic summarization methods are based on the up-to-date LSA method previously described: one taking as input only the news story to be summarized (*Simple*) and used as baseline; other taking as input only the selected background information (*Background only*); and, the last one, using both the news story and the background information (*Background + News*). The other two processes were human: extractive (using only the news story) and abstractive (understanding the news story and condensing it by means of paraphrase). Since the abstractive summaries had already been created, summary size was determined by their size (which means creating summaries using a compression rate of around 10% of the original size).

In what concerns the evaluation process, although ROUGE [26] is the most common evaluation metric for the automatic evaluation of summarization, since our approach might introduce in the summary information that is not present in the original input source, we found that a human evaluation was more adequate to assess the relevance of that additional information. A perceptual evaluation is also adequate to assess the perceived quality of the summaries and a better indicator of what is expected in a good summary. Each evaluator was given, for each story, the transcription of the news story (without background information) and five summaries, corresponding to the five different methods presented before. The evaluation procedure consisted in identifying the best summaries and in the classification of each summary (1–5, 5 is better) according to its content and readability (which covers issues like grammaticality, existence of redundant information, or entity references [44]).

---

[1]http://www.gnu.org/software/gsl/

Table 13.3 Differences between the transcriptions
used in the two experiments. Automatic data used in
Experiment 1 was generated using an older version
of the SSNT system, hence the higher WER and the
higher SER

Experiment	WER	SER	Stories
1	19.5	90.2	15
2	16.5	81.5	17

In order to assess the influence of speech-related phenomena and better support our claims, we performed trials (Experiment 1 and 2) in two different versions of the same speech corpus described in Sect. 13.5.2. The differences, detailed in Table 13.3, are caused by processing that corpus with two different releases of the SSNT system described in Sect. 13.5.1.

As mentioned before, the whole summarization process begins with the selection of the background information. Using the threshold estimated as described in Sect. 13.4.2 and the method described in Sect. 13.4.1 to compute similarity between sentence-like units, in the first experiment, no background information was selected for 11 of the 26 news stories of the test corpus. For the remaining 15 news stories, summaries were generated using the three automatic summarization strategies described before. The evaluation was performed by a heterogeneous group of 16 people, whose characterization is presented in Fig. 13.3. To measure the agreement level among the human evaluators we applied several methods. To analyse the agreement concerning the identification of the best summary, we use the well-known measures Krippendorff's α [23] and Fleiss' Kappa [14], both adequate for multiple observers and nominal data. We obtained similar values for both measures: $\alpha = 0.215$ and $\kappa = 0.215$. α shows an agreement above chance ($\alpha = 0$ indicates that the preferences of the evaluators were equally distributed; $\alpha = 1$ indicates perfect agreement), but does not reach 0.667, the minimum needed to consider that there is strong agreement. The same happens with κ, which reaches the level of fair agreement [24] (with $\kappa < 0$ meaning poor agreement; and, $\kappa = 1$ meaning complete agreement). Nevertheless, that value is still far from what is considered a good agreement: a κ above 0.61. These low values are to a certain extent justified by the good performance of the proposed model (see Fig. 13.4): its summaries were several times preferred over the human summaries. To analyse the dispersion from the average scores of content and readability, we used the standard deviation. The values of standard deviation are analysed alongside the discussion of the results.

In Experiment 2, as can be seen in Table 13.3, the WER and SER were smaller than in the first experiment and the number of news stories used in the evaluation is larger. This larger number of news stories results from two aspects: in this trial, the SSNT system generated a larger number of story segments and the process of background information selection was able to select background information for two more news stories. To evaluate the summaries generated for the 17 news stories, we set up an Internet page and publicized its existence in different forums (for this

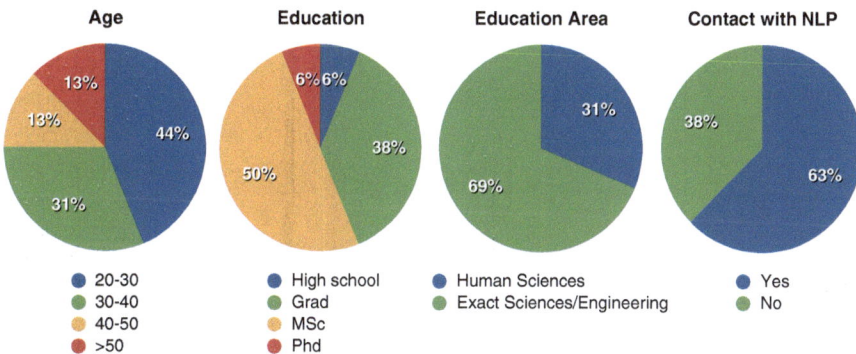

Fig. 13.3 Characterization of the human judges that participated in experiment 1

reason, in this experiment we do not have a characterization of the evaluators). We obtained an average of 9.5 answers per story. Concerning the inter-evaluator agreement, for experiment 2, it was only possible to calculate Krippendorff's α (which is able to deal with missing data), since not all evaluators completed the evaluation. We obtained an $\alpha = 0.12$, which is even lower than the α obtained for experiment 1. A reason for this decrease, beyond the good performance of the summarizers based on our model, is that, in fact, evaluators had more difficulty in selecting the best summary: the average number of summaries selected as best summary for each story was, in this experiment, 3.3, while in experiment 1, it was 2.8. This difficulty can be explained by the lower WER of the automatically transcribed data (see Table 13.3): automatic summaries with a higher WER are easily discarded. Note, for instance, that the summaries generated by the LSA baseline were voted as best summary in a larger number of stories than in experiment 1. Standard deviation for content and readability average values is again analysed further ahead in this section.

Concerning the overall results, it is surprising (see Figs. 13.4 and 13.5) that, in general, the extractive human summaries were preferred over the abstractive ones. According to Niggemeyer [12], in a specific scenario (in text summarization) in the case, summarizing from the WWW for physicians specializing in bone marrow transplantation users may sometimes prefer extractive summaries over abstractive ones. However, after a careful analysis of the human abstractive summaries, we believe that the main reason for the observed preference is that these summaries tend to mix both informative and indicative styles of summarization.

Beyond the natural preference for the human summaries, the automatically generated ones using background information (exclusively or not) were also selected as best summary (over the human created ones, especially in the first experiment) a non-negligible number of times. The poorest performance was attained, as expected, by the simple LSA summarizer. The summaries generated in this manner were only preferred in two news stories (for which all summaries were very similar) in the first experiment and on five stories in the second experiment. This growth from the

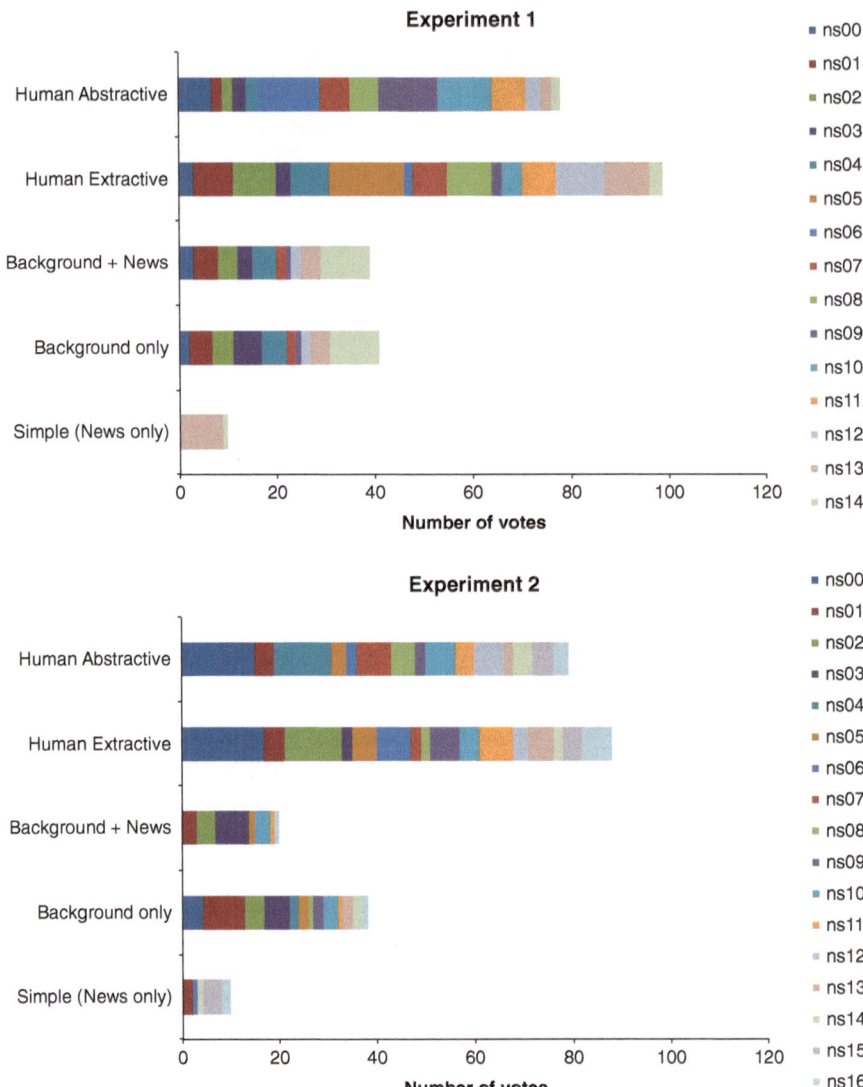

Fig. 13.4 Overall results for each summary creation method (nsnn identifies a news story). The cumulative number of votes as best summary in each news story is shown for each summarization method

first to the second experiment was also expected: the decrease of both the WER and the SER in this second trial helped improving the quality of the summaries based only on the news story. The results of the two approaches using background information were very similar in the first experiment. A result that can be explained by the fact that the summaries generated by these two approaches were equal for

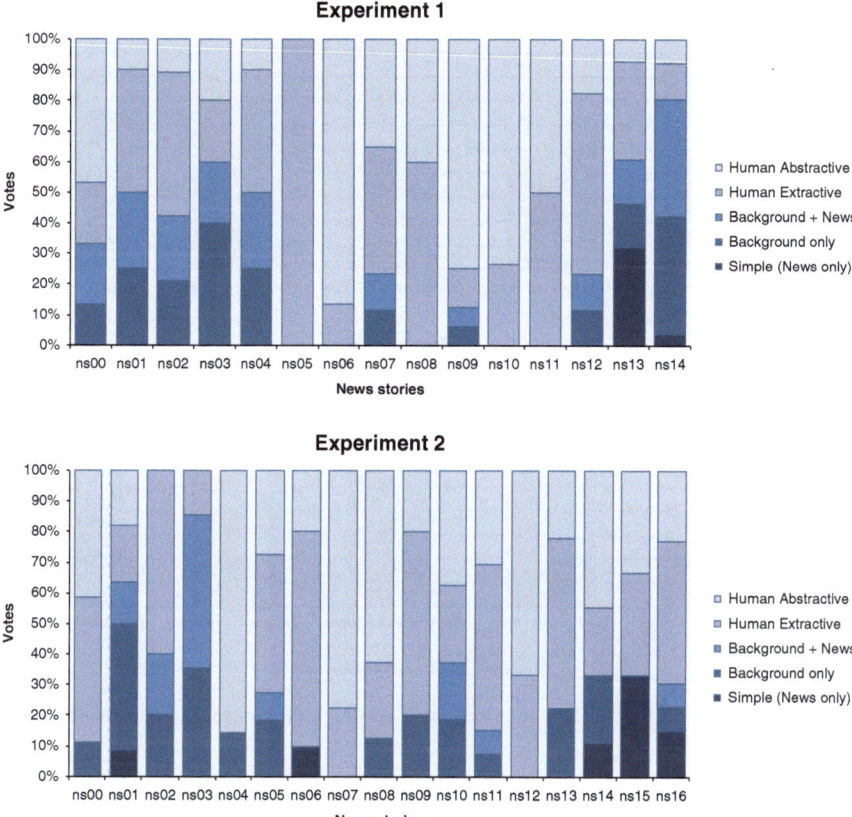

Fig. 13.5 Relative results for each news story (ns*n n* identifies a news story). Each column shows the distribution of the votes (in percentage) by the summarization methods for each news story

11 of the 15 news stories (in the remaining 4, the average distribution was 31.25% from the news story versus 68.75% from the background information). On the other hand, in the second experiment the results of these approaches were significantly different, something that can be explained by the different composition of the mixed approach summaries: they were equal only for 6 of the 17 news stories (in 4 of the remaining stories, the summaries using both the background information and the news story were completely composed by background information; and, in the remaining 7 news stories, the average distribution was 57.15% from the news story versus 42.85% from the background information).

Figure 13.6 further illustrates the results in terms of content and readability.

Regarding content, our results suggest that the choice of the best summary is highly correlated with its content, as the average content scores mimic the overall ones of Fig. 13.4. In what concerns readability, the summaries generated using background information achieved results comparable to human performance,

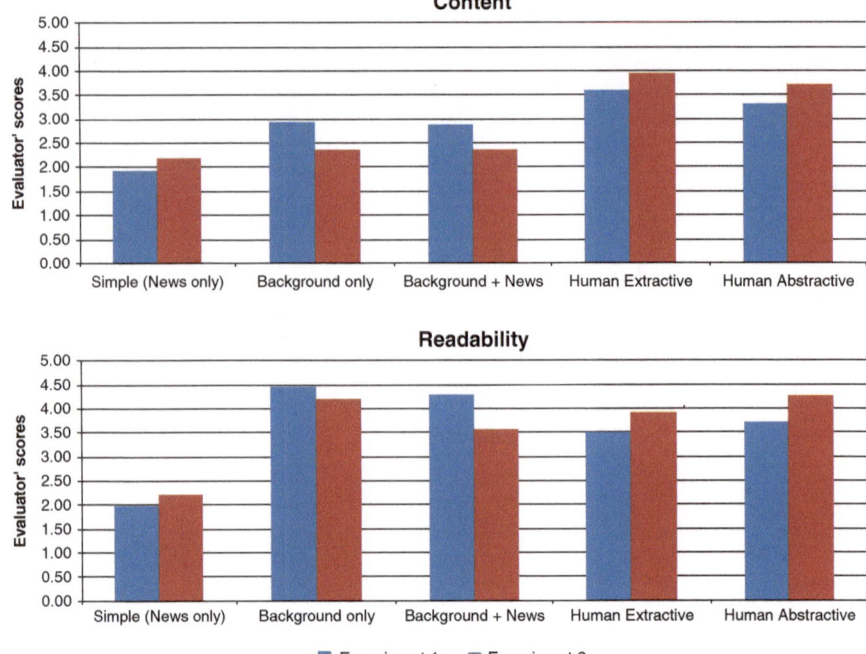

Fig. 13.6 Averages of the content and readability scores for each summary creation method

attaining the best results in the first experiment. The reasons underlying this outcome are that newspaper writing is naturally better planned than speech and that speech transcriptions are affected by the several problems described before (and the original motivation for the work), hence the idea of using written text as background information. Another interesting aspect is that, although in the first experiment, human abstractive summaries obtained worse results than the ones obtained by automatic generation using background information (both versions), which could suffer from coherence and cohesion problems, in the second experiment, results were similar to the ones obtained using exclusively background information and better that the ones obtained by mixing both the background information and the news story. Beyond differences due to a different evaluation group, the decrease in the mixed approach can be justified by the increase in the news story content in the composition of the summaries generated by this approach.

Figure 13.7 presents the standard deviation for content and readability scores. Concerning content, automatically generated summaries using background information achieved the highest standard deviation values. Those results are in part supported by some comments made by the human evaluators (on both experiments) on whether a summary should contain information that is not present in the input source. This aspect and the obtained results, suggest that this issue should be further analyzed, possibly using an extrinsic evaluation setup. On the other hand, standard

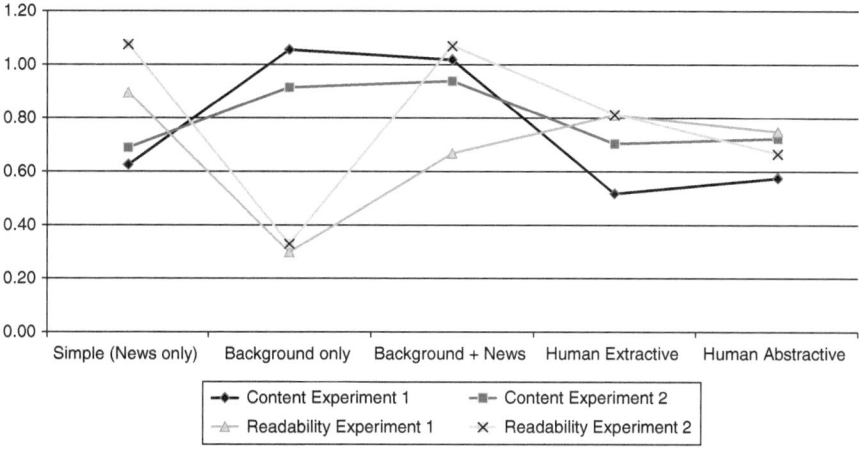

Fig. 13.7 Standard deviation of the content and readability scores

deviation values for readability show that there is a considerable agreement in what concerns this criterion. Note, however, the high standard deviation value for simple LSA, which can be explained by the occasional presence of sentences with a high WER. This can also justify the increase in the standard deviation value for readability of the mixed approach summaries, since, as previously mentioned, the percentage of content from the news story has also increased.

13.6 Conclusions

We present a new approach to speech summarization that closely integrates text and speech analysis, a strategy also suggested by other researchers [37, 45]. The main idea is the inclusion of related, solid background information to cope with the difficulties of summarizing spoken language and the use of multi-document summarization techniques, such as LSA, in single document speech-to-text summarization. In this work, we explore the possibilities offered by phonetic information to select the background information and conducted a perceptual evaluation to assess the relevance of the inclusion of that information.

The results obtained show that the human evaluators preferred human extractive summaries over human abstractive summaries. Moreover, simple LSA summaries had the poorest results, both in terms of content and readability, while human extractive summaries achieved the best performance (concerning content), and a considerably better performance than simple LSA (concerning readability). This suggests that it is still relevant to pursue new methods for relevance estimation. On the other hand, automatically generated summaries using background information were significantly better than simple LSA. This indicates that background information is a viable way to increase the quality of automatic summarization systems.

Acknowledgements We would like to thank Fernando Batista for his help with the speech corpus; Joana Paulo Pardal for her help with the web evaluation form; and, all the human judges for their invaluable contribution. We would also like to thank the insightful comments of the anonymous reviewers.

This work was partially supported by FCT (INESC-ID multiannual funding) through the PIDDAC Program funds.

References

1. Amaral, R., Trancoso, I.: Improving the topic indexation and segmentation modules of a media watch system. In: Proceedings of the 8th International Conference on Spoken Language Processing (INTERSPEECH 2004 – ICSLP), Jeju Island (2004)
2. Amaral, R., Meinedo, H., Caseiro, D., Trancoso, I., Neto, J.P.: Automatic vs. manual topic segmentation and indexation in broadcast news. In: Proceedings of the IV Jornadas en Tecnologia del Habla, Saragoza (2006)
3. Amaral, R., Meinedo, H., Caseiro, D., Trancoso, I., Neto, J.P.: A prototype system for selective dissemination of broadcast news in European Portuguese. EURASIP J. Adv. Signal Process. **2007**, 037507 (2007)
4. Batista, F., Caseiro, D., Mamede, N.J., Trancoso, I.: Recovering punctuation marks for automatic speech recognition. In: Proceedings of the 8th Annual Conference of the International Speech Communication Association (INTERSPEECH 2007), Antwerp, pp. 2153–2156. ISCA (2007)
5. Batista, F., Mamede, N.J., Trancoso, I.: The impact of language dynamics on the capitalization of broadcast news. In: Proceedings of the 9th Annual Conference of the International Speech Communication Association (INTERSPEECH 2008), Brisbane, pp. 220–223. ISCA (2008)
6. Charniak, E., Johnson, M.: Edit detection and parsing for transcribed speech. In: Proceedings of the 2nd Conference of the North American Chapter of the ACL, Pittsburgh, pp. 1–9. Association for Computational Linguistics (2001)
7. Chatain, P., Whittaker, E.W.D., Mrozinski, J.A., Furui, S.: Topic and stylistic adaptation for speech summarisation. In: Proceedings of the 2006 IEEE International Conference on Acoustics, Speech and Signal Processing, Toulouse, pp. 977–980. IEEE (2006)
8. Chen, Y.T., Chen, B., Wang, H.M.: A Probabilistic Generative Framework for Extractive Broadcast News Speech Summarization. IEEE Trans. Audio Speech Lang. Process. **17**(1), 95–106 (2009)
9. Christensen, H., Gotoh, Y., Kolluru, B., Renals, S.: Are extractive text summarisation techniques portable to broadcast news? In: Proceedings of the IEEE Workshop on Automatic Speech Recognition and Understanding (ASRU '03), St. Thomas, pp. 489–494. IEEE (2003)
10. Edmundson, H.P.: New methods in automatic abstracting. J. Assoc. Comput. Mach. **16**(2), 264–285 (1969)
11. Endres-Niggemeyer, B.: Summarizing Information. Springer, Berlin (1998)
12. Endres-Niggemeyer, B.: Human-style WWW summarization. Tech. rep., University for Applied Sciences, Department of Information and Communication (2000)
13. Endres-Niggemeyer, B., Hobbs, J.R., Spärck Jones, K. (eds.): Summarizing Text for Intelligent Communication. Dagstuhl-Seminar-Report, vol. 79. IBFI, Wadern (1995)
14. Fleiss, J.L., Levin, B., Paik, M.C.: The measurement of interrater agreement. In: Statistical Methods for Rates and Proportions. Wiley Series in Probability and Statistics, 3rd edn., pp. 598–626. John Wiley & Sons, Inc., Hoboken, NJ, USA (2004)

15. Furui, S.: Recent advances in automatic speech summarization. In: Proceedings of the 8th Conference on Recherche d'Information Assistée par Ordinateur (RIAO), Pittsburgh. Centre des Hautes Études Internationales d'Informatique Documentaire (2007)
16. Golub, G.H., van Loan, C.F.: Matrix analysis. Matrix Computations. Johns Hopkins Series in the Mathematical Sciences 3rd edn., pp. 48–86. The Johns Hopkins University Press, Baltimore (1996)
17. Gong, Y., Liu, X.: Generic text summarization using relevance measure and latent semantic analysis. In: SIGIR 2001: Proceedings of the 24st Annual International ACM SIGIR Conference on Research and Development in Information Retrieval, New Orleans, pp. 19–25. ACM (2001)
18. Hirohata, M., Shinnaka, Y., Iwano, K., Furui, S.: Sentence-extractive automatic speech summarization and evaluation techniques. Speech Commun. **48**, 1151–1161 (2006)
19. Hori, T., Hori, C., Minami, Y.: Speech summarization using weighted finite-state transducers. In: Proceedings of the 8th EUROSPEECH – INTERSPEECH 2003, Geneva, pp. 2817–2820. ISCA (2003)
20. Hovy, E.: Text summarization. In: Mitkov, R. (ed.) The Oxford Handbook of Computational Linguistics, pp. 583–598. Oxford University Press, Oxford/New York (2003)
21. Kessler, B.: Phonetic comparison algorithms. Trans. Philol. Soc. **103**(2), 243–260 (2005)
22. Kikuchi, T., Furui, S., Hori, C.: Two-stage automatic speech summarization by sentence extraction and compaction. In: Proceedings of the ISCA & IEEE Workshop on Spontaneous Speech Processing and Recognition (SSPR-2003), Tokyo, pp. 207–210. ISCA (2003)
23. Krippendorff, K.: Reliability. Content Analysis: An Introduction to Its Methodology, 2nd edn., pp. 211–256. Sage Publications, Thousand Oaks (2004)
24. Landis, J.R., Kosh, G.G.: The measurement of observer agreement for categorical data. Biometrics **33**, 159–174 (1977)
25. Lavrenko, V., Croft, W.B.: Relevance models in information retrieval. In: Croft, W.B., Lafferty, J. (eds.) Language Modeling for Information Retrieval. The Information Retrieval Series, vol. 13. Kluwer Academic Publishers, Dordrecht, The Netherlands (2003)
26. Lin, C.Y.: ROUGE: a package for automatic evaluation of summaries. In: Moens, M.F., Szpakowicz S. (eds.) Text Summarization Branches Out: Proceedings of the ACL-04 Workshop, Barcelona, pp. 74–81. Association for Computational Linguistics, East Stroudsburg (2004)
27. Lin, S.H., Chen, B.: A risk minimization framework for extractive speech summarization. In: Proceedings of the 48th Annual Meeting of the Association for Computational Linguistics, Uppsala, pp. 79–87. Association for Computational Linguistics (2010)
28. Lin, S.H., Yeh, Y.M., Chen, B.: Extractive speech summarization – from the view of decision theory. In: Proceedings of the 11th Annual Conference of the International Speech Communication Association (INTERSPEECH 2010), Chiba, pp. 1684–1687. ISCA (2010)
29. Liu, F., Liu, Y.: Using spoken utterance compression for meeting summarization: a pilot study. In: 2010 IEEE Workshop on Spoken Language Technology, Berkeley, pp. 37–42 (2010)
30. Liu, Y., Xie, S.: Impact of automatic sentence segmentation on meeting summarization. In: 2008 IEEE International Conference on Acoustics, Speech, and Signal Processing, Las Vegas, pp. 5009–5012. IEEE (2008)
31. Liu, Y., Shriberg, E., Stolcke, A., Hillard, D., Ostendorf, M., Harper, M.: Enriching speech recognition with automatic detection of sentence boundaries and disfluencies. IEEE Trans. Speech Audio Process. **14**(5), 1526–1540 (2006)
32. Luhn, H.P.: The automatic creation of literature abstracts. IBM J. Res. Dev. **2**(2), 159–165 (1958)
33. Maskey, S.R., Hirschberg, J.: Comparing lexical, acoustic/prosodic, strucural and discourse features for speech summarization. In: Proceedings of the 9th EUROSPEECH – INTERSPEECH 2005, Lisbon (2005)
34. Maskey, S.R., Rosenberg, A., Hirschberg, J.: Intonational phrases for speech summarization. In: Proceedings of the 9th Annual Conference of the International Speech Communication Association (INTERSPEECH 2008), Brisbane, pp. 2430–2433. ISCA (2008)

35. McKeown, K.R., Radev, D.: Generating summaries of multiple news articles. In: Fox, E.A., Ingwersen, P., Fidel R. (eds.) SIGIR 1995: Proceedings of the 18th Annual International ACM SIGIR Conference on Research and Development in Information Retrieval, Seattle, pp. 74–82. ACM (1995)

36. McKeown, K.R., Barzilay, R., Evans, D., Hatzivassiloglou, V., Klavans, J.L., Nenkova, A., Sable, C., Schiffman, B., Sigelman, S.: Tracking and summarizing news on a daily basis with Columbia's newsblaster. In: Marcus, M. (ed.) Proceedings of the Second International Conference on Human Language Technology Research (HLT 2002), San Diego, pp. 280–285. Morgan Kaufmann (2002)

37. McKeown, K.R., Hirschberg, J., Galley, M., Maskey, S.R.: From text to speech summarization. In: Proceedings of 2005 IEEE International Conference on Acoustics, Speech, and Signal Processing, Pennsylvania, vol. V, pp. 997–1000. IEEE (2005)

38. Meinedo, H., Souto, N., Neto, J.P.: Speech recognition of broadcast news for the european portuguese language. In: Proceedings of the IEEE Workshop on Automatic Speech Recognition and Understanding (ASRU '01), Madonna di Campiglio. IEEE (2001)

39. Meinedo, H., Caseiro, D., Neto, J.P., Trancoso, I.: AUDIMUS. Media: a broadcast news speech recognition system for the European Portuguese language. In: Computational Processing of the Portuguese Language: 6th International Workshop, PROPOR 2003, Faro, 26–27 June 2003. Proceedings. Lecture Notes in Computer Science (Subseries LNAI), vol. 2721, pp. 9–17. Springer (2003)

40. Meinedo, H., Viveiros, M., Neto, J.P.: Evaluation of a live broadcast news subtitling system for portuguese. In: Proceedings of the 9th Annual Conference of the International Speech Communication Association (INTERSPEECH 2008), Brisbane, pp. 508–511. ISCA (2008)

41. Mohri, M.: Finite-state transducers in language and speech processing. Comput. Linguist. **23**(2), 269–311 (1997)

42. Murray, G., Renals, S., Carletta, J.: Extractive summarization of meeting records. In: Proceedings of the 9th EUROSPEECH – INTERSPEECH 2005, Lisbon (2005)

43. Murray, G., Renals, S., Carletta, J., Moore, J.: Incorporating speaker and discourse features into speech summarization. In: Proceedings of the Human Language Technology Conference of the North American Chapter of the ACL, New York, pp. 367–374. Association for Computational Linguistics (2006)

44. Nenkova, A.: Summarization evaluation for text and speech: issues and approaches. In: Proceedings of INTERSPEECH 2006 – ICSLP, Pittsburgh, pp. 1527–1530. ISCA (2006)

45. Ostendorf, M., Favre, B., Grishman, R., Hakkani-Tür, D., Harper, M., Hillard, D., Hirschberg, J., Ji, H., Kahn, J.G., Liu, Y., Maskey, S., Matusov, E., Ney, H., Rosenberg, A., Shriberg, E., Wang, W., Wooters, C.: Speech segmentation and spoken document processing. IEEE Signal Process. Mag. **25**(3), 59–69 (2008)

46. Paulo, S., Oliveira, L.C.: Multilevel annotation Of speech signals using weighted finite state transducers. In: Proceedings of the 2002 IEEE Workshop on Speech Synthesis, Santa Monica, pp. 111–114. IEEE (2002)

47. Penn, G., Zhu, X.: A critical reassessment of evaluation baselines for speech summarization. In: Proceeding of ACL-08: HLT, Columbus, pp. 470–478. Association for Computational Linguistics (2008)

48. Radev, D.R., Otterbacher, J., Winkel, A., Blair-Goldensohn, S.: NewsInEssence: summarizing online news topics. Commun. ACM **48**(10), 95–98 (2005)

49. Ribeiro, R., de Matos, D.M.: Extractive summarization of broadcast news: comparing strategies for European Portuguese. In: Matoušek, V., Mautner, P. (eds.) Text, Speech and Dialogue – 10th International Conference, TSD 2007, Pilsen, 3–7 September 2007. Proceedings. Lecture Notes in Computer Science (Subseries LNAI), vol. 4629, pp. 115–122. Springer (2007)

50. Ribeiro, R., de Matos, D.M.: Mixed-source multi-document speech-to-text summarization. In: Coling 2008: Proceedings of the 2nd workshop on Multi-source Multilingual Information Extraction and Summarization, Manchester, pp. 33–40. Coling 2008 Organizing Committee (2008)

51. Ribeiro, R., de Matos, D.M.: Using prior knowledge to assess relevance in speech summarization. In: 2008 IEEE Workshop on Spoken Language Technology, Holiday Inn Goa, pp. 169–172. IEEE (2008)
52. Spärck Jones, K.: Automatic summarising: the state of the art. Inf. Process. Manag. **43**, 1449–1481 (2007)
53. Wan, X., Yang, J., Xiao, J.: CollabSum: exploiting multiple document clustering for collaborative single document summarizations. In: SIGIR 2007: Proceedings of the 30th Annual International ACM SIGIR Conference on Research and Development in Information Retrieval, Amsterdam, pp. 143–150. ACM (2007)
54. Zechner, K., Waibel, A.: Minimizing word error rate in textual summaries of spoken language. In: Proceedings of the 1st conference of the North American chapter of the ACL, Seattle, Washington, USA, pp. 186–193. Morgan Kaufmann (2000)
55. Zhang, J.J., Chan, R.H.Y., Fung, P.: Extractive speech summarization using shallow rhetorical structure modeling. IEEE Trans. Audio Speech Lang. Process. **18**(6), 1147–1157 (2010)
56. Zhu, X., Penn, G.: Summarization of spontaneous conversations. In: Proceedings of INTERSPEECH 2006 – ICSLP, Pittsburgh, pp. 1531–1534. ISCA (2006)

Chapter 14
Multi-Document Summarization Techniques for Generating Image Descriptions: A Comparative Analysis

Ahmet Aker, Laura Plaza, Elena Lloret, and Robert Gaizauskas

Abstract This paper reports an initial study that aims to assess the viability of multi-document summarization techniques for automatic captioning of geo-referenced images. The automatic captioning procedure requires summarizing multiple Web documents that contain information related to images' location. We use different state-of-the art summarization systems to generate generic and query-based multi-document summaries and evaluate them using ROUGE metrics [24] relative to human generated summaries. Results show that query-based summaries perform better than generic ones and thus are more appropriate for the task of image captioning or generation of short descriptions related to the location/place captured in the image. For our future work in automatic image captioning this result suggests that developing the query-based summarizer further and biasing it to account for user-specific requirements will prove worthwhile.

14.1 Introduction

Retrieving textual information related to a location shown in an image has many potential applications. It could help users gain quick access to the information they seek about a place of interest just by taking its picture. Such textual information could also, for instance, be used by a journalist who is planning to write an article

A. Aker (✉) · R. Gaizauskas
University of Sheffield, Regent Court, 211 Portobello, Sheffield, S1 4DP, UK
e-mail: a.aker@dcs.shef.ac.uk; r.gaizauskas@dcs.shef.ac.uk

L. Plaza
Universidad Complutense de Madrid, C/ José García Santesmases, s/n, 28040 Madrid, Spain
e-mail: lplazam@fdi.ucm.es

E. Lloret
University of Alicante, Apdo. de correos, 99, E-03080 Alicante, Spain
e-mail: elloret@dlsi.ua.es

T. Poibeau et al. (eds.), *Multi-source, Multilingual Information Extraction and Summarization 11*, Theory and Applications of Natural Language Processing, DOI 10.1007/978-3-642-28569-1__14, © Springer-Verlag Berlin Heidelberg 2013

about a building, or by a tourist who seeks further interesting places to visit nearby. In this paper we aim to generate such textual information automatically by utilizing multi-document summarization techniques, where documents to be summarized are Web documents that contain information related to the image content. We focus on geo-referenced images, i.e. images tagged with coordinates (latitude and longitude) and compass information, that show things with fixed locations (e.g. buildings, mountains, etc.).

Attempts towards automatic generation of image-related textual information or captions have been previously reported. Deschacht and Moens [8] and Mori et al. [31] generate image captions automatically by analyzing image-related text from the immediate context of the image, i.e. existing image captions, surrounding text in HTML documents, text contained in the image, etc. The authors identify named entities and other noun phrases in the image-related text and assign these to the image as captions. Other approaches create image captions by taking into consideration image features as well as image-related text [3, 32, 56]. These approaches can address all kinds of images, but focus mostly on images of people. They analyze only the immediate textual context of the image on the Web and are concerned with describing *what* is in the image only. Consequently, background information about the objects in the image is not provided. Our aim, however, is to have captions that inform users' specific interests about a location, which clearly includes more than just image content description. Multi-document summarization techniques have the potential to enable image-related information to be included from multiple documents. However, the challenge lies in being able to summarize unrestricted Web documents effectively.

Various multi-document summarization tools have been developed such as those described in [25, 34, 40, 42], to mention just a few. These systems generate generic and query-based summaries. Generic summaries address a broad readership whereas query-based summaries aim to support specific groups of people aiming to gain knowledge about specific topics quickly [28]. The performance of these tools has been reported for DUC tasks.[1] As Sekine and Nobata [49] note, although DUC tasks provide a common evaluation standard, they are restricted in topic and are somewhat idealized. For our purposes the summarizer needs to create summaries from unrestricted Web input, for which there are no previous performance reports.

For this reason we evaluate the performance of these systems in generic and query-based mode on the task of image captioning or generation of short descriptions related to the location/place captured in an image. We hypothesize that a query-based summarizer will better address the problem of creating summaries tailored to users' needs. This is because the query itself may contain important hints as to what the user is interested in. A generic summarizer generates summaries based on the topics it observes in the documents supplied to it and cannot take user specific input into consideration. Using the four systems mentioned above, we generate both generic and query-based multi-document summaries of image-related documents

---

[1] http://www-nlpir.nist.gov/projects/duc/index.html

obtained from the Web. We use a social Web site to obtain our model summaries against which we evaluate the automatically generated ones.

The paper is organized as follows. Section 14.2 provides a comprehensive related work concerning text summarization. Section 14.3 describes the image collection we use for evaluation, the model summaries and the image related Web documents. In Sect. 14.4 we describe the four different systems we use in our experiments. Section 14.5 discusses the results, and Sect. 14.6 concludes the paper and outlines directions for future work and improvements.

14.2   Text Summarization: An Overview

Text Summarization is a very active research area, despite being more than 50 years old [27]. Taking into account different factors concerning the input, output or purpose of the summaries [50], summarization approaches can be characterized according to many features. Although it has traditionally been focused on text, the input to the summarization process can also be multimedia information, such as images [11], video [19] or audio [58], as well as on-line information or hypertexts [52]. Furthermore, we can talk about *single-document summarization* [53], i.e. summarizing only one document, or *multi-document summarization*, i.e. summarizing multiple documents about the same topic [5]. Regarding the output, a summary may be an *extract* [16], i.e. when a selection of "significant" sentences of a document is shown, or an *abstract* [13], when the summary can serve as a substitute to the original document and new vocabulary is added. It is also possible to distinguish between *generic* summaries [4] and *query-based* summaries (also known as user-based or topic-based) [18]. The first type of summaries can serve as surrogate of the original text as they may try to represent all relevant facts of a source text. In the latter, the content of a summary is biased towards a user need, query or topic. Concerning the style of the output, a broad distinction is normally made between two types of summaries. *Indicative* summaries are used to indicate what topics are addressed in the source text. As a result, they can give a brief idea of what the original text is about. The other type, *informative* summaries, are intended to cover the topics in the source text and provide more detailed information. Methods for producing this kind of summaries are specifically studied in [22].

In recent years, new types of summaries have emerged. For instance, the birth of the Web 2.0 has encouraged new types of textual genres, containing high degree of subjectivity, thus allowing the generation of *sentiment-based or opinion-oriented summaries* [54]. Another example of new summary type are *update summaries* [7], which assume that the user has already a background and he/she needs only the most recent information about a topic. Finally, concerning the language of the summary, it can be distinguished between *mono-lingual* [9], *multi-lingual* [20] and *cross-lingual* [44] summaries, depending on the number of languages dealt with. The cases where the input and the output language is the same lead to mono-lingual summaries. However, if different languages are involved, the summarization approach is considered multi-lingual or cross-lingual.

As far as the methods used for addressing the summarization task is concerned, a great number of techniques have been proven to be effective for generating summaries automatically. Such approaches include statistical techniques, for instance, term frequency [26], discourse analysis [29], graph-based methods [34], language models [1], and machine learning algorithms [48].

Furthermore, although most work in text summarization has traditionally focused on newswire [17], scientific documents [21] or even legal documents [6], these are not the unique scenarios in which text summarization approaches have been tested on. Other domains have been recently drawn special attention. For instance, literary text [30], patents [55], or image captions [36]. In this work we apply text summarization to the image description generation task.

14.3 Data

This section describes the image collection we use for evaluation, including the model summaries and the image-related Web documents.

14.3.1 Image Collection

Our image collection has 308 different images which are toponym-referenced, i.e. assigned with toponyms (place names). The subjects of our toponym-referenced images are locations around the world such as *Parc Guell, the London Eye, Edinburgh Castle*, etc. For each image we manually generated model summaries as described in the following section.

14.3.2 Model Summaries

For each image we generated up to four model summaries based on image captions taken from *VirtualTourist*[2] [2]. VirtualTourist is one of the largest online travel communities in the world containing 3 million photos with captions (in English) of more than 58,000 destinations worldwide.

VirtualTourist uses a tree structured schema for organizing the descriptions. The tree has *world* at the root and the *continents* as the direct children of *world*. The *continents* contain the *countries* which have the *cities* as direct children. The leaves in the tree are the places or objects visited by travelers.

We selected from these structure a list of popular cites such as *London, Edinburgh, New York, Venice*, etc., assigned different sets of cities to different human subjects and asked them to collect up to four model summaries for each

---

[2]www.VirtualTourist.com

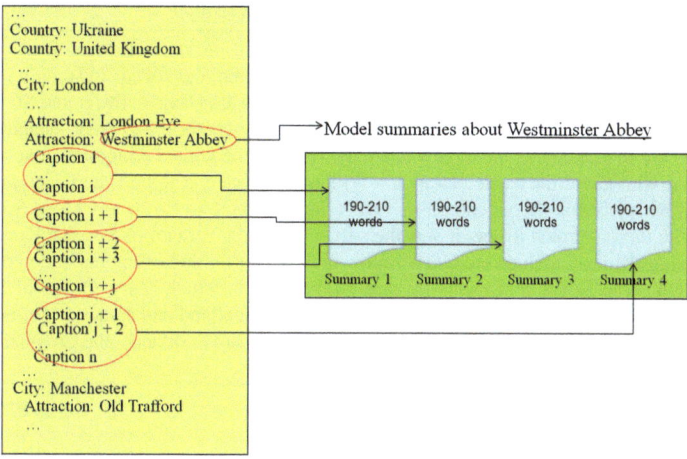

Fig. 14.1 Model summary collection

object from their descriptions with length ranging from 190 to 210 words (see Fig. 14.1).

During the collection it was ensured that the summaries did not have personal information and that they did genuinely describe a place, e.g. *Westminster Abbey, Edinburgh Castle, Eiffel Tower*. If the descriptions contained personal information this was removed. In case a description did not have enough words, i.e. the number of words was less than 190, more than one description was used to build a model summary. While doing this it was also ensured that the resulting summary did not contain redundant information. In addition, a manually written sentence based on directions and address information which is given by VirtualTourist users in form of single terms after each description, was optionally added to the model summary. However, this was only done if the description contained less than 190 words. If the description contained more than 210 words we deleted the less important information. What information is considered less important is subjective and depends on the person collecting the descriptions. Some VirtualTourist descriptions contain sentences recommending what one can do when visiting the place. These sentences usually have the form "you can do X". We allowed our model summary collectors to retain these kinds of sentences as they contain relevant information about the place. Finally, some descriptions contain sentences which refer to their corresponding images. We asked our summary gatherers to delete any sentences which refer to images.

The number of images with four model summaries is 170. Forty-one images have three model summaries, 33 have two and 63 have only one model summary. An example model summary about the *Edinburgh Castle* is shown in Table 14.1.

Table 14.1 Example model summary about the Edinburgh Castle

Edinburgh Castle stands on an extinct volcano. The Castle pre-dates Roman Times and bears witness to Scotland's troubled past. The castle was conquered, destroyed, and rebuilt many times over the centuries. The only two remain original structures are David's Tower and St. Margaret's Chapel. Edinburgh Castle – now owned and managed by Historic Scotland – stands 2nd only to the Tower of London as the most visited attraction in the United Kingdom. Take note of the two heros who guard the castle entrance – William Wallace and Robert the Bruce their bronze statues were placed at the gatehouse in 1929 a fitting tribute to two truely Great Scots. Inside the Castle, there is much to see. It was the seat (and regular refuge) of Scottish Kings, and the historical apartments include the Great Hall, which houses an interesting collection of weapons and armour. The Royal apartments include a tiny room in which Mary, Queen of Scots gave birth to the boy who was to become King James VI of Scotland and James I of England upon the death of Queen Elizabeth in 1603. The ancient Honours of Scotland – the Crown, the Sceptre and the Sword of State – are on view in the Crown Room.

14.3.3   Image Related Web Documents

For each image we used its associated toponym as a query to collect relevant documents from the Web. We passed the toponym to the Yahoo! Search engine[3] and the 100 best search results were retrieved, from which only 30 were taken for the summarization process. Before we selected the 30 documents we filtered from the 100 original documents those documents where we could not access the content. In addition to this, multiple hyperlinks belonging to the same domain were ignored as it is assumed that the content obtained from the same domain would be similar. From the remaining document list we took the first 30 documents to summarize. Each of these 30 documents is crawled to obtain its content (raw text).

The Web-crawler downloads only the content of the document residing under the hyperlink, which was previously found as a search result, and does not follow any other hyperlinks within the document. The content obtained by the Web-crawler encapsulates an HTML structured document. We further process this using an HTML parser[4] to select the *pure* text, i.e. text consisting of sentences.

The HTML parser removes advertisements, menu items, tables, java scripts, etc. from the HTML documents and keeps sentences which contain at least four words. This number was chosen after several experiments. The resulting data is passed on to the multi-document summarization systems which are described in Sect. 14.4.

14.4   Summarization Systems

In this section we describe the four summarization systems we use to generate both generic and query-based or query-biased summaries. Each summary (generic or query-based) contains sentences extracted from the Web documents and does not

---

[3]http://search.yahoo.com/

[4]http://htmlparser.sourceforge.net/

exceed 200 words in length. The queries used in the query-based mode are toponyms as described in Sect. 14.3.

14.4.1 SUMMA

SUMMA[5][42, 43] is a set of language and processing resources to create and evaluate summarization systems (single-document, multi-document, multi-lingual). The components can be used within GATE[6] to produce summarization applications. SUMMA can produce both generic and query-based summaries.

In the case of generic summarization SUMMA uses a single cluster approach to summarize n related documents which are given as input. Using GATE, SUMMA first applies sentence detection and sentence tokenisation to the given documents. Then each sentence in the documents is represented as a vector in a vector space model [45], where each vector position contains a term (word) and a value which is a product of the *term frequency (TF)* in the document and the *inverse document frequency (IDF)*, a measurement of the term's distribution over the set of documents [46].

Then it extracts features for each sentence such as:

- *centroidSimilarity*: Sentence similarity to the centroid (cosine similarity over the vector representation of the sentence and the centroid which is derived from the cluster).
- *sentencePosition*: Position of the sentence within its document. The first sentence in the document gets the score 1 and the last one gets $\frac{1}{n}$ where n is the number of sentences in the document.
- *leadSimilarity*: Sentence similarity to the lead part of the document where the sentence come from (cosine similarity over the vector representation of the sentence and the lead part).

The cosine similarity [47] that is used to compute similarity between two S_i and S_j sentences is given in the following formula:

$$Cos(S_i, S_j) = \frac{(TF*IDF\ of\ Words\ In S_i) * (TF*IDF\ of\ Words\ In S_j)}{\sqrt{\sum(TF*IDF\ of\ Words\ in S_i)^2} * \sqrt{\sum(TF*IDF\ of\ Words\ in S_j)^2}}$$
(14.1)

In the sentence selection process, each sentence in the collection is ranked individually, and the top sentences are chosen to build up the final summary. The ranking of a sentence depends on its distance to the centroid, its absolute position in its document and its similarity to the lead-part of its source document.

---

[5]http://www.dcs.shef.ac.uk/~saggion/summa/default.htm
[6]http://gate.ac.uk

In the case of the query-based approach, SUMMA extracts an additional feature. For each sentence, cosine similarity to the given query is computed. Finally, the sentences are scored by summing all features according to the following formula:

$$Sentence_{score} = \sum_{i=1}^{n} feature_i * weight_i \qquad (14.2)$$

After the scoring process, SUMMA starts selecting sentences for summary generation. In both generic and query-based summarization, the summary is constructed by first selecting the sentence that has the highest score, followed by the sentence with the second highest score, etc. until the compression rate is reached. However, before a sentence is selected a similarity metric for redundancy detection is applied to each sentence which decides whether a sentence is distinct enough from those sentences already selected to be included in the summary or not. SUMMA uses the following formula – a sort of weighted Jaccard similarity coefficient over n-grams – to compute the similarity between two sentences:

$$NGramSim(S_1, S_2, n) = \sum_{j=1}^{n} w_j * \frac{|ngrams(S_1, j) \bigcap ngrams(S_2, j)|}{|ngrams(S_1, j) \bigcup ngrams(S_2, j)|} \qquad (14.3)$$

where n specifies maximum size of the n-grams to be considered, $ngrams(S_X, j)$ is the set of j-grams in sentence X and w_j is the weight associated with j-gram similarity. Two sentences are similar if they are above a threshold, i.e. $NGramSim(S_1, S_2, n) > \alpha$. In this work n is set to 4 and the threshold α to 0.1. For j-gram similarity, weights $w_1 = 0.1$, $w_2 = 0.2$, $w_3 = 0.3$ and $w_4 = 0.4$ are selected. These values are coded in SUMMA as defaults.

14.4.2 MEAD

MEAD[7] is a publicly available toolkit for multi-document and multi-lingual summarization and evaluation. MEAD [39] produces extractive summaries using a linear combination of features to rank the sentences in the source documents. To avoid redundancy, it employs a 'sentence reranker' that ensures that the chosen sentences are not too similar to one another in terms of lexical unit overlap. MEAD allows users to customize the sentence selection criteria for various summarization tasks, including query-based summarization. In this work, MEAD has been used to create both a generic and a query-based multi-document summarizer.

For generic summarization, we use MEAD default parameters, which means that three different features are computed and combined to score and rank the sentences for the summary [38]:

---

[7]http://www.summarization.com/mead/

- **Centroid (C):** The centroid score is a measure of the centrality of a sentence to the overall topic of the cluster of documents. The centroid feature is computed as the cosine similarity between the vector representation of the sentence (TF×IDF of words in the sentence) and the centroid of the documents' cluster (TF×IDF of words of the cluster).
- **Position (P):** The position score assigns a value to each sentence that decreases linearly as the sentence gets farther from the beginning of a document.
- **Length (L):** This is a cutoff feature. Any sentence with a length shorter than *Length* is automatically given a score of 0, regardless of its other features.

The three features above are normalized in the interval [0,1]. An overall score for each sentence is then calculated using a linear combination of these features, as stated in (14.4). The combination of weights used in this work is $weight_c = 1$, $weight_p = 1$ and $weight_l = 100$ which were selected experimentally.

$$Score(S_i) = C_i \times weight_c + P_i \times weight_p + L_i \times weight_l \qquad (14.4)$$

For query-based summarization, MEAD provides three further features: *QueryCosine*, *QueryCosineNoIDF* and *QueryWordOverlap*, which compute the various measures of similarity between the sentence and the given query [41]. In this work, we use the **QueryCosineNoIDF (QC)** feature, and combine this query-based feature with the three generic features using (14.5), where the weight for the new feature $weight_{qc}$ is set to 10. These weights were again selected experimentally.

$$Score(S_i) = C_i \times weight_c + P_i \times weight_p + L_i \times weight_l + QC_i \times weight_{qc} \qquad (14.5)$$

14.4.3   COMPENDIUM

COMPENDIUM is a summarization framework able to produce either generic or query-focused, as well as single- or multi-document extractive summaries. It mainly relies on four stages for generating summaries: (1) redundancy detection; (2) similarity to the query; (3) relevance detection; and (4) summary generation. Prior to these stages, basic pre-processing, comprising sentence segmentation, tokenization, part-of-speech tagging, and *stopword* removal, is carried out in order to prepare the text for further processing. Once redundant information has been removed, a sentence is given two weights, one indicating its relevance within the text, and the other one its similarity with respect to the query. These weights will give the final sentence score in the last stage of the summarization process, thus determining which sentences are to be selected and extracted. The effectiveness of such modules for summarization has been shown in previous research [25, 26]. Each of them is explained in more detail below.

Redundancy Detection The aim of this stage is to avoid repeated information in the summary. Textual entailment is employed to meet this goal. A textual entailment relation holds between two text snippets when the meaning of one text snippet can be inferred from the other [15]. If such entailment relation can be identified automatically then it is possible to identify which sentences within a text can be inferred from others, as to avoid incorporating into summaries the sentences whose meaning is already contained in other sentences in the summary. In other words, the main idea here is to obtain a set of sentences from the text with no entailment relation, and then keep this set of sentences for further processing. To compute such entailment relations we have used the textual entailment approach presented in [12].

Query Similarity In order to determine the sentences potentially related to the query, we compute the cosine similarity between each sentence and the query using the Text Similarity package,[8] obtaining a query similarity weight later used for computing the overall relevance of the sentence. Furthermore, since sentences can contain pronouns referring to the image itself, before accounting for the similarity, we attempt to resolve the anaphora in the texts, using JavaRap[9] [37]. Anaphora resolution helps this module since the cosine similarity is more precise when computing the similarity of each sentence with respect to the query, specifically for those sentences containing pronouns. For instance, given the query *"Euston Railway Station"*, and the sentence *"It is located in London"*, the cosine similarity will not detect any similarity between them unless we are able to associated the pronoun *"it"* to the entity *"Euston Railway Station"*. Equation 14.6 shows how the query similarity weight is calculated. The cosine similarity is computed according to (14.1).

$$qSim_{s_i} = CosineSimilarity(S_i, Query) \qquad (14.6)$$

Relevance Detection The relevance detection module assigns a weight to each sentence, depending on how relevant it is within the text. This weight is based on the combination of two features: **term frequency** and the **code quantity principle**. On the one hand, concerning term frequency, it is assumed that the more times a word appears in a document, the more relevant become the sentences that contain this word, following Luhn's idea [27]. On the other hand, the code quantity principle [14] is a linguistic theory which states that the less predictable information will be given more coding material. In other words, the most important information within a text will contain more lexical elements, and therefore it will be expressed by a high number of units (for instance, syllables, words or phrases). Noun-phrases within a document are flexible coding units that can vary in the number of elements depending on the level of detail desired for the information. Therefore, it is assumed that sentences containing longer noun-phrases are more important. The way the relevance of a sentence is computed is shown in (14.7).

$$r_{s_i} = \frac{1}{\#NP_i} \sum_{w \in NP} |tf_w| \qquad (14.7)$$

---

[8]http://www.d.umn.edu/~tpederse/text-similarity.html

[9]http://aye.comp.nus.edu.sg/~qiu/NLPTools/JavaRAP.html

where:

$\#NPi$ = number of noun-phrases contained in sentence i,

tf_w = frequency of word w that belongs to a noun-phrase NP.

Summary Generation At this stage, the final score of a sentence is computed, and consequently the most important sentences (i.e. the ones with highest scores) are selected and extracted to form the final summary up to a desired length. Having computed the two different weights for each sentence (its relevance r and its similarity with regard to the query $qSim$), the final score for a sentence (Sc) is calculated according to (14.8). It is worth stressing upon the fact that this formula is based on the F-measure formula. Since it provides a good way to combine precision and recall, it is also appropriate for the combination of the suggested weights for deciding the overall relevance of a sentence.

$$Sc_{s_i} = (1 + \beta^2)\frac{r * qSim}{\beta^2 * r + qSim} \tag{14.8}$$

β can be assigned different weights between 0 and 1, depending on whether we would like to give more importance to the relevance or to the query similarity weight when producing query-focused summaries. It was empirically established that the optimal value for β in this case was 0, therefore meaning that the sentences related to the query have an important value for the summary. As far as generic summaries are concerned, the query-similarity stage is not taken into account, and we take as the final score of each sentence its relevance weight.

14.4.4 SummGraph

SummGraph is a generic architecture for knowledge-driven text summarization, which makes use of a graph-based method and the knowledge in a lexical knowledge base in order to perform single document summarization. The SummGraph summarizer has been already presented in previous work and used to generate summaries of documents in two different domains: news items [35] and biomedical papers [34]. In this work, the summarizer has been adapted and used to create summaries from multiple Web documents containing information related to places or locations, both generic and query-based.

In order to deal with multi-document summarization, we simply merge all documents about the same topic or image into a single document, and run the summarizer over it. After producing the summary, we apply a textual entailment module (the same used in the COMPENDIUM summarizer and explained above in Sect. 14.4.3) to detect and remove redundancy [12].

The summarizer first applies shallow preprocessing over the document, including sentence detection, POS tagging and removing stopwords and high frequency terms.

Next, it translates the text in the document into concepts from a knowledge base. In this work, we use WordNet as the knowledge source. Moreover, the *Lesk* algorithm [23] is used to disambiguate the meaning of each term in the document according to its context. After that, the resulting WordNet concepts are extended with their hypernyms, building a graph representation for each sentence in the document, where the vertices represent distinct concepts in the sentence and the edges represent *is-a* relations.

The system then merges all the sentence graphs into a single document graph, which is extended with a semantic similarity relation, so that a new edge is added that links every pair of leaf vertices whose similarity (calculated in terms of WordNet concept gloss overlaps, using the *WordNet Similarity* package [33] and the *jcn* similarity measure) exceeds a 0.25 threshold. Each edge in the document graph is assigned a weight that is directly proportional to the depth in the hierarchy of the nodes that it links (that is, the more specific the concepts connected by a link are, the more weight is assigned to the link).

Once the document graph is built, the vertices are ranked according to their *salience* or prestige. The salience of a vertex is calculated as the sum of the weight of the edges connected to it. The top *n* vertices are called *Hub Vertices* and grouped into *Hub Vertices Sets (HVS)*, which represent sets of concepts strongly related in meaning. A *degree-based clustering* method [10] is then executed over the graph and, as a result, a variable number of clusters or subgraphs are obtained. The working hypothesis is that each of these clusters represents a different *subtheme* or topic within the document, and that the most central concepts in a cluster (the so called HVS or centroids) give the necessary and sufficient information related to its topic.

The process continues by calculating the similarity between all the sentence graphs and each cluster. To this end, a non-democratic vote mechanism [57] is used, so that each vertex of a sentence gives to each cluster a different number of votes depending on whether or not the vertex belongs to the HVS of that cluster. The similarity is computed as the sum of the votes given by all vertices in the sentence to each cluster. Finally, under the hypothesis that the cluster with more concepts represents the main theme in the document, and hence the only one that should contribute to the summary, the *N* sentences with greatest similarity to this cluster are selected. Alternative heuristics for sentence selection were explored in previous work [34].

In order to deal with query-based summarization, we modify the function for computing the weight of the edges in the document graph, so that if an edge is linked to a vertex that represents a concept which is also present in the query, then the weight of the edge is multiplied by 2. This weight is distributed through the graph and the vertices representing concepts from the query and those other concepts connected to them in the document graph are assigned a higher salience and ranked higher. As a result, the likelihood of selecting sentences containing concepts closely related in meaning to those in the query for inclusion for the summary is increased.

14.5   Results and Discussion

The model summaries were compared against the summaries generated automatically using the four summarizers, both generic and query-based, by calculating ROUGE-2 and ROUGE-SU4 recall metrics [24]. For all these metrics, ROUGE compares each automatically generated summary s pairwise to every model summary m_i from the set of M model summaries and takes the maximum $ROUGE_{Score}$ value among all pairwise comparisons as the best $ROUGE_{Score}$ score:

$$ROUGE_{Score} = argmax_i ROUGE_{Score}(m_i, s) \qquad (14.9)$$

ROUGE repeats this comparison M times. In each iteration it applies the Jackknife method and takes one model summary from the M model summaries away and compares the automatically generated summary s against the remaining $M - 1$ model summaries. In each iteration one best $ROUGE_{Score}$ is calculated. The final $ROUGE_{Score}$ is then the average of all best scores calculated in the M iterations. In particular, ROUGE-2 computes the number of bigrams that are present in the automatic and model summaries, while ROUGE-SU4 measures the overlap of "skip-bigrams" between a candidate summary and a set of reference summaries, allowing a skip distance of 4. The ROUGE metrics produce a value in [0,1], where higher values are preferred, since they indicate a greater content overlap between the peer and model summaries.

It should be noted, however, that ROUGE metrics do not account for text coherence, but merely assess the content of the summaries. An important drawback of ROUGE metrics is that they use lexical matching instead of semantic matching. Therefore, peer summaries that are worded differently but carry the same semantic information may be assigned different ROUGE scores. In contrast, the main advantages of ROUGE are its simplicity and high correlation with the human judges gathered from previous Document Understanding Conferences.

In this way, generic and query-based summaries from the different systems are evaluated. The results are given in Table 14.2. Our interest here is to compare the generic and query-based versions of each summarizer. Moreover, in order to assess the significance of the results, we ran a Wilcoxon signed-rank test. Significance is also shown in Table 14.2, using the following convention: *** = p <.0001, ** = p <.001 and no star indicates non-significance, where in each case significance is calculated between the query-based and generic version of the same summarizer under the same metric.

The results in Table 14.2 support our hypothesis that query-based summaries perform better than generic ones on image-related summaries. All ROUGE scores for the query-based summaries are greater than the generic summary scores. Moreover, for all summarizers except SUMMA, the query-based summaries are significantly better than the generic ones. However, the improvement achieved by using the query considerably differs across summarizers. While the improvement for MEAD and SummGraph is greater than 12% for ROUGE-2, COMPENDIUM and

Table 14.2 Comparison: generic summaries vs. query-based summaries

Summarizer	ROUGE-2	ROUGE-SU4
SUMMA (Generic)	0.06423	0.10919
SUMMA (Query-based)	0.06532	0.10946
MEAD (Generic)	0.08866	0.13769
MEAD (Query-based)	**0.10192*****	0.15353***
COMPENDIUM(Generic)	0.08551	0.13371
COMPENDIUM (Query-based)	0.08864**	0.13892***
SummGraph (Generic)	0.08950	0.14290
SummGraph (Query-based)	0.10075***	**0.15430*****

SUMMA only achieve an average improvement of 3.6% and 1.7%, respectively. We think the reason is that MEAD and SummGraph attach greater weight to the query feature than COMPENDIUM and SUMMA.

Table 14.2 also shows differences among the summaries generated by the four systems, both generic and query-based. In particular, it is surprising that SUMMA behaves significantly worse than the remaining summarizers. This may be due to the fact that SUMMA strongly relies on information about the frequency of the terms in the documents to select the sentences for the summaries. However, it frequently occurs that the most frequent information in the documents is not directly related with the main topic but can be considered as noisy information (e.g. nearby hotels and other tourist services, advertisements from the website that hosts the information...).

As far as the COMPENDIUM approach is concerned, the main problem lies also in the nature of the corpus. Most documents in the corpus contain sentences with a high number of noun phrases, but which are unrelated to the topic (e.g., Mahogany, Maple, crown mouldings, multiple Viking ovens, Sub-Zero refrigerators...). According to the code quantity principle feature, these kind of sentences are scored higher, thus being considered relevant to incorporate them to the summary. In these cases, the quality of the generated summaries is directly affected by these sentences.

Regarding SummGraph, the main problem is directly related to the type of the documents to summarize: in most of these documents, the salient information is concerned with proper nouns describing monuments, cities, beaches, etc., that are not likely to be found in WordNet (e.g. Sacre Coeur, Santorini or Ipanema). If no concept is found for these terms, the document graph will be inevitably losing essential information to identify the topics covered in the document.

Table 14.3 compiles the results of the four query-based summarizers. The table also shows an upper bound for the different ROUGE scores, which consists of the comparison of the different model summaries against each other. To achieve this, each human written summary was evaluated against the remaining human written ones for the same object or image. The corresponding results are shown in row *User To User*. As a baseline we generated summaries using the Wikipedia article describing each image, from which we select the first 200 words. We look at these summaries as "challenging target": first, it must be taken into account

Table 14.3 Comparison: query-based summaries vs. Wikipedia and human written summaries

Summarizer	ROUGE-2	ROUGE-SU4
SUMMA (Query-based)	0.06532	0.10946
MEAD (Query-based)	0.10192	0.15353
COMPENDIUM (Query-based)	0.08864	0.13892
SummGraph (Query-based)	0.10075	0.15430
Wikipedia	0.09632	0.14203
User To User	**0.11191**	**0.16655**

that these articles have been created by humans; second, the first paragraph in a Wikipedia article is usually just a summary of the entire document content; and third, Wikipedia articles almost exclusively contain salient information to the subject matter, and so do not present other information somehow related to the topic but not important (e.g. nearby hotels, restaurants, transport services, or even advertising). These results are shown in row *Wikipedia*.

It is important to mention that for some of the images the Wikipedia article is one of the ten input documents to the summarizers. However, despite it achieves good ROUGE results, our aim in this research work is to study and analyze to what extent summarization techniques are appropriate to generate image descriptions. On the one hand, not every place is found in Wikipedia, especially if we want to deal with other languages different from English. On the other hand, as it has been proven some summarization approaches (i.e., MEAD or SummGraph) perform better than Wikipedia. Therefore, we cannot rely always on the Wikipedia article. Moreover, in the cases where there is a Wikipedia article associated to the specific image, the relevant information it contains may be also selected by some of the summarizers, and this information can be enriched by additional relevant information found in other documents.

It can be seen from Table 14.3 that the best results are obtained using the MEAD and SummGraph summarizers. According to a pairwise Wilcoxon test performed with Bonferroni correction for multiple testing, MEAD and SummGraph produce better summaries than SUMMA, COMPENDIUM and Wikipedia (MEAD: $p = 0.0003$ for ROUGE-2 and ROUGE-SU4; SummGraph: $p = 0.0177$ for ROUGE-2, $p = 0.0003$ for ROUGE-SU4). However, no differences exist between SummGraph and MEAD. The performance of both summarizers is very close to that of humans, covering approximately 10% of the bigrams in the model summaries (ROUGE-2), compared to 11% for the model summaries (cf. row *User To User* in Table 14.3).

The results in Tables 14.2 and 14.3 confirm our hypothesis that query-based summaries will better address the aim of this research, which is to get summaries tailored to users' needs. A generic summary does not take the user query into consideration and generates summaries based on the topics it observes. For a set of documents containing mainly historical and little location-related information, a generic summary will probably contain a higher number of history-related than location-related sentences. This might satisfy a group of people seeking historical information, however, it might not be interesting for a group who want to look for location-related information. Besides, the documents being summarized, most of

Table 14.4 Generic and query-based MEAD summaries for *Baker beach*

Generic summary	Query-based summary
[1] Baker Beach is a public beach on the peninsula of San Francisco, California, U.S. [2] It is roughly a half mile long, beginning just south of Golden Gate Point where the Golden Gate Bridge connects with the peninsula , extending southward toward the Seacliff peninsula, the Palace of the Legion of Honor and the Sutro Baths. [3] The reasons I can state on Yelp are 1 ample free parking at least early in the morning 2 well the weather just happened to be perfect 3 sea glass! 4 clean sand 5 pelicans! 6 great climbing rocks over to secluded areas that were like being off on some far away beach 7 lots of naked men 8 killer view of the GG bridge 9 happy people 10 oh the possibilities Finding Baker Beach was like finding religion for me, having been a frequenter of Ocean Beach which was like an old religion I'm rejecting . [4] It is nice, and one of the cleaner beaches or area of beach in SF although today when I went I did pick up a few pieces of trash only feet from the garbage can it is really only worth going for the waves that is at high tide and when it is a nicer day than today rain, wind etc. you know, normal SF weather.	[1] Baker Beach is a public beach on the peninsula of San Francisco, California, U.S. [2] The beach lies on the shore of the Pacific Ocean to the northwest of the city. [3] It is roughly a half mile long, beginning just south of Golden Gate Point where the Golden Gate Bridge connects with the peninsula, extending southward toward the Seacliff peninsula, the Palace of the Legion of Honor and the Sutro Baths. [4] Baker Beach is part of the Presidio, which was a military base from the founding of San Francisco by the Spanish in 1776 until 1997. [5] When the Presidio was decommissioned as a U.S. Army base, it became part of the Golden Gate National Recreation Area, which is administered by the National Park Service. [6] From 1986 to 1990, the north end of Baker Beach was the original site of the Burning Man art festival. [7] In 1990, park police allowed participants to raise the traditional large statue but not to set it on fire, since the beach enforces a limit on the size of any campfires. [8] A shark attack occurred on Baker Beach on May 7, 1959 when 18-year old Albert Kogler Jr. was attacked by a great white shark while he was 15 feet deep in water.

which have been gathered from tourist information websites, contain a good deal of information which is secondary or unrelated to the image (e.g. nearby hotels and other tourist services, advertisements from the website that hosts the information, personal opinions and experiences of users and so on). Clearly, this information is not relevant for generating image descriptions. A query-based summarizer biases the information selected toward the users' needs and therefore is more appropriate for image-related summaries than a generic one. To illustrate this assertion, consider the summaries in Table 14.4 for the image of the *Baker beach*. These have been generated using MEAD, both with and without using query information. By comparing the two summaries, it may be observed that they have only two sentences in common, since sentences [1] and [2] in the generic summary match sentences [1] and [3] in the query-based summary. However, the generic summary includes two sentences ([3] and [4]) stating the particular opinion of two different users about the beach being described. This subjective information might be of interest to people seeking others' experiences when visiting the beach, but it is not likely to be included in an image description aiming to provide users with broader location-related information in just 200 words.

Table 14.5 SummGraph query-based and human written summaries for *Acropolis*

Query-based summary	Human written summary
[1] Acropolis (Gr akros, akron, edge, extremity + polis, city, pl acropoleis) literally means city on the edge (or extremity). [2] The Acropolis was designated as a UNESCO World Heritage site in 1987, for its, illustrating the civilizations, myths, and religions that flourished in Greece over a period of more than 1,000 years, the Acropolis, the site of four of the greatest masterpieces of classical Greek art – the Parthenon, the Propylaea, the Erechtheum, and the Temple of Athena Nike-can be seen as symbolizing the idea of world heritage. [3] The Acropolis of Athens, a hill c.260 ft (80 m) high, with a flat oval top c.500 ft (150 m) wide and 1,150 ft (350 m) long, was a ceremonial site beginning in the Neolithic Period and was walled before the sixth century BC by the Pelasgians. [4] Devoted to religious rather than defensive purposes, the area was adorned during the time of Cimon and Pericles with some of the world's greatest architectural and sculptural monuments. [5] This temple is the first building visitors see as they make their way up the Acropolis. [6] The first stone temple to Athena, the patron goddess and protector of the city, was built on the Acropolis at the beginning of the sixth century BC.	[1] Acropolis is Athen's premier and a world famous landmark. [2] It was first inhabited in neolithic times but the buildings seen today date mostly from the greek classical period of approximately 400 BC. [3] The Acropolis towers over the city of Athens and can serve as a navigation guide. [4] It houses a number of temples, statues, the Acropolis Museum, Theatre of Dionysos and many others. [5] The Theatre of Dionysos, named after Dionysus, son of Zeus, is the most important building on the southern slope of the Acropolis. [6] It is the place where the famous tragedies of Aeschylus, Sophocles and Euripedes and the comedies of Aristophanes were first performed in the fifth century BC. [7] The irregular shape of the theatre is a consequence of the site restrictions. [8] Fourteen staircases divide the auditorium into thirteen Cunei (segments). [9] The rows of seats consist of large carved blocks of limestone. [10] The most elaborate seat in the middle was reserved for the priest of Dionysos. [11] Admission tickets for the Acropolis is 12 Euros. [12] Between November and March free admission on Sundays, but bear in mind that it will be overcrowded. [13] The ticket is valid for the Archaeological Sites of Athens.

Even though the query-based summaries are more appropriate for our purposes, they are not completely satisfactory with regard to the human written summaries, since they only cover a part of the information the users included in the model summaries. This may be observed in Table 14.5, where an example of Summ-Graph query-based summary for the image *Acropolis* is shown, along with its corresponding human written summary. By examining the human made summary, we realize that it encodes the following types or categories of information: (1) information about the type of the object (sentences [1] and [6]); (2) information about when the object was built (sentence [2]); (3) information about where the object is located (sentence [3]); (4) more specific or background information about the object (sentences [4,5] and [7–10]); and (5) information about visiting times, prices, transportation, etc. (sentences [11–13]). By comparing with the automatic summary, we can see that the latter addresses information from categories (1), (2) and (4), but nothing is said about location or visiting information. One reason for this is that the query-based summarizer takes relevant sentences according to the

Table 14.6 SummGraph query-based and human written summaries for *Mount Floyen*

Query-based summary	Human written summary
[1] Floyen mountain (320 m height) is one of the most well-known sights of Norway. [2] It is the second largest city in Norway, which is hard to believe given its dramatic setting. [3] A new viewing platform was opened here in the summer of 2007, from which a more fascinating view on Bergen, fjords and mountains opens. [4] It is the only funicular cable car for passengers in Scandinavia. [5] The hotel is very well located near to old Bergen (the wooden house area) and to the Harbour, yards away. [6] The market area and the funicular railway are also close by. [7] The airport bus which runs every 15 min or so, has its last stop at the SAS Royal hotel, about 200 yards away. [8] The fare is 72 Kroner. [9] There was a huge gathering of motorcycle riders, two guys were getting handcuffed by the Politi and everyone else sitting around drinking beer and eating ice cream and watching all the action. [10] The continental breakfast at the hotel was very good–lots of variety. [11] The student town has great atmosphere with a bustling fish market, cosy cafs and bakeries (try out the local cinnamon rolls) and narrow, cobbled streets. [12] This proved useful as there is a decent small supermarket -Rema 1000-just a block away.	[1] Mount Floyen (320 m above sea level) is probably one of the most famous tourist attractions in Norway and Scandinavia as well. [2] Tourists from all over the world coming here and taking the unique funicular to the mountaintop. [3] The ride up Mount Floeny in a funicular is an absolute must for anyone who wants to boast they have seen Bergen. [4] There, from the top, you can see it in all its splendour. [5] The ride takes around 6 min, and both while you're on board and once at the top you'll have great views of central Bergen, with surrounding fjords and islands. [6] The ascent is rather steep, so if you are faint-hearted don't look down. [7] Otherwise, it's a great experience. [8] The walk back down to town is very pleasant, and will take around 30–45 min – both through woods and old parts of Bergen. [9] The funicular stops a number of times on the way, so you can walk part of the way. [10] There is a restaurant and a souvenir shop at the top, which seems to offer an even wider choice of souvenirs than the ones in Bryggen. [11] The funicular is accessible to people with disabilities: there is a lift there that can take you up straight to the first carriage.

query given to it (in our case just the toponym) and does not take into consideration more generally the information likely to be relevant to the user.

Similar information patterns can be found in human written summaries for other completely different objects, such as landforms, parks or gardens. This may be observed in Table 14.6, where the SummGraph query-based and human written summaries for the image *Mount Floyen* are shown. Again, the human made summary gathers information about the object type, location, background and visiting. However, another important category of information for this kind of object seems to be that related to the object's dimensions (e.g. the height of a mountain or the length of a river).

Therefore, humans appear to have a conceptual model of what is salient regarding a certain object type and this model informs their choice of what to say when describing an instance of this type. One way to improve the performance of the query-based summarizer is to give the summarizer the information that users typically associate with a particular object type as input, and bias the multi-document summarizer towards this information. To do this we plan to build conceptual models for different object types from the large number of existing image captions from Web resources, which we believe will improve the quality of automatically generated captions. A further improvement would be to learn the structure of the information flow (i.e. the order in which the different types of information appear in the human written summaries), which again might be a function of the type of object described.

Another weakness of the query-based summaries shown in Tables 14.5 and 14.6 is that they suffer from a lack of coherence resulting from 'dangling' anaphoric expressions (expressions whose antecedent has not been included in the summary). These must be avoided and so future work must investigate these using anaphora resolution techniques [51].

14.6 Conclusion

This paper has presented a comparison of four summarization systems' performance of producing two types of summaries, generic and query-based, for the image captioning task. The objective of this analysis was to account for the importance of the user's interests when generating this kind of summary. We showed that query-based summarizers perform better than generic on an image captioning task. However, the resulting summaries are not completely satisfactory when compared to human produced summaries. One conjecture as to why this might be so is that humans seem to have a conceptual model of what type of information should be included in a summary with respect to a specific object type (such as the location of the object, the year it was built, as well as some background information or attractions and places nearby) that cannot be successfully captured using only the information in the query.

Therefore, our future work will concentrate on extending the query-based summarizer to improve its performance in generating captions that match user expectations regarding specific image types. This will include collecting a large number of existing captions from Web sources and applying machine learning techniques for building models of the kinds of information that people use for captioning. Besides, although query-based summaries are more adequate than generic ones, they are still poor in readability. Consequently, further work also needs to be carried out on improving the readability of the summaries.

Acknowledgements This work was supported by the EU-funded TRIPOD project (IST-FP6-045335) and by the Spanish Government through the FPU program and the projects TIN2009-14659-C03-01, TSI 020312-2009-44, and TIN2009-13391-C04-01; by Conselleria d'Educació – Generalitat Valenciana (grant no. PROMETEO/2009/119 and ACOMP/2010/286); and the FPI program (BES-2007-16268) from the Spanish Ministry of Science and Innovation (project TEXT-MESS (TIN2006-15265-C06-01)). We would like to thank Horacio Saggion for his support with SUMMA. We are also grateful to Emina Kurtic for comments on the previous versions of this paper.

References

1. Aker, A., Gaizauskas, R.: Summary generation for toponym-referenced images using object type language models. In: Proceedings of the International Conference on Recent Advances in Natural Language Processing (RANLP), Borovets (2009)
2. Aker, A., Gaizauskas, R.: Model summaries for location-related images. In: Proceedings of the 7th International Conference on Language Resources and Evaluation (LREC), Valletta (2010)
3. Barnard, K., Duygulu, P., Forsyth, D., de Freitas, N., Blei, D., Jordan, M.: Matching words and pictures. J. Mach. Learn. Res. **3**, 1107–1135 (2003)
4. Bellare, K., Das Sarma, A., Loiwal, N., Mehta, V., Ramakrishnan, G., Bhattacharyya, P.: Generic text summarization using WordNet. In: Proceedings of the 4th International Conference on Language Resources and Evaluation (LREC), Lisbon (2004)
5. Carenini, G., Ng, R., Pauls, A.: Multi-document summarization of evaluative text. In: Proceedings of the 11th Conference of the European Chapter of the Association for Computational Linguistics (EACL), Trento (2006)
6. Cesarano, C., Mazzeo, A., Picariello, A.: A system for summary-document similarity in notary domain. In: Proceedings of the International Workshop on Database and Expert Systems Applications, Regensburg (2007)
7. Chowdary, C., Kumar, P.S.: Update summarizer using MMR approach. In: Proceedings of the Text Analysis Conference (TAC), Gaithersburg (2008)
8. Deschacht, K., Moens, M.: Text analysis for automatic image annotation. In: Proceedings of the 45th Annual Meeting of the Association for Computational Linguistics (ACL), Prague (2007)
9. El-haj, M., Hammo, B.: Evaluation of query-based arabic text summarization system. In: Proceedings of the International Conference on Natural Language Processing and Software Engineering, Beijing (2008)
10. Erkan, G., Radev, D.: LexRank: graph-based lexical centrality as salience in text summarization. J. Artif. Intell. Res. **22**, 457–479 (2004)
11. Fan, J., Gao, Y., Luo, H., Keim, D., Li, Z.: A novel approach to enable semantic and visual image summarization for exploratory image search. In: Proceedings of the 1st ACM International Conference on Multimedia Information Retrieval (MIR), Vancouver (2008)
12. Ferrández, O., Micol, D., Muñoz, R., Palomar, M.: A perspective-based approach for solving textual entailment recognition. In: Proceedings of the ACL-PASCAL Workshop on Textual Entailment and Paraphrasing, Prague (2007)
13. Fiszman, M., Rindflesch, T., Kilicoglu, H.: Abstraction summarization for managing the biomedical research literature. In: Proceedings of the HLT-NAACL Workshop on Computational Lexical Semantics, Boston (2004)
14. Givón, T.: Syntax: A Functional-Typological Introduction, vol. II. John Benjamins Publishing Company, Amsterdam/Philadelphia (1990)
15. Glickman, O.: Applied textual entailment. Ph.D. thesis, Bar Ilan University (2006)
16. Goldstein, J., Mittal, V., Carbonell, J., Kantrowitz, M.: Multi-document summarization by sentence extraction. In: Proceedings of the NAACL-ANLP Workshop on Automatic summarization, Seattle (2000)
17. Gotti, F., Lapalme, G., Nerima, L., Wehrli, E.: GOFAISUM: a symbolic summarizer for DUC. In: Proceedings of the Document Understanding Conference (DUC), Rochester (2007)
18. Gupta, S., Nenkova, A., Jurafsky, D.: Measuring importance and query relevance in topic-focused multi-document summarization. In: Proceedings of the 45th Annual Meeting of the Association for Computational Linguistics (ACL), Prague. Demo and Poster Sessions (2007)
19. He, L., Sanocki, E., Gupta, A., Grudin, J.: Auto-summarization of audio-video presentations. In: Proceedings of the Seventh ACM International Conference on Multimedia (MULTIMEDIA), Orlando (1999)
20. Hsin-Hsi, C., Chuan-Jie, L.: A multilingual news summarizer. In: Proceedings of the 18th Conference on Computational Linguistics (COLING), Saarbrücken (2000)

21. Jaoua, M., Ben Hamadou, A.: Automatic text summarization of scientific articles based on classification of extract's population. In: Proceedings of the 4th International Conference on Computational Linguistics and Intelligent Text Processing (CICLing), Mexico City (2003)
22. Kan, M.Y., McKeown, K., Klavans, J.: Domain-specific informative and indicative summarization for information retrieval. In: Proceedings of the Document Understanding Conference (DUC), New Orleans (2001)
23. Lesk, M.: Automatic sense disambiguation using machine readable dictionaries: how to tell a pine cone from a ice cream cone. In: Proceedings of SIGDOC, Toronto (1986)
24. Lin, C.Y.: ROUGE: a package for automatic evaluation of summaries. In: Proceedings of the Workshop on Text Summarization Branches Out (WAS), Barcelona (2004)
25. Lloret, E., Ferrández, O., Muñoz, R., Palomar, M.: A text summarization approach under the influence of textual entailment. In: Proceedings of the 5th Natural Language Processing and Cognitive Science Workshop, Barcelona (2008)
26. Lloret, E., Palomar, M.: A gradual combination of features for building automatic summarisation systems. In: Proceedings of the 12th International Conference on Text, Speech and Dialogue (TSD), Pilsen (2009)
27. Luhn, H.P.: The automatic creation of literature abstracts. IBM J. Res. Dev. **2**, 159–165 (1958)
28. Mani, I.: Automatic Summarization. John Benjamins Publishing Company, Amsterdam/Philadelphia (2001)
29. Marcu, D.: Discourse trees are good indicators of importance in text. In: Mani, I., Mayburg, M.T. (eds.) Advances in Automatic Text Summarization. MIT, Cambridge, MA (1999)
30. Mihalcea, R., Ceylan, H.: Explorations in automatic book summarization. In: Proceedings of the Joint Conference on Empirical Methods in Natural Language Processing and Computational Natural Language Learning (EMNLP-CoNLL), Prague (2007)
31. Mori, Y., Takahashi, H., Oka, R.: Automatic word assignment to images based on image division and vector quantization. In: Proceedings of RIAO 2000: Content-Based Multimedia Information Access, Paris (2000)
32. Pan, J.Y., Yang, H.J., Duygulu, P., Faloutsos, C.: Automatic image captionin. In: Proceedings of the IEEE International Conference on Multimedia and Expo (ICME), Taipei (2004)
33. Pedersen, T., Patwardhan, S., Michelizzi, J.: WordNet::Similarity – measuring the relatedness of concepts. In: Proceedings of the National Conference on Artificial Intelligence (AAAI), San Jose (2004)
34. Plaza, L., Díaz, A., Gervás, P.: Concept-graph based biomedical automatic summarization using ontologies. In: Proceedings of the 3rd Textgraphs Workshop on Graph-based Algorithms for Natural Language Processing, Manchester, pp. 53–56. (2008)
35. Plaza, L., Díaz, A., Gervás, P.: Automatic summarization of news using WordNet concept graphs. In: Proceedings of the Informatics IADIS International Conference, Algarve (2009)
36. Plaza, L., Lloret, E., Aker, A.: Improving automatic image captioning using text summarization techniques. In: Proceedings of the 13th International Conference on Text, Speech and Dialogue (TSD), Brno (2010)
37. Qiu, L., Kan, M.Y., Chua, T.S.: A public reference implementation of the RAP anaphora resolution algorithm. In: Proceedings of the 4th International Conference on Language Resources and Evaluation (LREC), Lisbon (2004)
38. Radev, D., BlairGoldensohn, S., Zhang, Z.: Experiments in single and multidocument summarization using MEAD. In: Proceedings of the Document Understanding Conference (DUC), New Orleans (2001)
39. Radev, D., Allison, T., Blair-Goldensohn, S., Blitzer, J., Çelebi, A., Dimitrov, S., Drabek, E., Hakim, A., Lam, W., Liu, D., Otterbacher, J., Qi, H., Saggion, H., Teufel, S., Topper, M., Winkel, A., Zhang, Z.: MEAD – a platform for multidocument multilingual text summarization. In: Proceedings of the 4th International Conference on Language Resources and Evaluation (LREC), Lisbon (2004)
40. Radev, D., Jing, H., Styś, M., Tam, D.: Centroid-based summarization of multiple documents. Inf. Process. Manage. **40**(6), 919–938 (2004)

41. Radev, D., Blitzer, J., Winkel, A., Allison, T., Topper, M.: Mead documentation v3.10. Tech. rep. URL http://www.summarization.com/mead/ (2006). Accessed June 2011
42. Saggion, H.: SUMMA: A robust and adaptable summarization tool. Rev. Trait. Automat. Lang. **49**(2), 103–125 (2008)
43. Saggion, H., Gaizauskas, R.: Multi-document summarization by cluster/profile relevance and redundancy removal. In: Proceedings of the Document Understanding Conference (DUC), Boston (2004)
44. Saggion, H., Teufel, S., Radev, D., Lam, W.: Meta-evaluation of summaries in a cross-lingual environment using content-based metrics. In: Proceedings of the 19th International Conference on Computational Linguistics (COLING), Taipei (2002)
45. Salton, G.: Automatic Text Processing. Addison-Wesley Longman Publishing Co., Inc., Boston, MA, USA (1988)
46. Salton, G., Buckley, C.: Term-Weighting Approaches in Automatic Text Retrieval. Pergamon, Tarrytown, NY, USA (1988)
47. Salton, G., Lesk, M.: Computer evaluation of indexing and text processing. ACM J. **15**(1), 8–36 (1968)
48. Schilder, F., Kondadadi, R.: FastSum: fast and accurate query-based multi-document summarization. In: Proceedings of the 46th Annual Meeting of the Association for Computational Linguistics (ACL): Human Language Technologies. Short Papers, Columbus (2008)
49. Sekine, S., Nobata, C.: A survey for multi-document summarization. In: Proceedings of the HLT-NAACL Workshop on Text Summarization, Edmonton (2003)
50. Spärck Jones, K.: Automatic summarizing: factors and directions. In: Mani, I., Mayburg, M.T. (eds.) Advances in Automatic Text Summarization, pp. 1–14. MIT, Cambridge, MA (1999)
51. Steinberger, J., Poesio, M., Kabadjov, M., Ježek, K.: Two uses of anaphora resolution in summarization. Inf. Process. Manage. **43**(6), 1663–1680 (2007)
52. Sun, J.T., Shen, D., Zeng, H.J., Yang, Q., Lu, Y., Chen, Z.: Web-page summarization using clickthrough data. In: Proceedings of the 28th Annual International ACM SIGIR Conference on Research and Development in Information Retrieval, Salvador (2005)
53. Svore, K., Vanderwende, L., Burges, C.: Enhancing single-document summarization by combining RankNet and third-party sources. In: Proceedings of the Joint Conference on Empirical Methods in Natural Language Processing and Computational Natural Language Learning (EMNLP-CoNLL), Prague (2007)
54. Titov, I., McDonald, R.: A joint model of text and aspect ratings for sentiment summarization. In: Proceedings of the 46th Annual Meeting of the Association for Computational Linguistics (ACL): Human Language Technologies, Columbus (2008)
55. Trappey, A., Trappey, C., Wu, C.Y.: Automatic patent document summarization for collaborative knowledge systems and services. J. Syst. Sci. Syst. Eng. **1**, 71–94 (2009)
56. Westerveld, T.: Image retrieval: content versus context. In: Proceedings of RIAO 2000: Content-Based Multimedia Information Access, Paris (2000)
57. Yoo, I., Hu, X., Song, I.Y.: A coherent graph-based semantic clustering and summarization approach for biomedical literature and a new summarization evaluation method. BMC Bioinformatics **8**(9) (2007)
58. Zechner, K., Waibel, A.: DiaSumm: flexible summarization of spontaneous dialogues in unrestricted domains. In: Proceedings of the 18th Conference on Computational Linguistics (COLING), Saarbrücken (2000)

Index